THE CORROSION OF DUPLEX STAINLESS STEELS:

A Practical Guide for Engineers

Roger Francis

NACE INTERNATIONAL

NACE International
The Worldwide Corrosion Authority

Printed in the United States of America

ISBN: 978-1-57590-369-9

NACE International
The Worldwide Corrosion Authority
15835 Park Ten Place
Houston, TX 77084
nace.org

Dedication

To all my friends and colleagues at Weir Materials and Foundries and Rolled Alloys for sharing your knowledge and expertise with me over the years.

Table of Contents

Preface

I first really became aware of duplex stainless steels with the launch of 2205 by Sandvik in the early 1980s. Within a few years, I was conducting crevice corrosion tests on a range of duplex stainless steels being considered for seawater piping systems. Since that time, I have carried out numerous corrosion tests on duplex stainless steels under a wide range of conditions for many industries. At the same time, I have attended numerous conferences where many people have presented the results of their work on duplex stainless steels. This class of stainless steel is now widely used in many industries, but I realized that there is no one place where data on the corrosion of duplex stainless steels is collected together. Hence, I conceived the idea of this book as a place to go for information on where the different grades of duplex will be satisfactory.

I am indebted to a number of friends for sharing their data with me, and for reviewing early drafts, in particular Carol Powell, Glenn Byrne, Geoff Warburton, Alan Turnbull, Bob Kain, Steve Clarke, Bop Phull, Les Boulton, Lena Wegrelius, and Mick Schofield. I would like to thank Rolled Alloys, Outokumpu Stainless AB, Industeel, and Arup for the use of some of their photographs and technical information. I would also like to thank my brother Ian Francis, who did the original line drawings. I hope that you find this document useful.

ROGER FRANCIS
June 2017

1 Introduction

Duplex stainless steels contain more than 19% chromium and have more than 30% of both ferrite and austenite. They were first invented in the early part of the 20th century, but not until the 1970s and 80s, with the introduction of argon oxygen decarburization (AOD) melting and the recognition of the benefits of nitrogen additions, that duplex stainless steels became attractive for widespread industrial use.

Modern duplex stainless steels are approximately 50/50 austenite and ferrite, and they combine the high strength of ferrite with the ductility of austenite. The addition of nitrogen enables better equi-partioning of elements between the two phases, higher strength in the austenite, better corrosion resistance, and improved weldability, to name but a few benefits.

Since the 1980s, a range of duplex stainless steels has been developed that have higher alloy content, and therefore greater corrosion resistance and strength. The commonly used duplex stainless steels have good corrosion resistance in a wide range of environments and excellent resistance to stress corrosion cracking (SCC). Their high strength and hardness also gives them excellent resistance to both erosion corrosion and cavitation. There are also cast versions of many of the duplex compositions and these have seen extensive use for pumps, valves, and other equipment.

Duplex stainless steels have been adopted by many industries to varying degrees. They are the workhorse corrosion resistant alloy (CRA) of the oil and gas industry, and are also widely used in the chemical and process industries for their SCC and corrosion resistance. Superduplex stainless steel has become the main alloy for piping, pumps, and valves in seawater reverse osmosis (SWRO) desalination plants, because of its resistance to crevice corrosion in seawater.

Duplex stainless steels have found applications in most industries, including oil and gas, marine, desalination, power, chemical and process, pulp and paper, and mineral processing. Figures 1.1–1.5 show some typical examples. Figure 1.1 is a 2205[(1)] subsea manifold for a North Sea gas field, while Figure 1.2 shows superduplex piping in an SWRO desalination plant. The pump in Figure 1.3 is Z100 superduplex for a flue gas desulfurization (FGD) plant at a United Kingdom power station, handling corrosive and erosive limestone slurry. The vessels in Figure 1.4 are 2304 lean duplex, and are a digester and an impregnation vessel

1. See Appendix for alloy generic names, UNS numbers, and nominal compositions.

in a Kraft pulp mill awaiting assembly. Figure 1.5 shows the mixed sulfides vessel in superduplex stainless steel being installed at a nickel laterite mineral processing plant. Further examples of the wide range of applications of duplex stainless steels are described in the chapters of this book relating to particular corrosion mechanisms.

FIGURE 1.1 A subsea well head in 2205 for a North Sea gas field. [Photo courtesy of Rolled Alloys]

FIGURE 1.2 Superduplex piping in an SWRO desalination plant. [Photo courtesy of Rolled Alloys]

FIGURE 1.3 A Z100 superduplex pump for an FGD plant in a UK power station. [Photo courtesy of Rolled Alloys]

FIGURE 1.4 A 2304 digester and impregnation vessel digester for a Canadian Kraft pulp mill awaiting assembly. [Photo courtesy of Steve Clarke]

FIGURE 1.5 Mixed sulfides vessel in Z100 superduplex being installed at a nickel laterite mineral processing plant.
[Photo courtesy of Rolled Alloys]

Tables 1.1–1.3 show some examples of the wide range of uses of the different grades of duplex stainless steel in the oil and gas, refining and petrochemical, and pulp and paper industries.

The next chapter reviews the various duplex alloy compositions, mechanical properties, and the design stresses for vessels and pipes to various codes. Chapter 3 covers the basics of welding duplex alloys, both to themselves and to other alloys, and their corrosion resistance. The next chapters cover nine different types of corrosion that may affect duplex stainless steels in service, presenting the available data and, in some cases, how to avoid problems. The final chapter covers the use of standards and specifications to procure good quality material for specific applications. It also covers the requirements over and above standard specifications necessary to obtain good quality material. Failure to appreciate this has led to a number of expensive service failures.

The alloys are referred to by their generic names, but the compositions to UNS standards[1] are used throughout and ASTM standards[2] are used to discuss the various product forms. The Appendix contains the nominal composition and generic names of all the cast and wrought stainless steels discussed in this book in Tables A.1, A.2, and A.3. The nominal composition of the nickel, copper, and titanium alloys mentioned in the text are in Tables A4, A5, and A6.

If this document is read from beginning to end, it will be noticed that there is a certain amount of repetition. This is because it is expected that most readers will dip into sections relevant to their current problem and they will not wish to keep following references to other chapters. Where the point at issue requires more lengthy discussion, reference to other chapters has been unavoidable.

TABLE 1.1 Examples of the uses of duplex stainless steels in oil and gas production.

AREA	APPLICATION	ALLOY
Topside	Process piping Process pumps Injection pumps Process valves Separator vessels Manifolds	2205/ Superduplex
	Seawater/waste water piping Seawater pumps Seawater valves Seawater filter vessels	Superduplex
Subsea	Flexibles	2003/ 2205
	Flowlines Gathering lines Manifolds Valves Hub connectors	2205/ Superduplex
	Risers	Superduplex
Downhole	OCTG Wireline Valves	2205/ Superduplex

TABLE 1.2 Examples of the uses of duplex stainless steels in refineries and petrochemical plants.

AREA	APPLICATION	ALLOY
Refinery	Crude desalters Amine towers lean/rich H/Ex Amine reclaimer units with caustic Filter drums (at risk of SCC) Fresh and brackish water coolers	2205 & Superduplex 2205 2205 2205 2205
Steam Cracker	Freshwater cooling towers/coolers Dilution steam generators Spent caustic oxidation vessels/piping Coolers for acidic process condensate	2101/ 2102/ 2205 2205 2205 & Superduplex 2205 & Superduplex
Aromatics	Feed fraction tower airfins/piping/vessels Aromatics/HC separation units, reboilers, H/Ex tubing	2205 2205
Polymers	Polypropylene piping, drums, dust collectors, dryers and rotating equipment Elastomer slurry tanks, vessels, piping and rotating equipment LLDPE[(A)] extruder feed shafts LDPE co-polymer piping Elastomer extruder barrels and screws	2205 2205 2205 2205 2205
Intermediates and other fluids	Corrosive service with organic acids, sulfuric acid, nitric acid, phosphoric acid, caustic, carbonyls etc., at high temperature and pressure Intermediate storage tanks	2205 2101/ 2102

(A) LLDPE= Linear low-density polyethylene

TABLE 1.3 Examples of applications of duplex stainless steels by the pulp and paper industry.

AREA	APPLICATION	ALLOY
Cooking Plant	Pre-steaming vessels Impregnation vessel Digester vessel Pulp discharge tank Digester flash tanks Liquor heaters High pressure feeder components Liquor piping	2205 2101, 2304, 2205 2101, 2304, 2205 2101, 2304, 2205 2205 2205 2205 2205/2101
Oxygen Delignification	O_2 Deliginification vessel Post O_2 Washers	2205 2101
Bleach Plant	Brownstock washers	2101/2205
Recovery	Evaporators White and green liquor storage Weak black liquor storage Heavy/strong black liquor storage	2101/2205 2101/2003/2202 2101 2205/superduplex
Paper Machine	Headbox Suction rolls	2205 2205

REFERENCES

1. SAE International, *Metals and Alloys in the Unified Numbering System* (Warrendale, PA: SAE International).
2. ASTM International (West Conshohocken, PA).

2 Alloys

Although some of the older duplex stainless steels are still available, this discussion will be confined to alloys that are commonly used in industrial plants.

2.1 COMPOSITIONS

The duplex stainless steels can be split up into four main groups: lean duplex, standard duplex, superduplex, and hyperduplex. The compositions of some common wrought alloys are shown in the Appendix, Table A.1, and those of some common austenitic alloys are shown in Table A.2 for comparison. Table A.3 shows the composition of some commonly used duplex cast alloys, with some common austenitic cast alloys for comparison.

Modern duplex stainless steels are mostly produced with a balance of approximately 50/50 austenite/ferrite phase. Any heat treatment or other process that takes the alloy too far from the nominal balance, or introduces third phases, can result in reduced mechanical properties, corrosion resistance, or both.

A common way of comparing stainless steels is by their Pitting Resistance Equivalent Number (PREN), where

$$\text{PREN} = \%\text{Cr} + 3.3(\%\text{Mo} + 0.5\times\%\text{W}) + 16\times\%\text{N}.$$

This is an empirical formula that relates to an alloy's resistance to pitting and crevice corrosion in chloride-containing solutions. The higher the PREN, the greater is the resistance to localized attack. The formula is widely used in industry and has been adopted by NACE MR0175/ISO 15156–the standard for materials used in sour oil and gas service[1]. Many standards, including the Norwegian oil and gas standard (NORSOK), insist on a PREN>40 for stainless steels and nickel alloys in seawater service[2].

For almost every austenitic alloy, there is a duplex stainless steel with similar corrosion resistance, higher strength, and often, lower cost. The 304L and 316L austenitic stainless steels are widely used in industry, and steel mills have developed a range of lean duplex alloys with low nickel and molybdenum to provide a higher strength alternative to 304L and 316L. The earliest lean duplex was 2304, which provides a corrosion resistance close to that of 316L and better than 304L. More recently, alloys with less nickel and more nitrogen have been developed to provide similar corrosion resistance and high strength at lower cost. The three alloys, 2101, 2102, and 2202 are all proprietary with similar properties. They are lower cost than 316L, but as of the time of writing, the factor holding up their wider use is the limited availability in product forms such as pipes, fittings, flanges, and bars.

Alloy 2003, with its PREN>30, is also proprietary and offers a greater corrosion resistance than 316L, but less corrosion resistance than that of 2205 duplex. This alloy has found uses in the oil and gas industry and has potential for the process and chemical industries, among others.

Alloy UNS S31803 is the original specification for 2205 duplex, and UNS S32205 was introduced to guarantee a consistently high level of corrosion resistance with higher minimum contents for chromium, molybdenum, and nitrogen. UNS S32205 is now the preferred composition, and most manufacturers dual certify their 2205 to both specifications. Alloy 2205 is very widely used in many industries, particularly where the SCC resistance of the 300 series austenitic alloys is inadequate. It offers a higher strength and lower cost than the austenitic alloy 904L, with a similar corrosion resistance in many fluids.

Ferralium is an older 25% Cr alloy that has been used in many applications by the process industries, particularly when acids are present. Nowadays it is not as widely available as the superduplex alloys, which can often be used as alternatives, but it has been written into alloy specifications for some processes.

The superduplex stainless steels offer a similar corrosion resistance or better when compared with the 6% Mo austenitic alloys, but at lower cost, because of their lower nickel and molybdenum contents. These alloys have found a wide variety of uses in aggressive process fluids, as well as seawater, and sour oil and gas fluids. Alloys 2507 and Z100 have found a wider use in industry than 52N+ and DP3W, but data for the latter is included if available.

Hyperduplex alloys have increased levels of chromium and nitrogen giving higher corrosion resistance to chlorides compared with superduplex, as shown by their high PREN values. They are currently only available as heat exchanger tube and seamless pipe.

The cast alloys are a little different, in that there is not such a wide variety to choose from. Only six alloys are listed in ASTM A995 (for pressure-containing castings)[3]. The compositions of some common cast alloys are shown in Table A.3.

Although no cast equivalents to the lean duplex alloys (such as 2101) are listed, the proprietary alloy Z21 LCD is shown, because it is a cast equivalent to 2003, with only minor compositional differences. To date, no cast versions of the leaner duplex alloys are known. ASTM A995 grade 4A is the cast version of 2205, and this alloy has been widely used for pumps, valves, and other cast components. CD4MCuN is a 25% Cr cast alloy that was originally developed for sulfuric acid service, but the addition of nitrogen gives this alloy a useful resistance to chlorides. Ferralium is listed in Table A.1 and it has been cast, but it is not listed in ASTM A995, although it is listed in ASTM A890 for non-pressure-containing castings. The cast composition is the same as the wrought alloy.

Of the superduplex alloys, Z100 (grade 6A) started life as a casting alloy and has been widely used for many cast components including pumps and valves. Care is required in casting when producing thick section components, but castings up to 5 tonnes finished weight have been produced successfully. The cast version of 2507 (grade 5A) has a slightly higher strength than Z100 because of its higher ferrite content. This can lead to cracking problems when trying to produce thicker section components[4].

There are newer methods available for manufacturing duplex alloys that are particularly suited to complex shapes. The first is hot isostatic pressing (HIP), in which fine powder is placed in a mild steel mold, which is then sealed and evacuated (ASTM A988[5]). The whole item is then heated at high temperature and pressure to produce a 100% dense solid that is anisotropic. The steel outer part can then be pickled off. The second method is 3-D printing, which currently is limited to small and medium size parts. It is known that HIP duplex products do not always have the same corrosion resistance as conventional material. At present, there is no published data on corrosion of 3-D printed duplex alloys. It is recommended that alloys manufactured by either of these methods be subjected to corrosion tests prior to use, because it is not safe to assume equivalent corrosion resistance to conventional wrought and cast products.

2.2 MECHANICAL PROPERTIES AND DESIGN STRESSES

The minimum mechanical properties at room temperature of some common wrought austenitic and duplex stainless steels are shown in Table 2.1. The mechanical properties of the three lean duplex alloys, 2101, 2102 and 2202, shown in Table 2.1, are the same. It can be seen that the duplex stainless steels are significantly stronger than their austenitic counterparts. The mechanical properties of 2507 vary with thickness and the product form in ASTM Standards. Thin products in 2507 often have a higher tensile strength than that of Z100, while thicker sections have the same or a lesser proof strength. The properties of 2507 have been shown as the same as Z100 for simplicity; anyone using thin sections (<5 mm) should consult the relevant ASTM standard. The mechanical properties of 52N+ and DP3W are similar to those of Z100.

The duplex castings are also substantially stronger than the cast austenitic alloys, as shown in Table 2.2. This means there is scope for significant wall thickness savings with duplex alloys for applications involving higher pressures and temperatures. Duplex castings are also much more resistant to hot tearing than are fully austenitic castings, and therefore much easier and less expensive to produce. The 6% Mo austenitic alloys suffer from molybdenum segregation at grain boundaries, which can reduce corrosion resistance unless the alloys are heat treated for extended periods. However, they are also subject to grain growth in extended heat treatments and short heat treatments are not always possible when they can compromise corrosion resistance.

The high strength of duplex stainless steels is reflected in the high design stresses for both vessels and pipes, as shown in Table 2.3. Again, the stresses for 2507 have been shown to be the same as for Z100, but may be slightly higher when using thin sections (<5 mm). It is interesting that the design stresses for vessels to the European code are much higher than under ASME VIII, Division 1. This is because ASME is an older code based on design stress as a fraction of the tensile strength. The European code uses a design stress based on a fraction of the 0.2% proof stress, which is high for duplex stainless steels. There are numerous vessels in 2205 and superduplex designed to EN13345 that have been in service for many years without problems. These high-design stresses can offer significant wall thickness savings for higher pressure applications, while not increasing the risk significantly for stress corrosion cracking. ASME Section VIII, Division 2 uses a reduced divisor on tensile strength compared with Division 1 and therefore, is much more desirable for high yield-to-tensile ratio alloys, such as duplex stainless steels.

TABLE 2.1 Minimum mechanical properties of some common wrought stainless steels at room temperature.

TYPE	UNS No.	GENERIC NAME	0.2% PROOF STRESS (MPa)	TENSILE STRENGTH (MPa)	ELONGn. (%)
Austenitic	S31603	316L	170	485	40
	N08904	904L	220	490	35
	S31254	6% Mo	310	655	35
Lean Duplex	S32304	2304	400	630	25
	S32101	2101	450	650	25
	S82011	2102	450	650	25
	S32202	2202	450	650	25
	S32003	2003	450	655	25
Standard Duplex	S32205	2205	450	655	25
Superduplex	S32750	2507	550*	750*	25
	S32760	Z100	550	750	25
Hyperduplex	S32707+	2707	700	920	25
	S33207+	3207	770	950	25

* See text.

\+ - For thicknesses ≤ 4mm

TABLE 2.2 Minimum mechanical properties of some common cast stainless steels at room temperature.

TYPE	ASTM+	GENERIC NAME	0.2% PROOF STRESS (MPa)	TENSILE STRENGTH (MPa)	ELONGn. (%)
Austenitic	CF8M	316	205	485	30
	CN7M	Alloy 20	170	425	35
	CK3MCuN	6% Mo	260	550	35
Lean Duplex	None	Z21 LCD	415	620	25
Standard Duplex	Grade 4A	2205	415	620	25
25% Cr Duplex	Grade 1B	CD4MCuN	485	690	16
Superduplex	Grade 5A	2507	515	690	18
	Grade 6A	Z100	450	690	25

\+ - ASTM properties taken from A351 for austenitic and A995 for duplex

TABLE 2.3 Design stresses at room temperature for some common wrought stainless steels.

TYPE	UNS No.	GENERIC NAME	DESIGN STRESS (MPa)		
			EN 13445 (Vessels)	ASME VIII Div 1 (Vessels)	ASME B31.3 (Pipes)
Austenitic	S31603	316L	140	113	113
	N08904	904L	167	140	147
	S31254	6% Mo	227	187	207
Lean Duplex	S32101	2101	271	186	217
	S32003	2003	273	187	218
Standard Duplex	S32205	2205	273	187	218
Superduplex	S32750	2507	313	214	250
	S32760	Z100	313	214	250

2.3 LIMITS OF USE

Because duplex stainless steels contain ~50% ferrite, it is sometimes assumed that they cannot be used to low temperatures without significant loss of toughness. However, duplex alloys are routinely tested and used down to -46°C without problems by the oil and gas industry. This is usually a requirement of a company specification or NORSOK[2]. Some projects have lower blow-down temperatures, and good impact toughness of Z100 superduplex has been obtained down to -70°C with modified processing[6].

A common requirement is an average Charpy Impact energy of 45 J with a single minimum of 35 J, and this has been shown to give adequate toughness. However, if a component is to be welded, the weld will have a lower toughness than the parent metal, and a minimum Charpy impact energy of 70 J average, 65 J minimum, has been recommended for parent metal so that the weld will still have adequate toughness[7].

In the chemical and process industries, there are some processes that involve cycling between high and low temperatures. Winnik and Fitzgerald[8] describe one such process with a cycle between +80° and -120°C. They required a special pump that was cast in Z100 superduplex in two halves. The impact toughness requirement was 0.38 mm minimum lateral expansion at -120°C. Lateral expansion is defined in ASTM A370[9] and E23[10], and a value of 0.38 mm has been found adequate for carbon steels for pressure vessels (e.g., ASME VIII); it has been applied to a number of CRAs as well. In a duplex stainless steel, 0.38 mm lateral expansion corresponds to an approximate impact energy of 40 J[11]. Although the parent castings met the impact toughness requirement, the welded joint with superduplex filler did not; a nickel alloy C-276 filler was used (UNS N10276), which gave satisfactory toughness and strength in the welded joint.

The lean duplex alloys have somewhat lower impact toughness resistance compared with 2205 duplex, because of the low nickel content, but it is still adequate for many applications where very low temperatures do not occur. Suitable impact toughness test criteria to assess material quality have yet to be determined.

When duplex stainless steels are heated, third phases can precipitate depending on the temperature[12]. In the temperature range 700–1,000°C phases such as sigma, chi, and nitrides may precipitate if the dwell time is too long. There is also the possibility of precipitating alpha prime phase in the temperature range 300–550°C. This fine intermetallic phase has little effect on corrosion, but causes a substantial loss of ductility. For this reason, duplex stainless steels are limited to a maximum temperature of 300°C. Some specifications use even lower maximum temperatures because of the known inclination of process engineers to increase temperatures to increase productivity.

For these reasons, duplex stainless steels cannot be given a full stress relief heat treatment, but alpha prime is slow to precipitate. The author has found that a short heat treatment at 300–320°C can significantly reduce peak stresses in some components. The heat treatment must be short: 2 or 3 hours maximum, followed by air cooling. It is important to test the impact toughness after stress relief to ensure it is still adequate for the specific application.

2.4 STANDARDS AND SPECIFICATIONS

There is a range of ASTM standards for various stainless steel products that include most, if not all, the duplex stainless steels. However, these usually only specify the composition limits, heat treatment temperature, and minimum mechanical properties. They do not discuss the heat treatment in detail, such as the time at temperature or how the quench should be done. These factors are very important to obtain a microstructure with the desired toughness and corrosion resistance. There have been some well-publicized failures of duplex stainless steel because of poor manufacturing procedures[13-16]. Francis and Hebdon[4] discussed in detail the requirements to obtain good quality duplex castings and these also apply to wrought products.

To ensure good quality material, it is common to ask for testing in addition to the mechanical tests required by ASTM. The following tests have been suggested for all wrought and cast superduplex products[17].

1. An ASTM G48[18] corrosion test at 50°C for 24 h. There must be no pitting and a weight loss less than 4 g/m^2.
2. A Charpy impact toughness test at –46°C with a minimum average energy of 70 J and a single minimum of 65 J.
3. A microstructure check at X500 after electrolytic etching in oxalic acid followed by electrolytic etching in NaOH or KOH.

The first test shows that there are insufficient third phases to affect the corrosion resistance. The weight loss criterion shows that there is no significant chromium depletion at the surface. The Charpy test also shows that there are insufficient third phases to significantly reduce the toughness. The microsection enables a determination of the phase balance and will also show which third phases are present if the material fails one or both of the other tests. The oxalic acid etch is required to show nitrides, which are not clearly shown in a caustic etch, except in large quantities. Figure 2.1 shows a double etched microsection where there are no nitrides in the ferrite, but the sigma is clearly shown as an orange precipitate. Figure 2.2 shows a double etched sample of superduplex where nitrides are clearly visible in the ferrite.

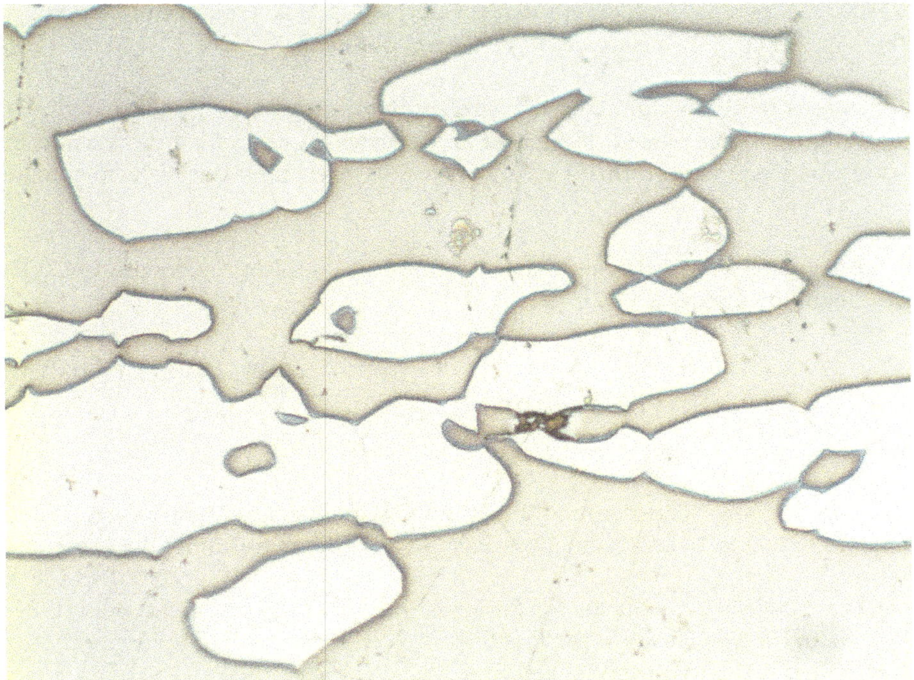

FIGURE 2.1 Microsection showing ferrite, austenite, and sigma after an oxalic/caustic electrolytic double etch.

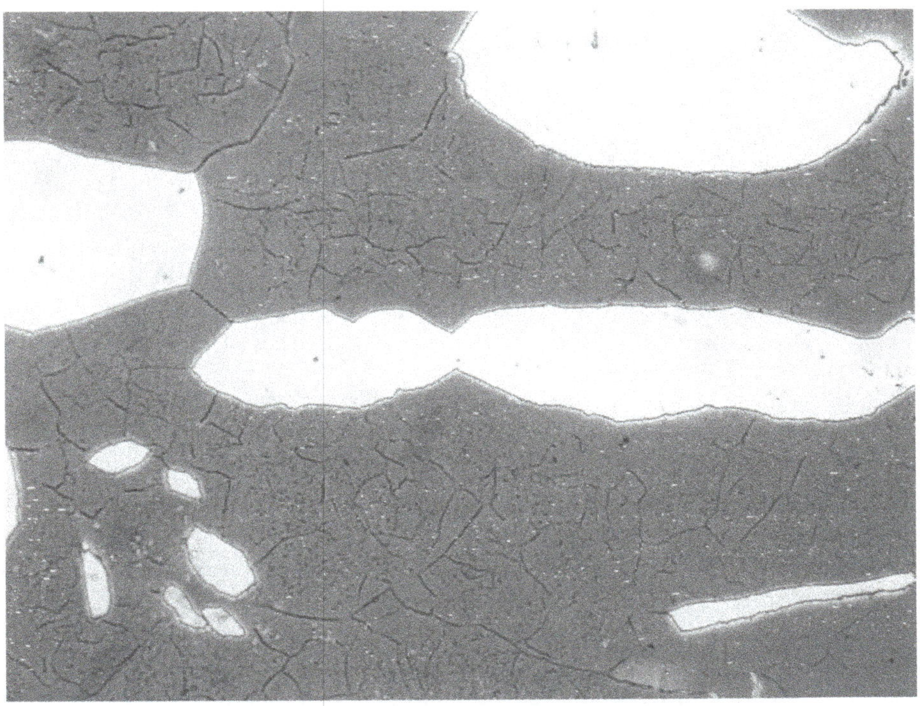

FIGURE 2.2 Microsection showing ferrite, austenite, and nitrides after an oxalic/caustic electrolytic double etch.

For 2205 duplex, the tests are the same except that the ASTM G48 test temperature is reduced to 25°C. Test conditions for the lean duplex alloys have yet to be clearly defined.

In addition to these tests by the manufacturer, it is possible to do further nondestructive tests on delivery to ensure that nothing has been missed. Francis et al. describe the use of the Feritscope to detect sigma/chi phases and describe some case histories where such tests prevented unsatisfactory material going into service[19].

The importance of using good procurement specifications and applying adequate QA/QC monitoring is discussed in more detail in Chapter 13.

REFERENCES

1. NACE MR0175/ISO 15156, "Petroleum and Natural Gas Industries–Materials for Use in H_2S-Containing Environments in Oil and Gas Production, Part 3 Corrosion Resistant Alloys" (Houston, TX: NACE International).
2. NORSOK M-630, "Material Data Sheets for Piping," Edition 5 (Oslo, Norway: Standards Norway, 2010).
3. ASTM A995, "Standard Specification for Castings, Austenitic-Ferritic (Duplex) Stainless Steel for Pressure-Containing Parts" (West Conshohocken, PA: ASTM International).
4. R. Francis and S. Hebdon, "The Selection of Stainless Steels for Seawater Pumps" CORROSION 2015, paper no. 5446 (Houston, TX: NACE International, 2015).
5. ASTM A988, "Standard Specification for Hot Isostatically Pressed Stainless Steel Flanges, Fittings, Valves and Parts for High Temperature Service" (West Conshohocken, PA: ASTM International).
6. G. Byrne, G. Warburton, Z. Schulz, and R. Francis, "ZERON 100 (UNS S32760 modified), Advanced Forging Process (AFP), Superduplex Stainless Steel for Increased Low Temperature Impact Toughness and Resistance to Hydrogen Induced Stress Corrosion Cracking (HISCC) Due to Cathodic Protection of API Forgings for Subsea Applications," Offshore Technology Conference Brasil, paper no. 130TCB-P-571-OTC (Rio de Janiero, Brazil: Offshore Technology Conference, 2013).
7. G. Byrne, R. Francis, G. Warburton, and M. Maligas, "The Selection of Superduplex Stainless Steel for Oilfield Applications," CORROSION 2004, paper no. 123 (Houston, TX: NACE International, 2004).
8. S. Winnik and B. Fitzgerald, "A User Perspective of Duplex Stainless Steel in Petrochemical Service," Duplex '94, paper no. 131 (Cambridge, United Kingdom: TWI, 1994).
9. ASTM A370, "Standard Test Methods and Definitions for Mechanical Testing of Steel Products" (West Conshohocken, PA: ASTM International).
10. ASTM E23, "Standard Test Methods for Notched Bar Impact Testing of Metallic Materials" (West Conshohocken, PA: ASTM International).
11. G. Warburton, Rolled Alloys, private communication to author (2016).
12. J. Charles, "Duplex Stainless Steels: Structure and Properties," Duplex Stainless Steels '91 (Les Ulis Cedex A: Les editions de physique, 1991): p. 3.
13. G. Rorvik, L.A. Marken, I.M. Kulbotte, M. Aursand, S. Olsen, and K.S. Karlsen, "Influence of Intermetallic Precipitates on the Mechanical Properties and Environmental Cracking Resistance of Duplex Stainless Steel Fittings–A Case History (Part 1)," Duplex Stainless Steels 2010 (Zutphen, Netherlands: KCI Publishing, 2010).
14. S. Mahajanam, H. Rincon, and D. McIntyre, "Metallurgical Examination of Defects in Duplex Stainless Steel Pipe Fittings," *MP*, 49, 4 (2010): p. 56.

15. E. Ryengen and C. Wintermark, "Lessons Learned from Heat Treatment of Components in 22Cr and 25Cr Duplex Stainless Steel (and Other Materials)," Duplex Stainless Steels 2010 (Zutphen, Netherlands: KCI Publishing, 2010).
16. E. Turbeville, F. Busschaert, S. Benum, and B. Hadsen, "A Summary of Recent Experiences on Designing and Fabricating Subsea Oil and Gas Production Systems with 22Cr and 25Cr Piping Materials," Duplex Stainless Steels 2010 (Zutphen, Netherlands: KCI Publishing, 2010).
17. S. Clarke and R. Francis, *The Selection of Materials for Mineral Processing Operations* (Houston, TX: NACE International, 2016).
18. ASTM G48, "Standard Test Methods for Pitting and Crevice Corrosion Resistance of Stainless Steels and Related Alloys by Use of Ferric Chloride Solution" (West Conshohocken, PA: ASTM International).
19. R. Francis, G. Warburton, G. Byrne, and J. Wilson, "QA/QC Tools to Ensure the Quality of Duplex Fittings and Other Components," Duplex Stainless Steels 2010 (Zutphen, Netherlands: KCI Publishing, 2010).

3 Welding

The combination of AOD melting and nitrogen additions mean that modern duplex stainless steels are fully weldable by all common arc welding processes. It is important to avoid the presence of third phases in either the weld metal or the HAZ, and care is needed to control both heat input and interpass temperatures. Manufacturers produce welding guidance documents for the various grades of duplex stainless steels.

All welding of duplex stainless steels requires shielding from the air. This is usually done with argon or an argon based gas mixture in both the torch and on the back face, or by use of fluxes as in SAW or SMAW welding.

3.1 AUTOGENOUS WELDING

Autogenous welding of duplex stainless steels is not generally recommended because the weld metal will be highly ferritic because of the fast cooling rate. However, if the weld is solution annealed after welding, the microstructure will be restored to a normal phase balance with good properties. Without post-weld heat treatment of autogenous welds, there will be a significant reduction in corrosion resistance and toughness. It is claimed that the use of nitrogen-containing welding gases and special welding techniques can maintain an adequate phase balance in the weld metal for single pass autogenous welds without post-weld heat treatment[1].

3.2 WELDING WITH FILLER

If duplex alloys are welded with matching composition fillers, then the weld will be ferritic, as with autogenous welding. However, a full post-weld solution anneal heat treatment will completely restore the microstructure and give good properties.

However, most welding of duplex stainless steels require them being used in the as-welded condition. The filler wires for this requirement are usually matching in composition with an extra 2–3% nickel. Common examples are 2209 (for 2205), 2509 (for 2507), and Z100X (for Z100). The extra nickel enables sufficient austenite to form during the fast cooling of the weld so that the 50/50 austenite/ferrite balance is maintained. With the superduplex alloys, ASME rules now permit welding of 2507 and Z100 with either of the two alloys' fillers.

However the weld is made and whatever filler is used, it is most important that the final weld has a good phase balance, ranging from 40–60% for preference, although some standards allow a phase balance up to 65%. This balance will ensure both good mechanical properties and corrosion resistance. In Chapter 6, there is an instance of a chloride SCC failure of a 2205 weld with a ferrite content ≥ 70%.

Manufacturers of alloys and welding products often have guidelines for joint design, heat inputs, and interpass temperatures for the various alloys, e.g., reference 2. A good overview of welding duplex alloys has been produced by the International Molybdenum Association (IMOA) and is free to download from their website[3].

The root pass of a duplex weld can sometimes be a little ferritic, but reheating during the second pass of a multipass weld will cause further transformation of ferrite to austenite. This usually ensures a good phase balance in the root, which is the side of the weld normally in contact with corrosive fluids. The capping pass is not reheated and may sometimes be a little ferritic, but this is not usually exposed to corrosive fluids. If it is, the corrosion resistance can be improved by pickling (see below). Figure 3.1A shows a typical root microstructure of a multipass weld, while Figure 3.1B shows a slightly ferritic capping pass. The ferrite is the darker phase.

It is often stated that welds are not as corrosion resistant as parent metal, but if correctly made, duplex welds are often similar in performance. Walker and Gooch[4] concluded that a filler wire with a PREN (see Chapter 2) about 2 points higher than that of the parent metal (2205) would give welds with similar corrosion resistance. Typical commercial welding fillers for 2205 usually meet this requirement, particularly if this is requested, because of the extensive use of 2205 by industry.

Where concern exists about the corrosion resistance of the weld, or if the correct filler is not available, it is common to weld with a higher alloy duplex filler. Thus, lean duplex alloys are often welded with 2209 filler, and 2205 has been welded with superduplex filler. For the superduplex alloys, it would be necessary to consider a nickel alloy filler for higher corrosion resistance. Alloy 625 (UNS N06625) is *not* suitable because segregation in the weld metal reduces its corrosion resistance. Superduplex has been successfully welded with C-276 (UNS N10276), C-22 (UNS N06022), and alloy 686 (UNS N06686) fillers, which have high chromium and molybdenum contents. Note that these nickel alloy fillers may not match the strength of the superduplex parent metal.

Gooch[5] commented that the threshold stress for SCC of duplex welds is no lower than that for parent metal and the fact that weld metal is usually somewhat stronger than parent metal means that welds are not particularly susceptible to SCC.

Sedriks[6] commented that welds may be more susceptible to SCC if they are poorly welded such that third phases, such as sigma, are precipitated, or the ferrite content is particularly high. In 2005, the latter caused the SCC of an offshore vessel when seawater was trapped and heated to around 100°C under highly absorbent insulation.[7] The SCC resistance of duplex welds is thoroughly reviewed by McIntyre and Francis

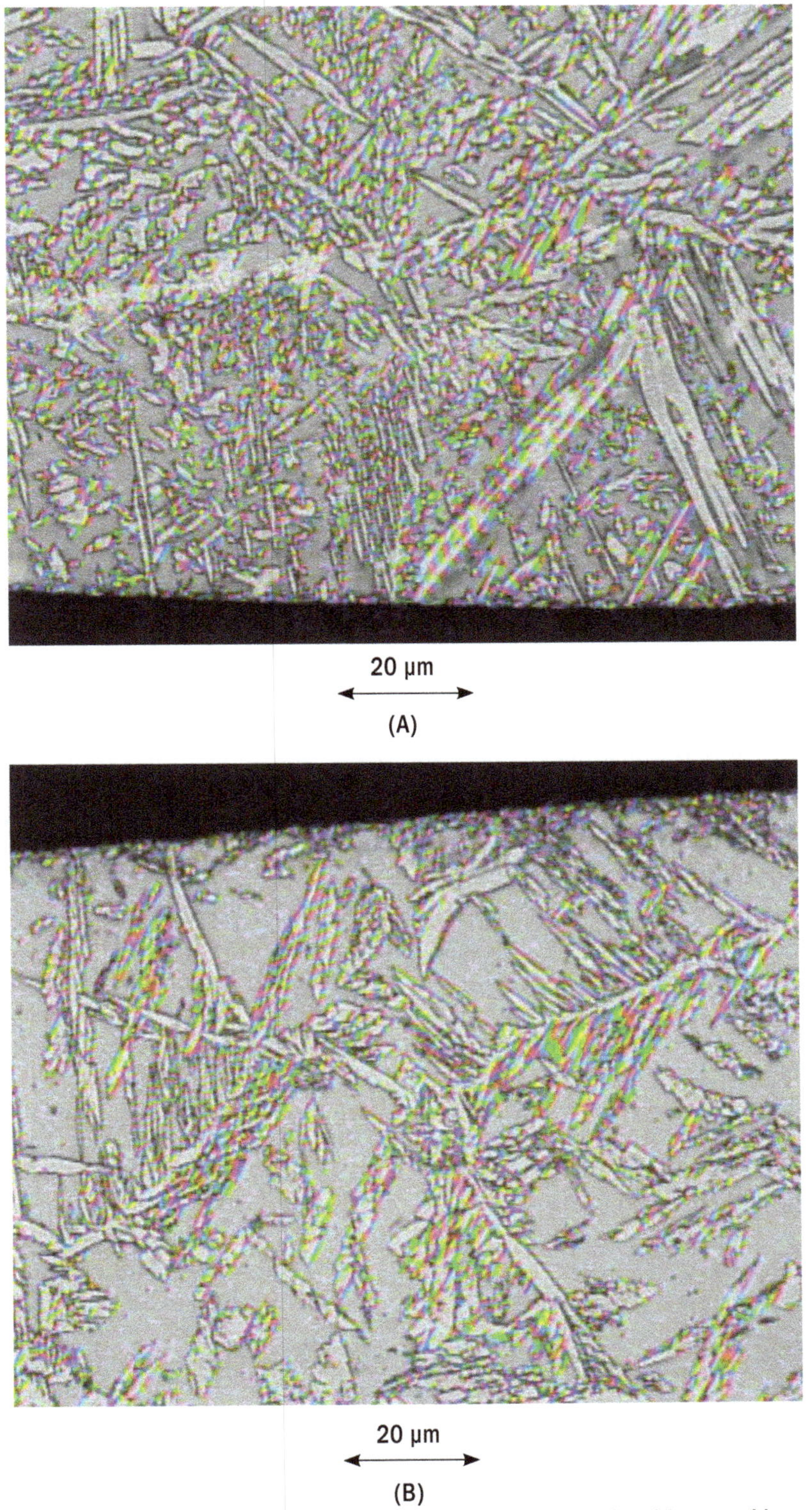

FIGURE 3.1 Typical microstructures of duplex stainless steel multipass welds. Weld root, B) Weld cap

and they show that properly made welds are not particularly susceptible to cracking compared with parent metal[8].

It is well known that if superduplex stainless steel is incorrectly welded, sigma and/or chi phase may precipitate in the HAZ, and the chromium and molybdenum depleted zone around the particles is then susceptible to corrosion. Francis and Warburton modelled the depleted zone and calculated the critical sigma content for corrosion to occur in warm, chlorinated seawater[9]. The calculation showed that for 2 μm diameter sigma particles, the threshold concentration was ~4%, and this was confirmed by exposure tests of Z100 welds with different sigma contents. The threshold was very susceptible to the third phase particle diameter and fitness for purpose tests were recommended to determine whether a particular weld qualification was acceptable. The ASTM G48[10] test at 35°C has been recommended as a suitable qualification test, with a pass being no pitting and a weight loss <4 g/m^2, e.g., reference 11. The whole issue of appropriate tests for duplex welds is discussed in Chapter 13.

The author has carried out a lot of corrosion tests of welded duplex stainless steels, and generally welds are as corrosion resistant as parent metal in many fluids. This was confirmed by Gunn for some common fluids[12]. Numerous publications show that cleaning welds improves their corrosion resistance compared with the as-welded condition. Gunn showed some data for both 2205 and superduplex welds, including manual wire brushing, grinding, blasting, and pickling with pastes or gels[12]. The results clearly show pickling to give the greatest increase in the critical pitting temperature (CPT).

Figure 3.2 shows the CPT of various welds in synthetic seawater at a potential of +600 mV SCE, typical of the potential of stainless steel in chlorinated seawater. The tests were with 2205 welded with either the regular 2209 filler or with superduplex filler Z100X, and Z100 welded with its Z100X filler. The pickling was carried out with commercial paste based on nitric/hydrofluoric acids. The results show that the CPT of all the welds can be improved by pickling, although the least advantage was gained by pickling 2205 welded with superduplex filler.

It has been pointed out that when duplex stainless steels are welded with pure argon backing gas and all the air is flushed out, it is possible to lose nitrogen from the weld root, which reduces corrosion resistance[13]. Several authors carried out tests using shielding and backing gases containing nitrogen; some of this data was reviewed by Gunn[12]. The results show significant increases in the CPT when nitrogen is present in the shielding gas for both 2205 and superduplex. There was a greater increase in CPT for superduplex with nitrogen in the shielding gas. Byrne et al. showed that the biggest improvement in weld corrosion resistance was obtained with argon with 2% nitrogen as the shielding gas and nitrogen plus 10% hydrogen as the backing gas[13]. Figure 3.3 shows the effect of gas composition on the CPT in the ASTM G48 ferric chloride test for both 2205 and Z100 superduplex. The beneficial effect of nitrogen in the welding gas on the pitting resistance of both alloys is clearly shown.

Francis and Byrne showed that the resistance of superduplex stainless steel welds to pitting in seawater (and the crevice corrosion resistance of parent metal) can be improved by a "soft" startup[14]. A typical start up regime is the following:

-start in cold seawater for 2 days (minimum)
-run in cold chlorinated seawater for 5 days (minimum)
-turn on heat exchangers and operate normally.

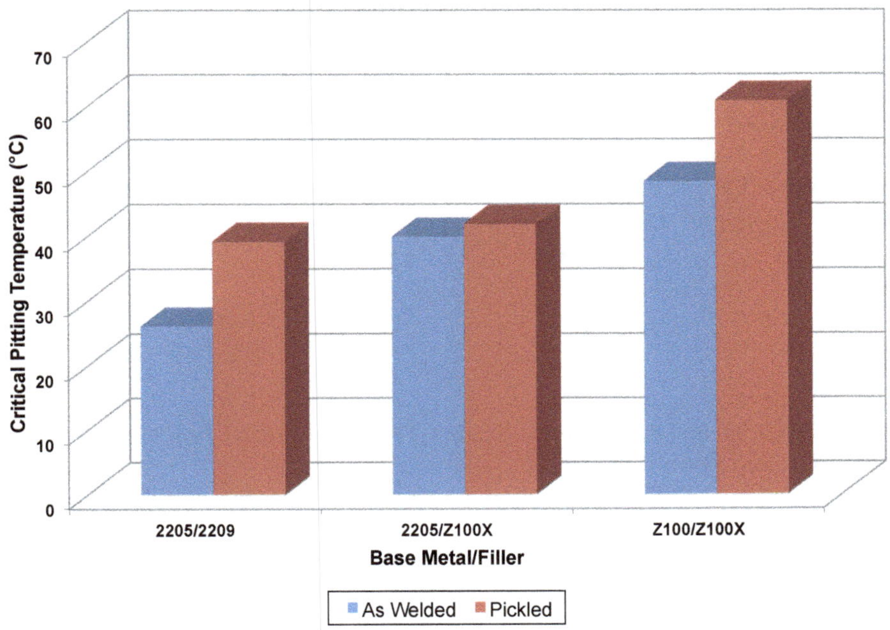

FIGURE 3.2 CPT for some duplex stainless steel welds in synthetic seawater at +600 mV SCE.

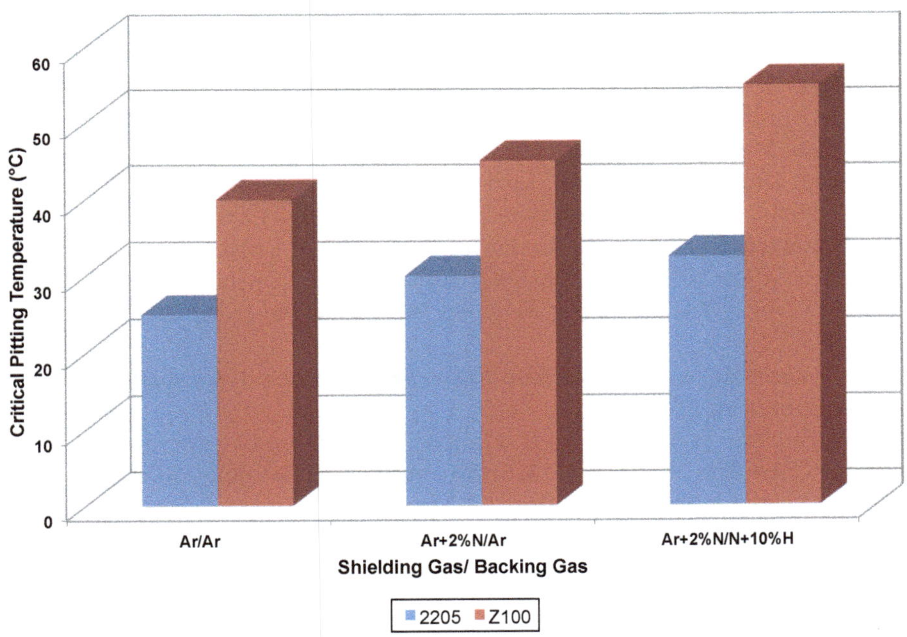

FIGURE 3.3 Typical CPT of duplex welds in ASTM G48 ferric chloride test with different backing gases[13].

Francis and Byrne gave an example where superduplex seawater piping was used with a titanium gas cooler for several months with a seawater discharge temperature of ~20°C[14]. The gas flow was then increased and the discharge temperature rose to around 50–60°C. When inspected after about 6 months operation at the higher temperature, there was no pitting at welds or crevice corrosion at flanged joints. This shows how the limits of use can be extended by a soft start-up.

Additional testing and QA/QC requirements, over and above those in ASME for duplex stainless steel welds, are discussed in detail in Chapter 13.

REFERENCES

1. A.W. Stevenson, "Weir Materials Technical Report TN906," Internal report (Manchester, United Kingdom: Weir Materials and Foundries, 1992).
2. "Welding Guidelines for ZERON 100," Rolled Alloys Publication No. 105 (Temperance, MI: Rolled Alloys, 2009).
3. IMOA, *Practical Guidelines for the Fabrication of Duplex Stainless Steels*, 3rd Edition (London, United Kingdom: International Molybdenum Association, 2014).
4. R.A. Walker and T.G. Gooch, *Brit. Corr. J*, 26, 1 (1991): p. 51.
5. T.G. Gooch, "Corrosion Resistance of Welds in Duplex Stainless Steel," Duplex Stainless Steels '91 (Les Ulis Cedex A: Les editions de physique, 1991): p. 325.
6. J. Sedriks, *Corrosion of Stainless Steels*, 2nd Edition (New York, NY: J. Wiley and Sons, 1996).
7. I. Oystetun, K. Johansson, and O.B. Andersen, "Stress Corrosion Cracking of Duplex SS, UNS S31803, on the Gyda HP Separator," Offshore Technology Conference, paper no. 7207 (Houston, TX: Offshore Technology Conference, 1993).
8. D.R. McIntyre and R. Francis (eds), "Environmental Cracking," MTI Project 216 (St. Louis, MO: Materials Technology Institute, 2016).
9. R. Francis and G.R. Warburton, "A Model for the Corrosion of the Depleted Zones Around Sigma Precipitates Produced During Welding of Superduplex Stainless Steel," Stainless Steel World Conference 1999, paper no. 99006 (Zutphen, Netherlands: KCI Publishing, 1999).
10. ASTM G48, "Standard Test Methods for Pitting and Crevice Corrosion Resistance of Stainless Steels and Related Alloys by Use of Ferric Chloride Solution" (West Conshohocken, PA: ASTM International).
11. S. Clarke and R. Francis, *The Selection of Materials for Mineral Processing Operations* (Houston, TX: NACE International, 2016).
12. R. Gunn (ed), *Duplex Stainless Steels; Microstructure, Properties and Applications* (Cambridge, United Kingdom: Abington Publishing, 1997).
13. G. Byrne, G. Warburton, J. Wilson, and R. Francis, "Fabrication of Superduplex Stainless Steel for Optimum Seawater Corrosion Resistance," World Congress on Desalination and Water Reuse (Topsfield, MA: International Desalination Association, 2011).
14. R. Francis and G. Byrne, "Experiences with Superduplex Stainless Steel in Seawater," *Stainless Steel World*, Vol. 16 (Zutphen, Netherlands: KCI Publishing, June 2004): p. 53.

4 General Corrosion

Corrosion resistance of duplex stainless steels is mostly conferred by a passive film rich in chromium. The corrosion resistance and repassivation kinetics, when the film breaks down, are boosted by the presence of molybdenum and nitrogen. In near neutral pH solutions, the general dissolution rate is usually very low (<0.01 mm/y), but the presence of aggressive ions, such as halides, can result in localized breakdown of the passive film and the initiation of localized corrosion, such as pitting and crevice corrosion. In solutions with very high concentrations of chloride at high temperatures, the lower alloy duplex stainless steels can lose passivity and undergo some general corrosion[1]. However, in solutions of very low or high pH without chlorides, the passive film can break down completely and then general dissolution occurs. This chapter describes the resistance of duplex stainless steels to corrosion in some common acids and alkalis.

4.1 SULFURIC ACID

4.1.1 Pure Sulfuric Acid

Sulfuric acid is one of the most commonly used chemicals in the world and it can vary from a reducing acid to an oxidizing acid depending on its concentration. A common way to assess the performance of alloys in acids is by the iso-corrosion curve. This plots the points where the corrosion rate is constant against acid concentration and temperature. A common value is 0.1 mm/y because this is a useful design criterion.

Figure 4.1 shows the iso-corrosion curves in pure acid for some of the lower alloy duplex stainless steels with the curve for 316L–a commonly used austenitic stainless steel. It has been assumed that alloys 2102 and 2202 are similar in performance to 2101. It can be seen that 2101 is superior to 316L at acid concentrations up to 15 wt.%, and 2304 and 2205 are significantly more resistant. An iso-corrosion curve for 2003 was not available, but it is assumed to lie somewhere between those of 2101 and 2205 because of its alloy content.

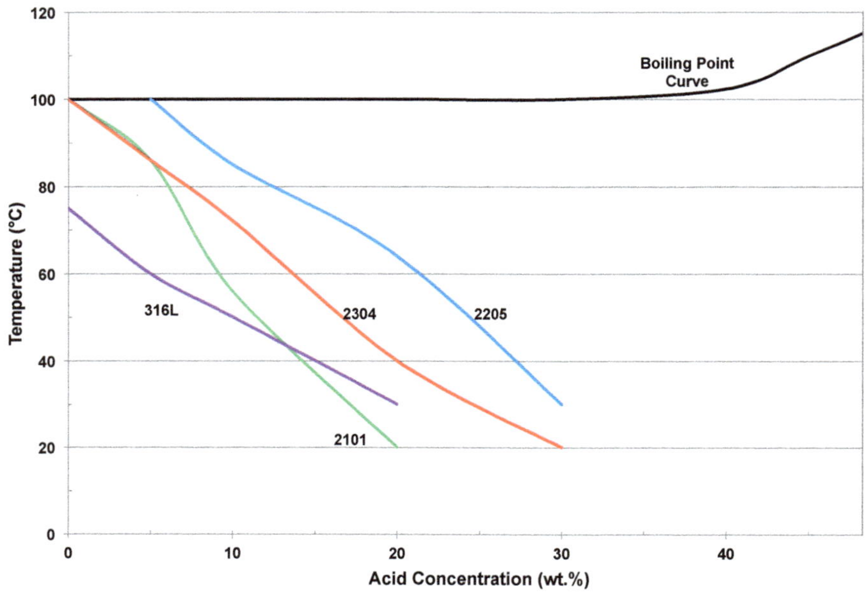

FIGURE 4.1 Iso-corrosion (0.1 mm/y) curves for some duplex stainless steels in pure, aerated sulfuric acid.

Figure 4.2 shows the iso-corrosion curves for some higher alloy austenitic and duplex stainless steels, again with that of 316L for comparison. It is clear that the superduplex stainless steels are superior to 2205, but it is also clear that alloys like 904L and alloy 20 (UNS N08020) are superior to 6% Mo austenitic (UNS S31254). This is because 904L and alloy 20 were developed for sulfuric acid service and they contain deliberate additions of copper, which is known to improve the performance of stainless steels in sulfuric acid.

This also explains the differences between the different grades of superduplex stainless steel and this is shown more clearly in Figure 4.3. Alloy 2507 has no deliberate copper addition, while that of Z100 is about 0.7% and that of Ferralium is ~2%. It is clear that increasing copper content improves the corrosion resistance, particularly in the mid-range acid concentration. In very dilute acids, there is no real difference between the grades, and in concentrated acid it has been shown that a little copper is beneficial, such as in Z100, but increasing it further has no significant benefit.

Cast alloys perform similarly to their wrought counterparts in sulfuric acid. Figure 4.4 shows the iso-corrosion curves for some cast alloys. The composition of cast alloys can be adjusted slightly with each heat to enhance particular properties and Z100AG, as shown in Figure 4.4, is standard Z100 with the copper content increased to 1%, the maximum in the UNS composition specification. This was to improve the corrosion resistance in sulfuric acid for a specific application and it can be seen in Figure 4.4 that the corrosion resistance in the midrange acid concentrations was significantly improved. It is clear that under many conditions, there are duplex stainless steels that offer superior performance to austenitic grades.

The high resistance of Z100 superduplex stainless steel to corrosion in hot, concentrated (96–100 wt.%) acid has led to its use in sulfuric acid plants. Applications include heat exchanger tubes and orifice plates, where it offers a cost effective alternative to high silicon austenitic alloys.

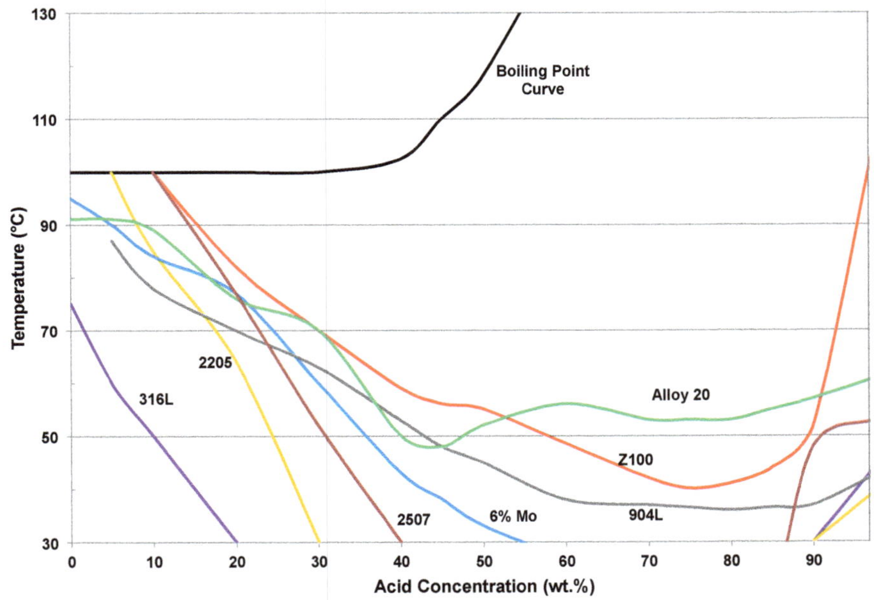

FIGURE 4.2 Iso-corrosion (0.1 mm/y) curves for some higher alloy stainless steels in pure, aerated sulfuric acid.

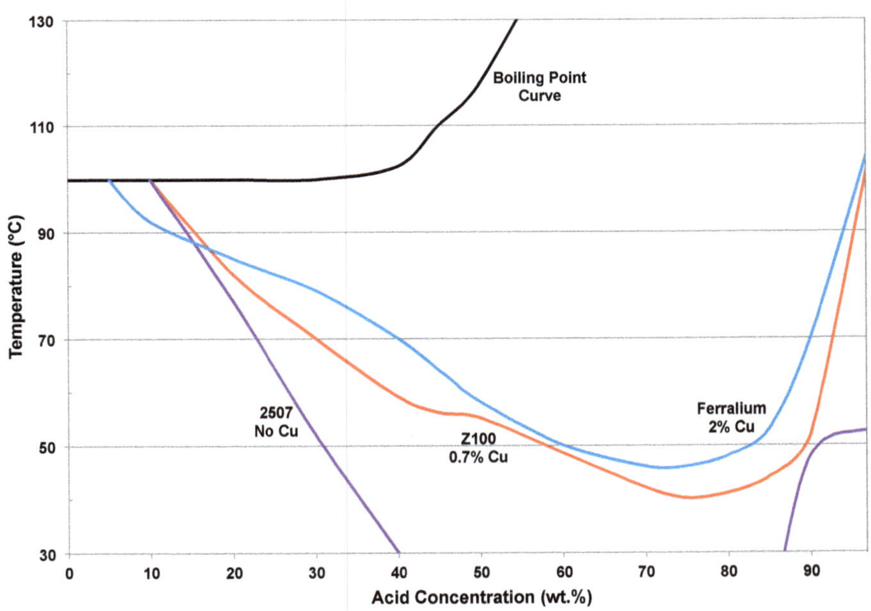

FIGURE 4.3 Iso-corrosion curves (0.1 mm/y) of some superduplex stainless steels in sulfuric acid.

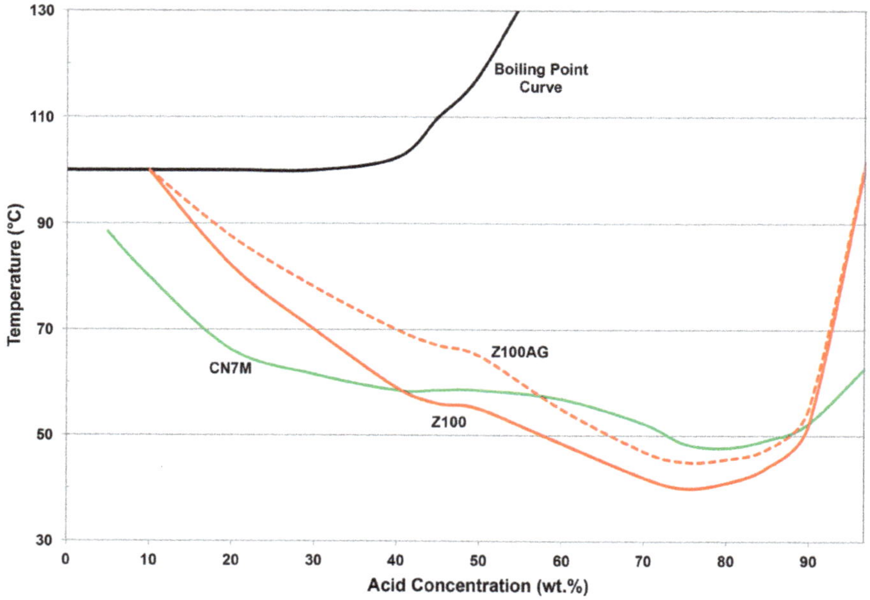

FIGURE 4.4 Iso-corrosion curves (0.1 mm/y) for some cast stainless steels in sulfuric acid.

4.1.2 Effect of Contaminants

The above section discussed the performance of stainless steels in commercially pure sulfuric acid, but in industrial service, the acid is rarely pure and it can contain a wide variety of other species that can have a significant effect on corrosion resistance. One of the most common is chloride, which is present in most water used in commercial solutions. The concentration may be anything from a few 10s of milligrams per liter to 1000s of milligrams per liter.

Figure 4.5 shows a schematic diagram of a polarization curve for stainless steel in an acidic environment. While the alloy is in the passive zone corrosion rates are low, but as temperatures and acid concentrations increase, the size of the passive zone decreases and alloys tend to move into the active zone. Here, the passive film has broken down and general dissolution is occurring. If the solution becomes oxidizing and the potential moves into the transpassive zone, a number of possible reactions may occur. In the absence of chlorides, the general corrosion rate will increase–the higher the molybdenum content of the alloy, the greater the dissolution rate, because molybdenum salts are soluble under these conditions. If halides are present, pitting and/or SCC may also occur.

Figure 4.6 shows the iso-corrosion curves for some stainless steels in sulfuric acid with 200 mg/L chloride. Unfortunately, there is no data for most duplex stainless steels with 200 mg/L chloride, and only limited data for 2205. However, if the curves are compared with those in Figures 4.1 and 4.2, it can be seen that even this concentration of chloride significantly decreases the resistance of many stainless steels. One of the biggest decreases is for alloy 20, which was developed for sulfuric acid service, but has a low tolerance for chlorides.

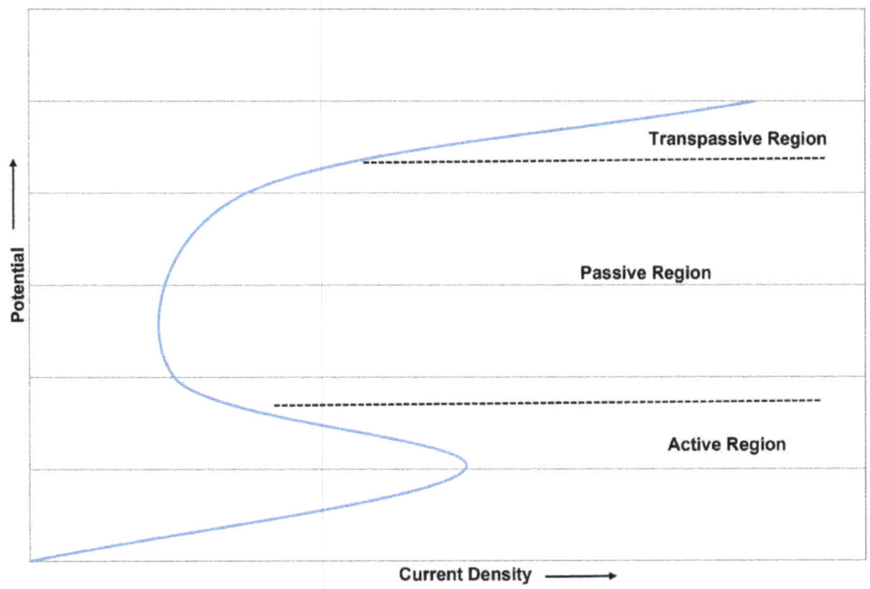

FIGURE 4.5 Schematic polarization curve for stainless steels in acids.

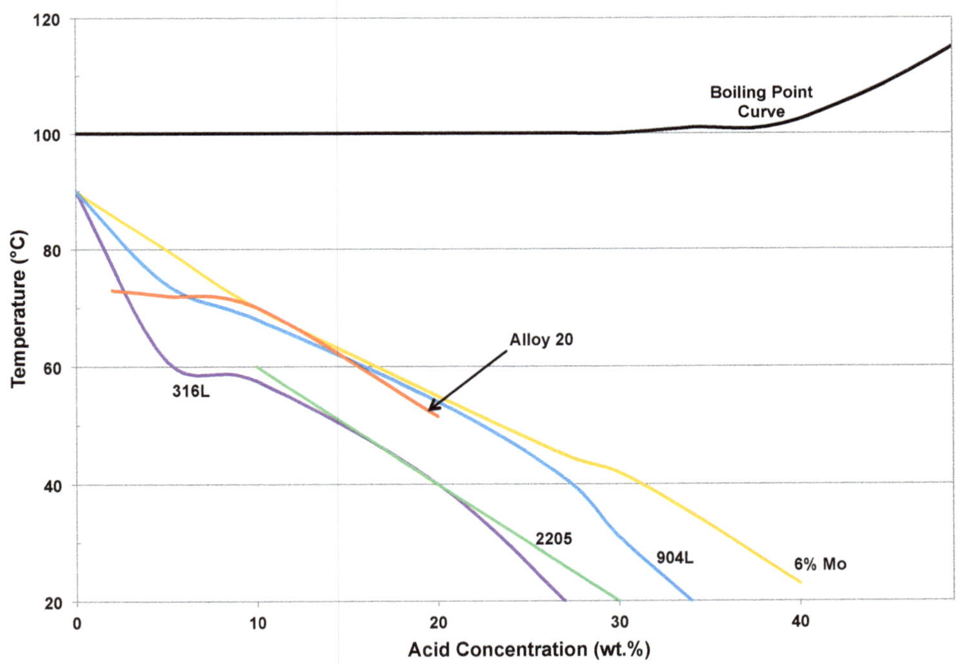

FIGURE 4.6 Iso-corrosion curves (0.1 mm/y) for some stainless steels in sulfuric acid with 200 mg/L chloride.

Figure 4.7 shows the iso-corrosion curves in sulfuric acid with 2,000 mg/L chloride, and this time there is more data for duplex alloys. The striking thing is that although the performance of the austenitic alloys decreases significantly with 2,000 mg/L chloride, that of the superduplex alloys decreases only slightly. Figure 4.8 shows the iso-corrosion curves for Z100 superduplex with various concentrations of chloride, and it can be seen that the alloy can tolerate high chloride concentrations before the performance decreases significantly. This has led to the use of this alloy in acid leach mineral processing projects, where high chlorides are common[2]. Although chloride is a common contaminant of acid solutions, other halides— particularly fluorides—will have a similar effect if the concentration is high enough.

Another type of contaminant that is often present in commercial fluids containing sulfuric acid is oxidizers. These may be metallic ions, such as Cu^{2+} and Fe^{3+}, as found in many acid leach mining fluids, or they may be present as catalysts, or they may be oxidizing chemicals such as nitric acid, or the oxidizer may be a high pressure of oxygen. All of these will raise the redox potential of the fluid, as discussed by Francis and McIntyre[3]. If the alloy is in the active region, the potential may move into the passive region and extend the limits of use. However, if the potential is raised into the transpassive region, increased rates of dissolution can occur, and pitting or SCC may also occur if halides are also present[3]. Note that if oxidizers are present and are preventing rapid corrosion, they must also be present in the solution in crevices, otherwise these may corrode rapidly.

Unfortunately there is very limited data on the effect of oxidizers on the corrosion resistance of stainless steels. Figure 4.9 shows the effect of small additions of ferric ions and cupric ions on the iso-corrosion (0.1 mm/y) curves in sulfuric acid for both 316L austenitic and Z100 superduplex. It is evident that both oxidizing species significantly increase the corrosion resistance of both alloys. Ferric ions seem to be more effective with superduplex, while there seems to be little difference between the two for 316L austenitic.

It is important to note that these concentrations are in the milligrams-per-liter range, but in industrial fluids, concentrations of grams per liter are common. In such cases, there is increased likelihood of alloys going transpassive, and corrosion testing is advised before relying on oxidizing additions to keep corrosion rates low. Francis reported the use of Z100 superduplex in an acid recovery unit (70 wt.% acid) at 125°C, where the presence of a small amount of nitric acid enabled the alloy to be suitable[4]. The corrosion rates were

$$\text{70 wt.\% sulfuric acid} \gg 1 \text{ mm/y}$$
$$\text{70 wt.\% acid} + 0.15\% \text{ nitric acid} = 0.029 \text{ mm/y}$$

In some acid solutions, the local water is often present and such waters can contain chlorides from ~10–20,000 mg/L if seawater is being used. If oxidizers are also present, then pitting or SCC is possible if the alloy goes transpassive. There is almost no data on the performance of stainless steels in such mixtures, but Clarke and Francis present some useful data for Z100 superduplex[2].

Figure 4.10 shows the iso-corrosion (0.1 mm/y) curves for Z100 superduplex in sulfuric acid with 2 g/L chloride and various concentrations of copper ions. Even with 2 g/L chloride, the presence of copper ions up to 1 g/L improved the corrosion resistance. These tests were at atmospheric pressure and no tests were conducted above the boiling point. It is likely that ferric ions will have a similar effect.

Figure 4.11 shows the effect of chloride on the iso-corrosion curve for Z100 superduplex in 5 wt.% sulfuric acid with different copper concentrations. While copper was beneficial at 2 g/L chloride, the corrosion resistance decreased at higher concentrations and pitting occurred at 20 g/L chloride and 1 g/L copper, while at 50 g/L chloride pitting occurred at all copper concentrations.

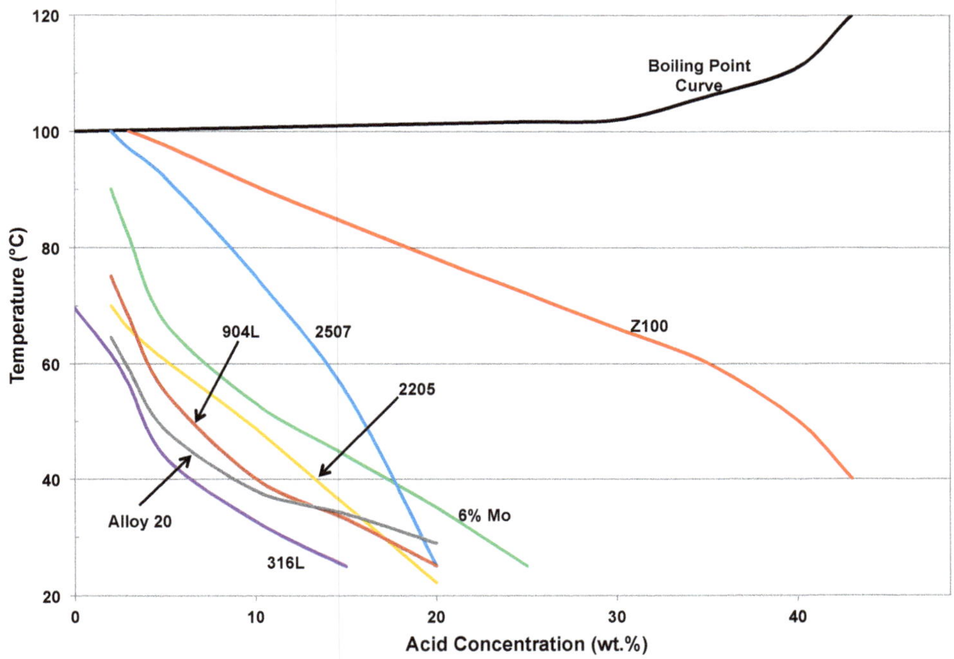

FIGURE 4.7 Iso-corrosion curves (0.1 mm/y) for some stainless steels in sulfuric acid with 2,000 mg/L chloride.

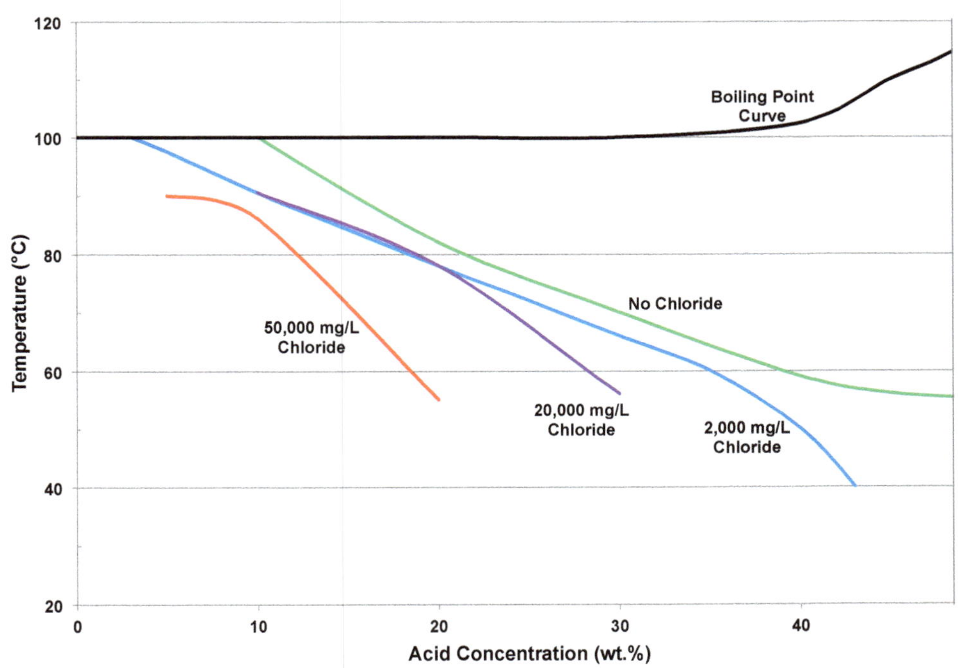

FIGURE 4.8 Iso-corrosion curves (0.1 mm/y) for Z100 superduplex in sulfuric acid with various concentrations of chloride.

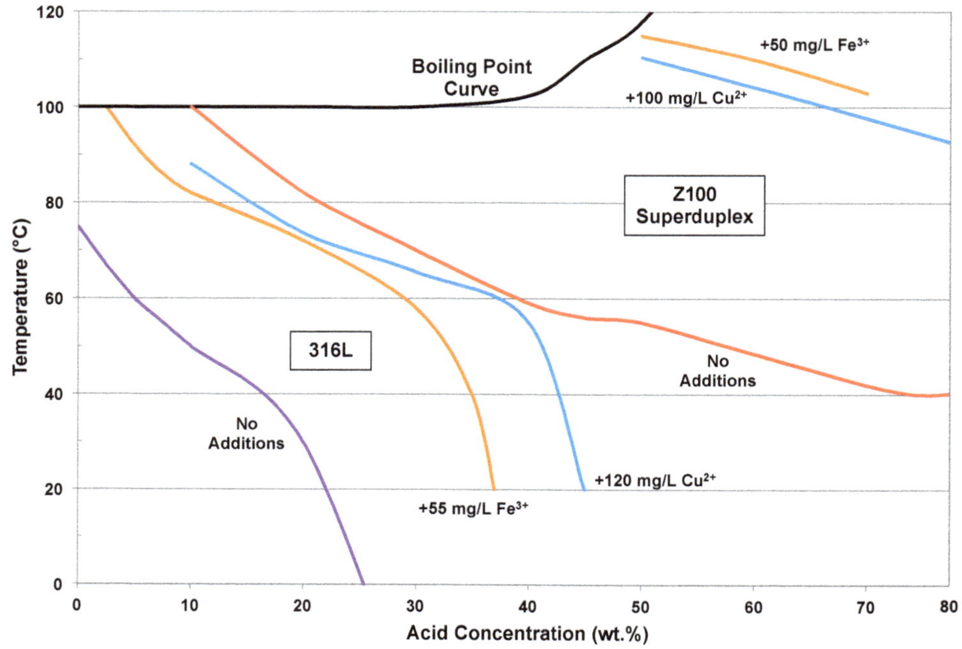

FIGURE 4.9 Effect of ferric and cupric ions on the iso-corrosion (0.1 mm/y) curves for 316L and Z100 superduplex.

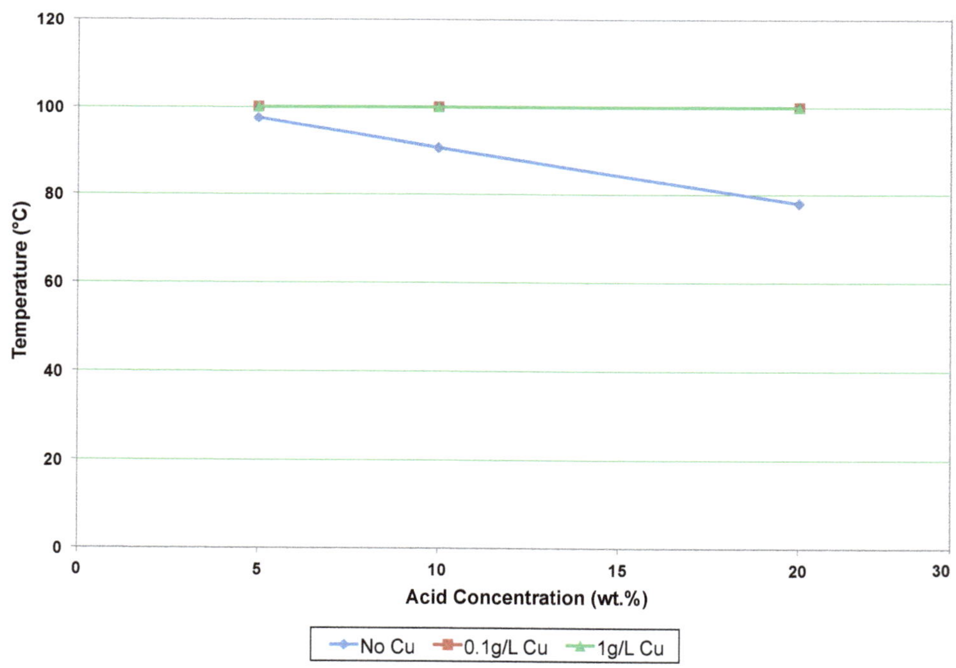

FIGURE 4.10 Effect of copper on the iso-corrosion curve for Z100 superduplex in sulfuric acid plus 2 g/L chloride[2]. Note that the green line is superimposed over the red line.

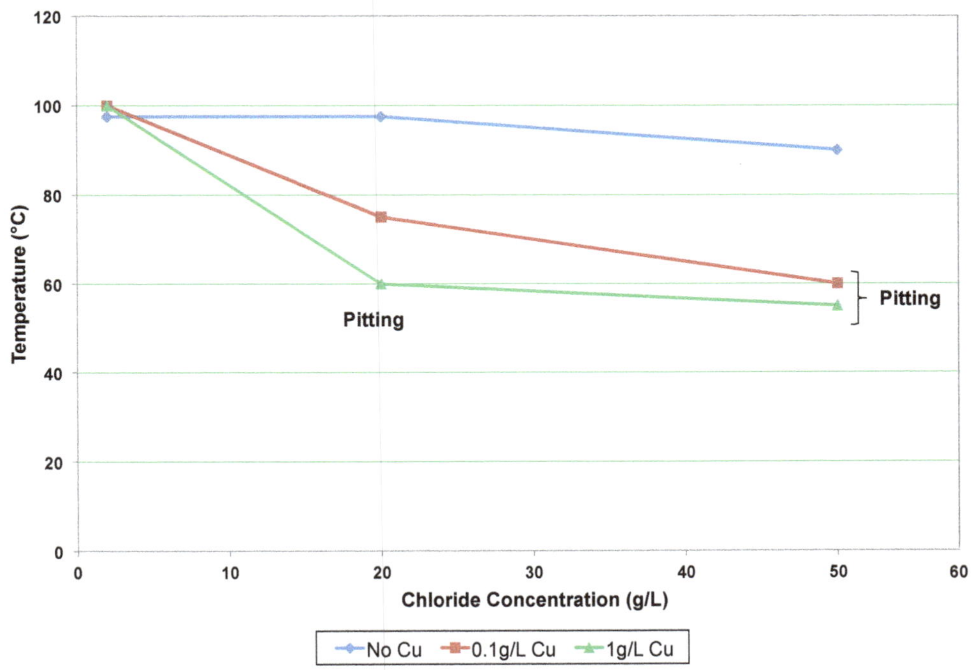

FIGURE 4.11 Effect of copper and chloride on the iso-corrosion curve for Z100 superduplex in 5 wt.% sulfuric acid[2].

Table 4.1 shows the temperatures at which the corrosion rate was 0.1 mm/y as a function of chloride, and copper content for Z100 superduplex in sulfuric acid up to 20 wt.%. The results show that chloride and copper concentrations are more important than acid concentration in this range.

TABLE 4.1 Temperature for 0.1 mm/y corrosion rate in sulfuric acid plus chlorides and copper for Z100 superduplex[2].

ACID CONC. (wt%)	CHLORIDE (g/L)	TEMPERATURE (°C) Copper (g/L)		
		0	0.1	1.0
5	50	90	65	55
	20	97.5	75	60
	2	97.5	100	100
10	50	86	55	55
	20	90.5	80	55
	2	90.5	100	100
20	50	55	55	45
	20	75	80	50
	2	78	100	100

Francis and McIntyre describe a failure of nickel alloy C-276 (UNS N10276) in a raffinate solution (dilute sulfuric acid) containing chlorides and ferric ions at 80°C[3]. The solution composition was

Chloride	0.2 g/L
Copper	13 g/L
Iron	50 g/L
Sulfuric acid	66.6 g/L
Zinc	4.9 g/L
Arsenic	0.4 g/L

The nickel alloy went transpassive and suffered severe pitting. The solution was to use Z100 superduplex, because its high chromium content gives it a wider passive range under these conditions, and there was test data to show it would be suitable.

4.2 HYDROCHLORIC ACID

Because of the high chloride concentration in hydrochloric acid and the fact that it is strongly reducing, stainless steels have limited use, and are confined to low acid concentrations. Hydrochloric acid is so aggressive that only the duplex alloys containing a minimum of 25% chromium can be considered.

Figure 4.12 shows the iso-corrosion (0.1 mm/y) curves for some of the higher alloy stainless steels with 316L austenitic for comparison. The duplex alloys are clearly superior to the 6% Mo alloy, and it might appear that the copper content of the superduplex alloys is having an effect on performance in hydrochloric acid, as

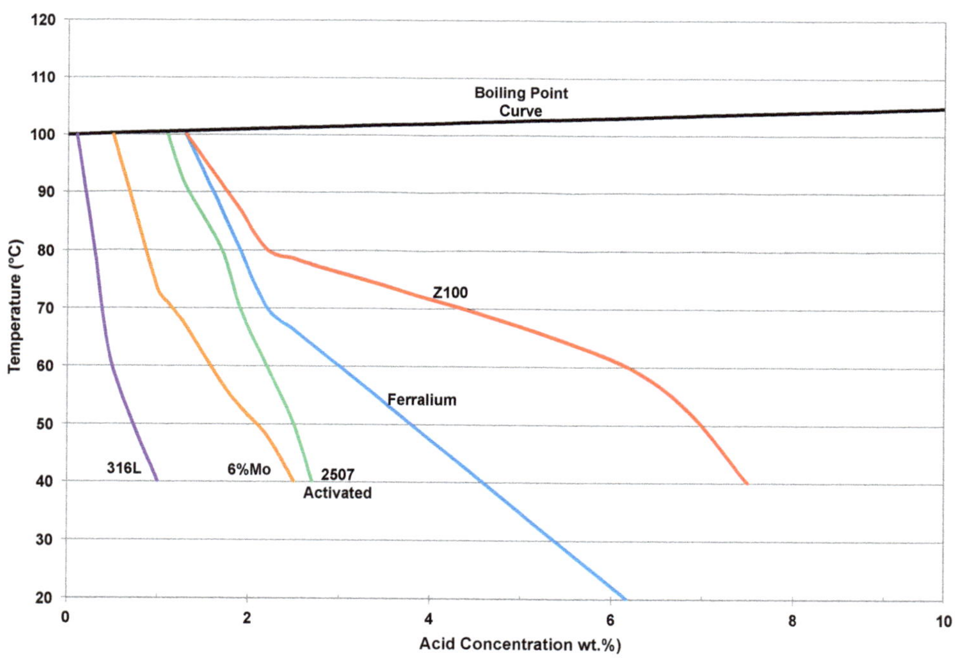

FIGURE 4.12 Iso-corrosion curves (0.1 mm/y) for some stainless steels in hydrochloric acid.

was observed for sulfuric acid, in Section 4.1.1. However, the data for 2507 was determined by first activating the alloy with a zinc rod, while the data for the other alloys was determined by a straight immersion in the test solution. Kivisakk tested several superduplex stainless steels in hydrochloric acid and showed that they had similar corrosion resistance in the activated state[5]. He did not derive an iso-corrosion curve for 2507 in the nonactivated state, but the results reported suggest that the corrosion resistance of 2507 would be similar to that of Z100.

If oxidizers are present, such as ferric ions as found in pickling baths, then the alloys will probably go transpassive and pitting is likely. However, Francis reported the successful use of Z100 superduplex in the spent acid lines of a rutile plant (TiO_2) using the chloride route[6]. The spent acid contained dilute hydrochloric acid at elevated temperature and the alloy performed well. Alloy Z100 was also tried for the hooks lifting carbon steel items in and out of hot, 5% hydrochloric acid pickling tanks, but the alloy suffered pitting and high corrosion rates, as might be expected from the discussion above on the effect of ferric ions.

4.3 NITRIC ACID

Nitric acid is very oxidizing and at higher temperatures and acid concentrations, stainless steels can be moved into the transpassive zone, as discussed in Section 4.1.2. If contaminants such as chlorides are also present, pitting or SCC might occur.

Figure 4.13 shows the iso-corrosion (0.1 mm/y) curve for 316L austenitic stainless steel in pure nitric acid. Alloys that have molybdenum and go transpassive will have high dissolution rates, as discussed in Section 4.2, so alloys high in chromium with low or zero molybdenum will perform best. Hence, the austenitic alloy 310 (UNS S31008) and lean duplex alloy 2304 should outperform 2205 and superduplex in strong, hot nitric acid.

Where nitric acid is present in small quantities with other chemicals, it is best to consider corrosion in the main constituent chemical(s) with the nitric acid being an oxidizer, and moving the potential towards or into the transpassive region. For alloys that would be in the active zone, the presence of nitric acid could be beneficial by moving the alloy into the passive region.

4.4 PHOSPHORIC ACID

Pure phosphoric acid is not particularly aggressive and Figure 4.14 shows the iso-corrosion curves for some austenitic and duplex stainless steels. It can be seen that at acid concentrations greater than 60 wt.% and temperatures exceeding 100°C, the austenitic alloy 904L has greater corrosion resistance than Z100 superduplex. The reason for this is not known. There are some general data for 2304 lean duplex that suggests the performance is similar to that of Z100 superduplex at acid concentrations from 50–86 wt.%[7].

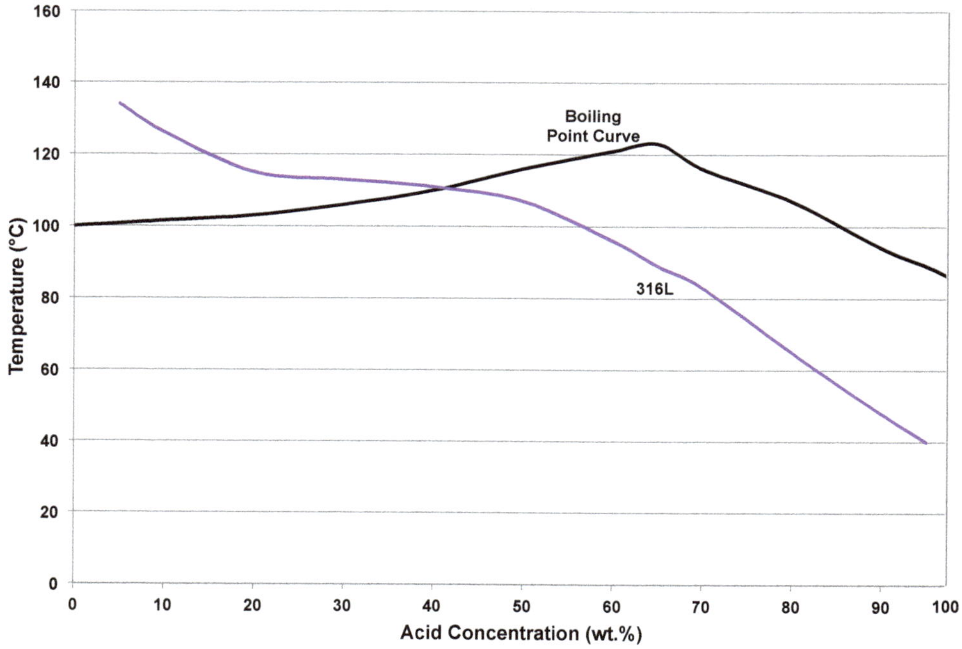

FIGURE 4.13 Iso-corrosion (0.1 mm/y) curve for 316L austenitic in pure nitric acid.

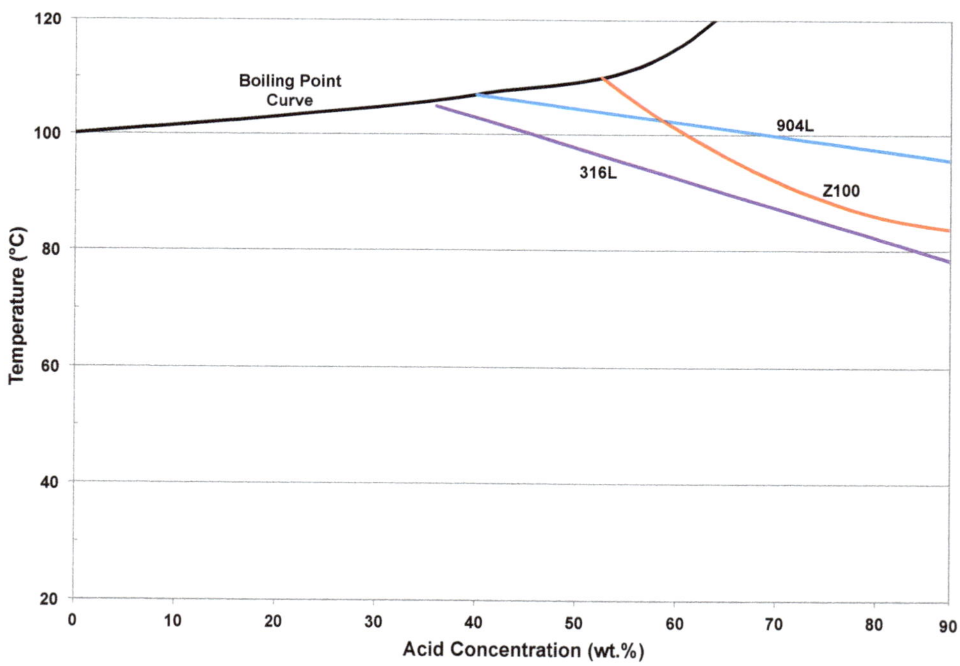

FIGURE 4.14 Iso-corrosion (0.1 mm/y) curve for some stainless steels in pure phosphoric acid.

4.5 COMMERCIAL PHOSPHORIC ACID

Commercial phosphoric acid is a complex fluid that contains many impurities, particularly fluorides and chlorides. Figure 4.15 shows a simplified schematic diagram of the production process. Crushed phosphate rocks and sulfuric acid are added to the primary reactor (~80°C), which is stirred to improve the reaction rate. The reacted slurry then goes to a separator where the gypsum ($CaSO_4$) is separated from the phosphoric acid. The acid is then heated and concentrated to around 54% (as P_2O_5), from where it goes to storage. A detailed description of the process and areas of concern for corrosion are given by Audouard and Soulignac[8].

The ore can contain variable amounts of fluoride and chloride, which may also be present in the water. In addition, there are significant quantities of iron and aluminum, which it has been suggested will reduce the corrosiveness of the mixture. Some data in reference 7 in simulated wet phosphoric acid (WPA) fluids suggests that this is correct. The iron tends to move the stainless steels into the passive zone and the aluminum tends to form complexes with the fluorides.

The composition of phosphate ores varies considerably around the world so simulated WPA fluids can also be quite varied. Hence, there can be a disparity in the results presented by different authors. Table 4.2 shows the composition of simulated WPA fluids by Avesta[7], Audouard and Soulignac[8], and Francis[9]. The solution used in reference 8 was derived by consideration of a number of ores processed in France at the time, while that of reference 9 was derived from a number of ores being processed in the United States, some of which were from North Africa, where chlorides and fluorides are high. The source of the fluid composition in the Avesta work is unknown, but its major difference is the phosphoric acid concentration (54% as P_2O_5), which represents the concentrated acid rather than the more aggressive and more dilute acid produced in the primary reactor. The Avesta publication does not quote corrosion rates, only ranges. However, its general results are in agreement with those in references 8 and 9.

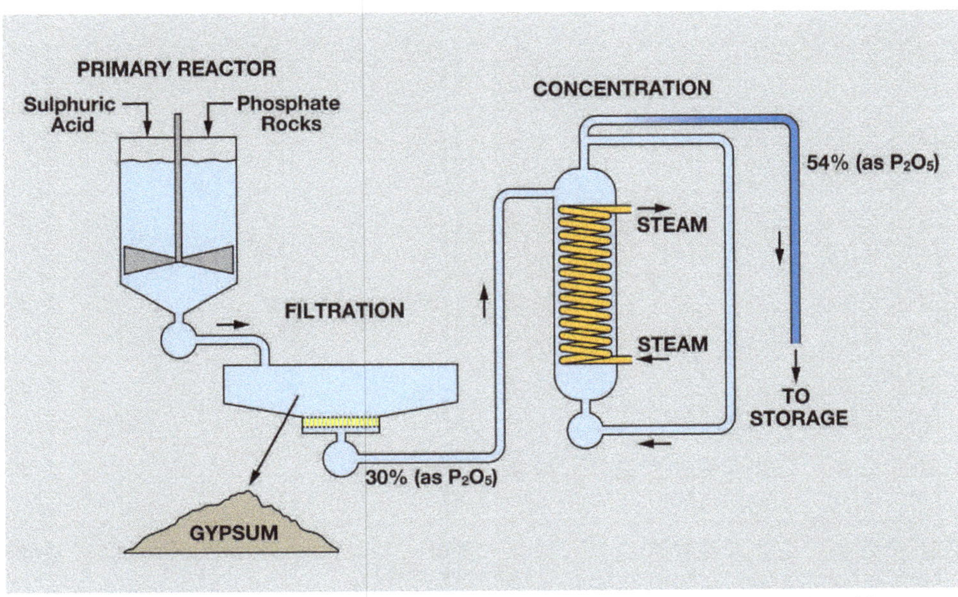

FIGURE 4.15 Simplified schematic diagram of the production of phosphoric acid.

TABLE 4.2 Composition of simulated WPA fluids used in references 7-9.

CHEMICAL	UNIT	CONCENTRATION		
		Ref 7	Ref 8	Ref 9
Phosphoric acid	wt% as P_2O_5	54	30	30
Sulfuric acid	wt%	4	2	3
Fluosilicic acid	wt%	0.2	1.5	2
Fluoride	wt%	0.2-1.0	0.2	0.2-0.8
Chloride	mg/L	200-1,000	600-3,000	300-3,000
Iron	wt%	0.14-0.17	0.35	0.35
Aluminium	wt%	0.1/-0.5	0.26	0.4

Figure 4.16 shows the corrosion rate of some stainless steels as a function of chloride content at 80°C from reference 8. Figure 4.17 shows similar data generated by Francis[9]. It is noticeable that Francis's data appears to show lower corrosion rates than in Figure 4.16, but the only alloy in common between the two is superduplex stainless steel, and there the results are similar. At a chloride concentration of 1,000 mg/L and 0.2 wt.% fluoride, both 2205 and Z100 or 52N+ superduplex are satisfactory. Figure 4.18 shows that at higher fluoride concentrations, superduplex stainless steel is a better choice.

The fluid in the primary reactor will also contain solids and will therefore be abrasive, so alloys with good resistance to erosion resistance are required. Audouard and Soulignac reported the successful use of 52N+ superduplex stainless steel in an agitator in the primary reaction vessel, where the performance was supe-

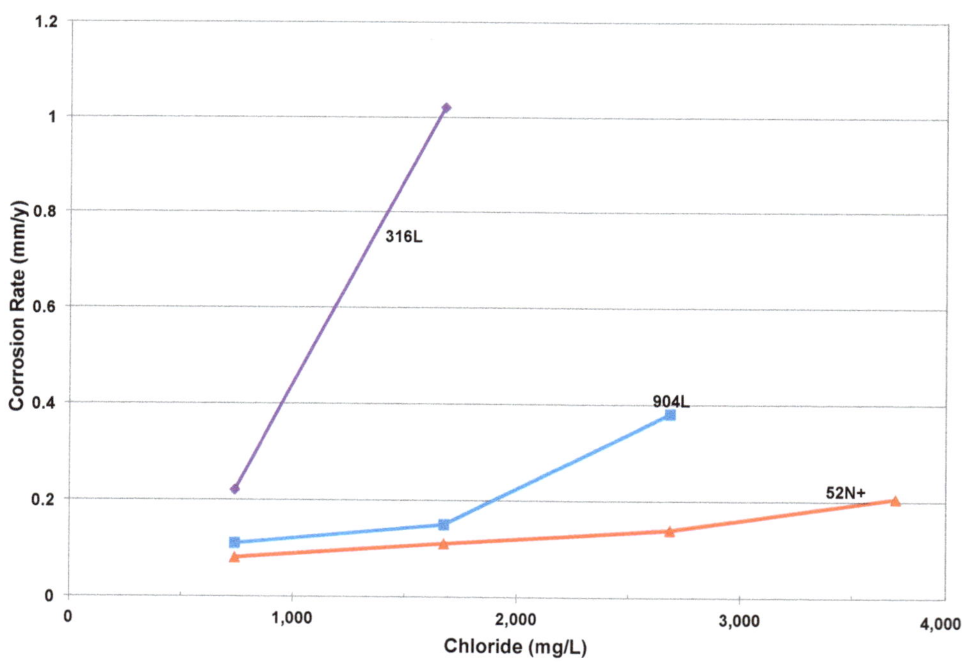

FIGURE 4.16 Corrosion of some stainless steels in simulated WPA fluid as a function of chloride, with 0.2 wt.% fluoride at 80°C[8].

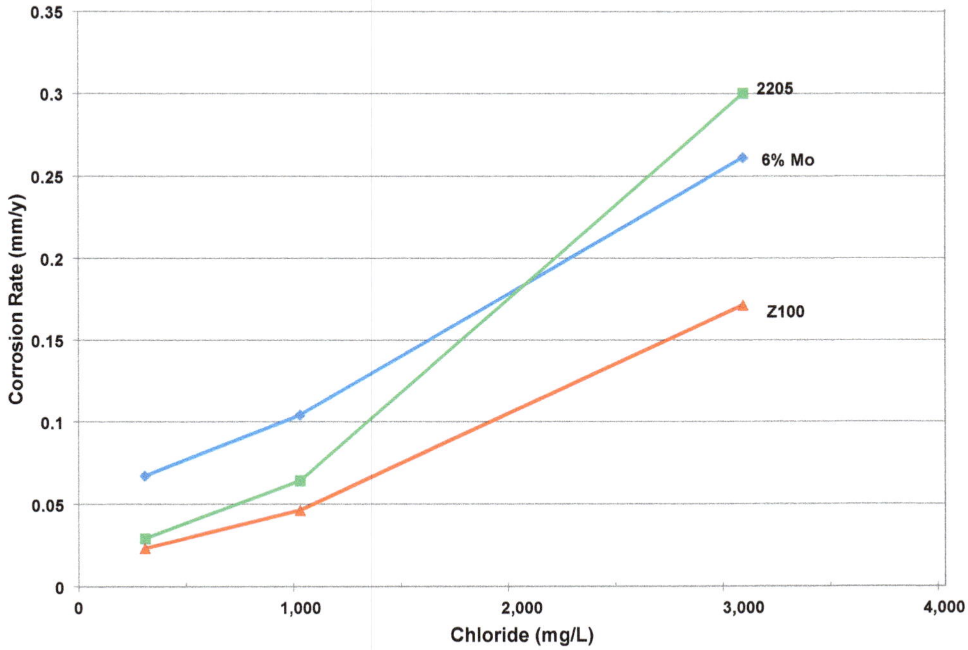

FIGURE 4.17 Corrosion of some stainless steels in simulated WPA fluid as a function of chloride, with 0.2 wt.% fluoride at 80°C[9].

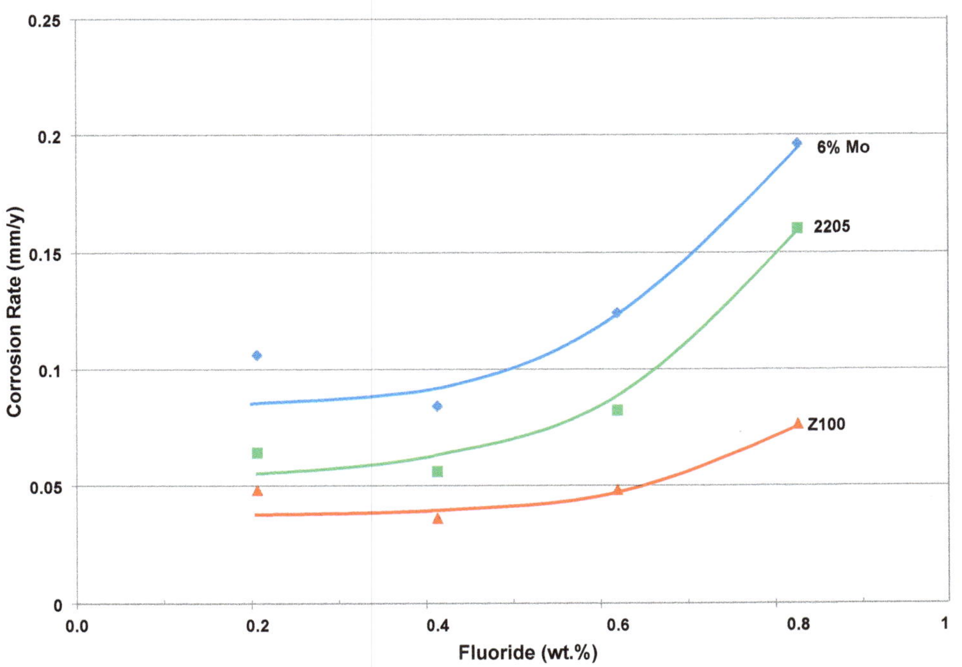

FIGURE 4.18 Corrosion of some stainless steels in simulated WPA fluid as a function of fluoride, with 1,000 mg/L chloride at 80°C[9].

rior to that of both a 6% Mo austenitic alloy and the nickel alloy 625 (UNS N06625)[8]. Francis reported the successful use of Z100 superduplex in phosphate reactor recirculation pumps because of the good resistance to erosion corrosion[9].

In the evaporator, the dilute acid is heated to concentrate it to 54% (as P_2O_5), and not only is the acid less corrosive at this concentration, the chloride and fluoride contents are less than in the primary reactor[8]. Audouard and Soulignac showed the safe chloride concentrations as a function of temperature for some stainless steels in 54% phosphoric acid (Figure 4.19). Unfortunately, they did not include a 6% Mo austenitic or a superduplex alloy. Because these should be more corrosion resistant than Ferralium, they should have some useful corrosion resistance up to ~120°C with low chlorides. This has been confirmed by the successful use of Z100 superduplex for the heating coil in a phosphoric acid concentrator, making food grade acid, which is very low in impurities, including fluoride and chloride.

Although the production of phosphoric acid using sulfuric acid is the most common route, there are some chemical plants that react phosphate ores with nitric acid to produce fertilizers. The reaction produces nitrophosphoric acid and calcium nitrate. The solution is cooled to crystallize the calcium nitrate, which can be separated, while the liquor is concentrated as required. The calcium nitrate may be reacted with ammonia and carbon dioxide to produce calcium ammonium nitrate, which is also used as a fertilizer. Although there appears to be no data on the performance of duplex stainless steels in phosphoric acid with nitric acid, the corrosion rate might be expected to be better than with sulfuric acid, because the nitric acid will ensure the stainless steels are in the passive zone.

Francis reported the successful use of Z100 superduplex for recirculation pipework on the primary reactor in a Scandinavian nitro-phosphate plant because of its good resistance to erosion corrosion, while 316L had failed[9]. The first stage sedimentation tank was also in superduplex stainless steel, because of its resistance to erosion corrosion.

4.6 ORGANIC ACIDS

4.6.1 Carboxylic Acids

The most common organic acids used commercially are the carboxylic acids and the shorter the chain, the more aggressive is the acid. Hence, formic acid is the most aggressive followed by acetic acid and then propionic acid.

Figure 4.20 shows the iso-corrosion curves for some stainless steels in pure formic acid. It is clear that the corrosion resistance of the lean duplex 2304 is similar to that of 904L austenitic and it is also greater than that of 316L. Superduplex works very well in all concentrations up to high temperatures.

Pure acetic acid is less corrosive than formic acid, and the low alloy stainless steels can handle it up to the boiling point. However, in commercial processes acetic acid is often present with other chemicals that can increase corrosion rates.

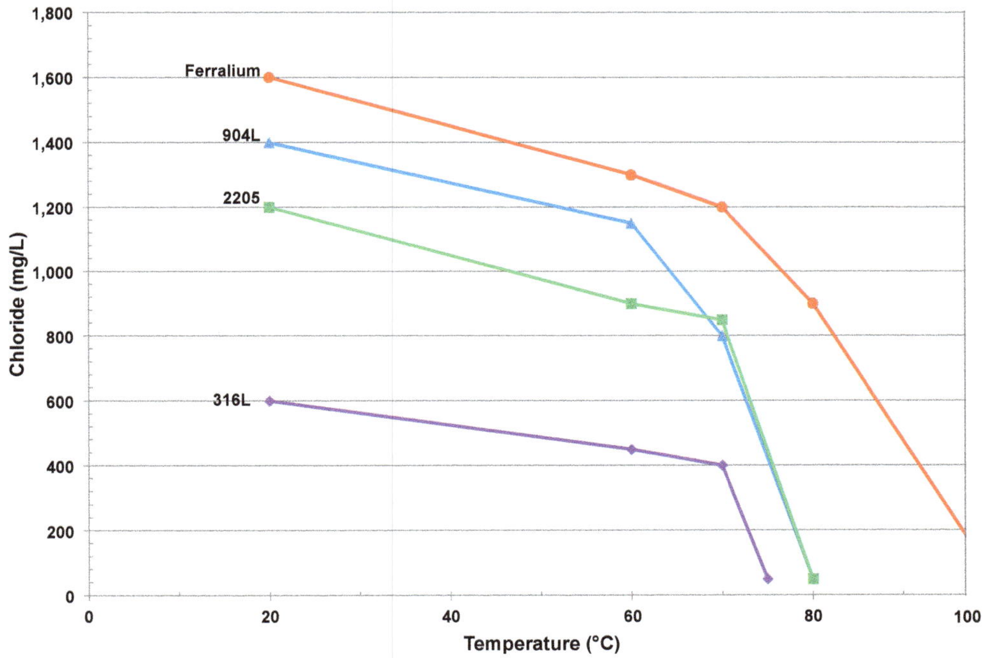

FIGURE 4.19 The temperature/chloride limits for some stainless steels in 54% phosphoric acid (as P_2O_5) [8].

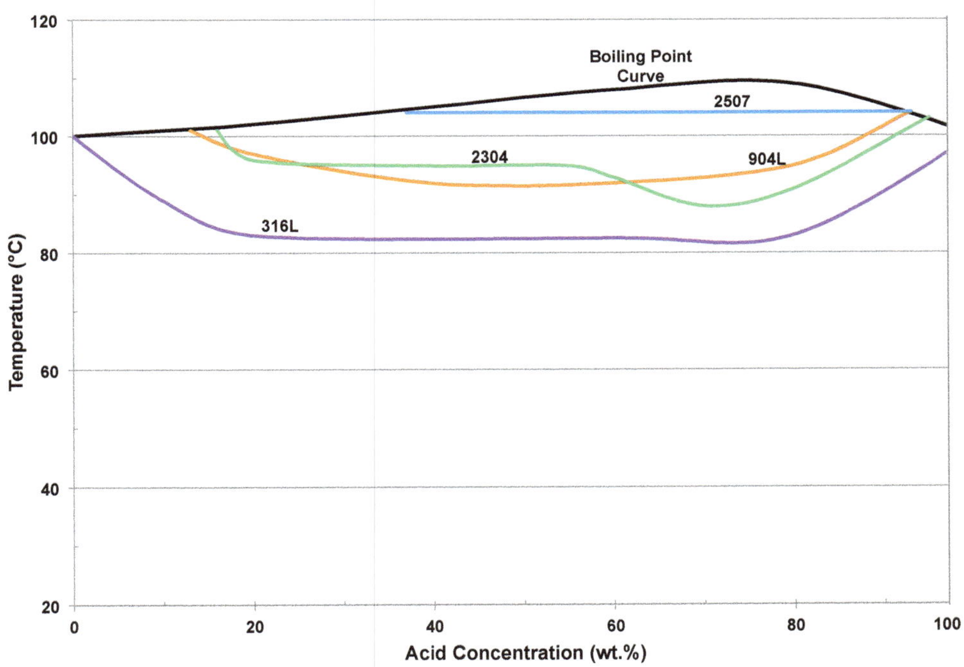

FIGURE 4.20 Iso-corrosion curves (0.1 mm/y) for some stainless steels in pure formic acid.

In the production of acetic acid, it is common for some formic acid to also be produced. Figure 4.21 shows the resistance of some stainless steels to boiling 50 wt.% acetic acid with various concentrations of formic acid. There were no data for 904L, but 317L (UNS S31703; Fe/18Cr/13Ni/3Mo) is a slightly lower alloyed stainless steel that has been used in plants producing acetic acid. The data show that 2205 has useful corrosion resistance in these mixtures and superduplex stainless steel was totally unattacked at all formic acid concentrations up to 25 wt.%.

Francis reported the results of tests on some alloys in a United Kingdom acetic acid plant in the following fluid[4]:

Acetone	2%	Water	19%
Formic Acid	10%	Acetic Acid	42%
Propionic Acid	8%		
Temperature	188°C	Pressure	680 psi

There had been corrosion problems with valves in nickel alloys C-276 (UNS N10276) and alloy 20 (UNS N08020), and the following corrosion rates were measured after extended plant exposure under oxidizing conditions:

Alloy C-276	0.100 mm/y
Alloy 20	0.075 mm/y
Z100	0.045 mm/y

The Z100 superduplex stainless steel valves subsequently performed well. When acetic acid is produced by oxidation processes, peracids or peroxides can be formed, which although unstable, can profoundly affect the corrosion behavior of many alloys. Testing is advised when selecting alloys for these kinds of conditions.

Acetic acid is widely used as a feedstock when manufacturing other chemicals, especially plastics. One example is the manufacturing of terephthalic acid (TPA) and dimethyl terephthalate (DMT). Francis reported corrosion tests in 80 wt.% acetic acid containing chemicals used as catalysts at 90°C[4]. Copper is an oxidant so helps to keep stainless steel in the passive zone. The results in Figure 4.22 show the superior resistance of superduplex stainless steel compared with both 2205 and 6% Mo austenitic. Kivisakk and Nicolio show similar data in 80 wt.% acetic acid at 90°C with 2 g/L chloride (Figure 4.23)[10].

A UK chemical company carried out autoclave tests simulating some of the steps in TPA/DMT production[4]. The results in Table 4.3 show a significantly reduced corrosion rate for Z100 superduplex compared with 317L stainless steel (UNS S31703). Kivisakk and Nicolio showed similar data for 2205 and 2507 under similar conditions[10], again showing the excellent resistance of duplex stainless steels to corrosion in this environment.

TABLE 4.3 The results of corrosion tests in simulated TPA/DMT fluids[4].

ALLOY	CORROSION RATE (mm/y)	
	Fluid 1*	Fluid 2+
317L	0.44	0.67
Welded 317L	0.47	0.68
Z100	0.004	0.011
Welded Z100	0.006	0.016

* Fluid 1: 75% 75 wt% Acetic acid with traces of Br, Cu, and Mn at 175°C
\+ Fluid 2: 75% 96 wt% Acetic acid with traces of Br, Cu, and Mn at 150°C

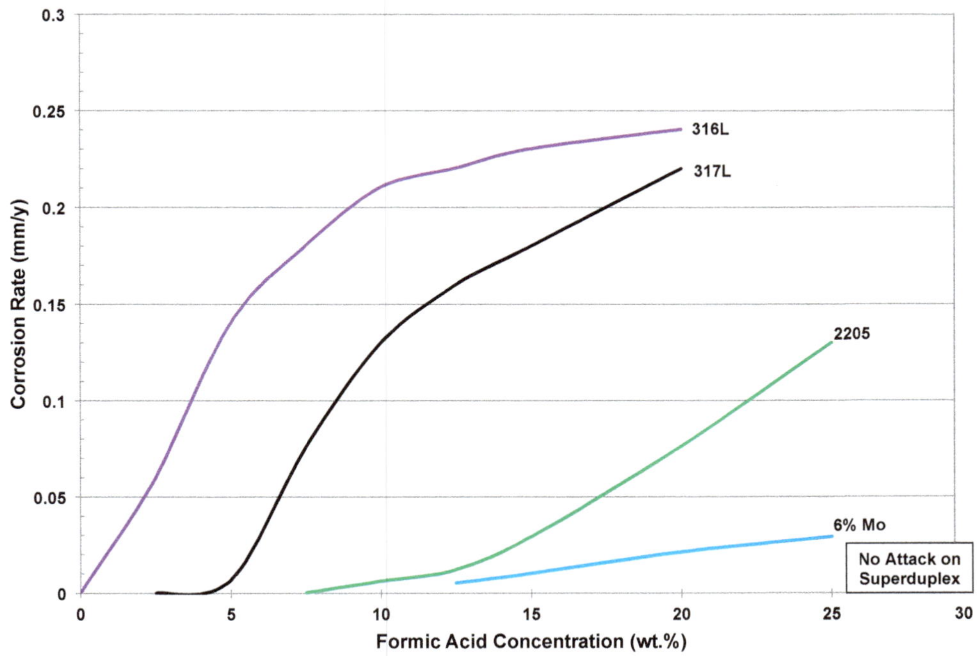

FIGURE 4.21 Iso-corrosion curves (0.1mm/y) for some stainless steels in boiling 50 wt% acetic acid as a function of the formic acid content.

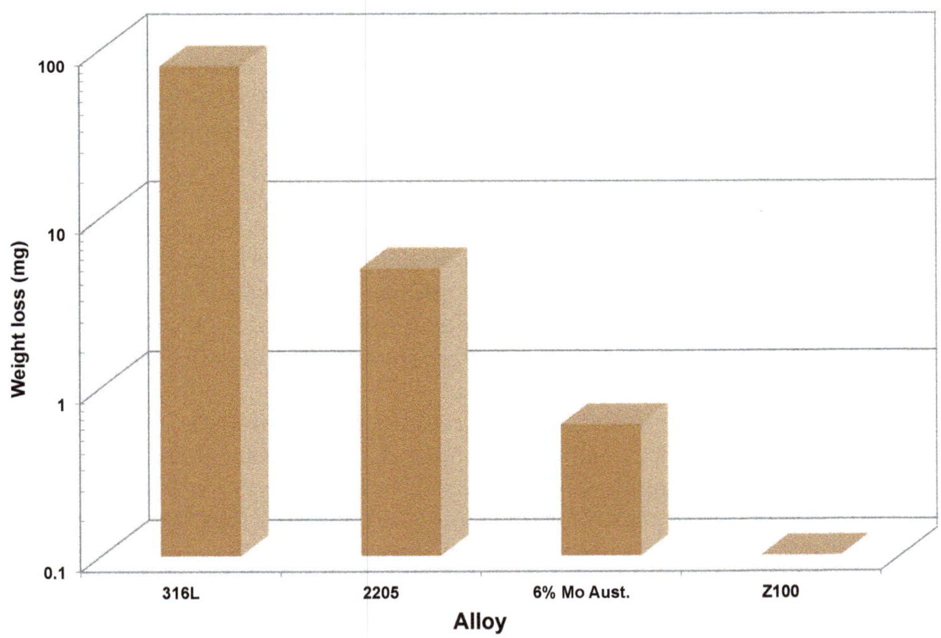

FIGURE 4.22 Weight loss of some stainless steels in 80 wt.% acetic acid plus 1.5 g/L bromide and 0.5 g/L manganese at 90°C[4].

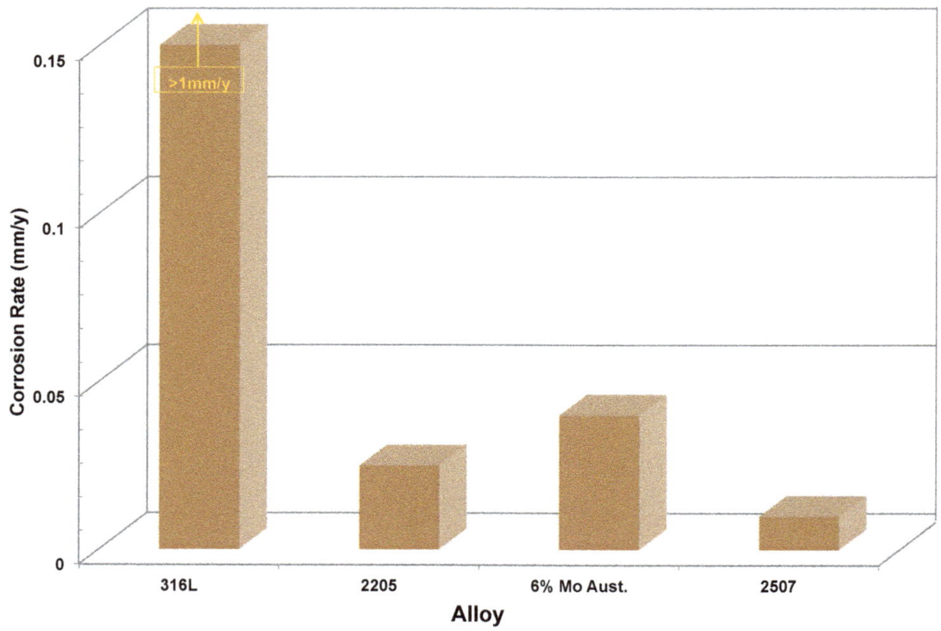

FIGURE 4.23 Corrosion of some stainless steels in 80 wt.% acetic acid with 2 g/L chloride at 90°C[10].

The carboxylic acids with longer chains, such as propionic acid, are far less aggressive and are mostly found in small quantities as process byproducts, as described above. The chloroacetic acids (monochloroacetic acid, dichloroacetic acid, etc.) are stronger than acetic acid; 100 times for mono chloroacetic acid and 3,000 times for dichloroacetic acid. However, corrosion data in these acids is hard to find. There is a little data in reference 7, which suggests that 2507 will suffer crevice corrosion in 80% monochloroacetic acid at 60°C, and will corrode at >1 mm/y at 80°C at the same concentration.

The hydroxycarboxylic acids, such as citric, tartaric, and lactic acids are not very aggressive and only cause corrosion of low alloy stainless steels at high temperatures and concentrations[7]. Hence, they can usually be handled as industrial fluids by 316L or one of the lean duplex stainless steels. These acids can become more aggressive at elevated temperatures if oxidizers are present and/or halides. There is little published data under such conditions, and corrosion testing is advised before selecting alloys for such applications.

One organic acid that can be very aggressive is oxalic acid. It is often used at ~10 wt.% concentration to electrolytically etch stainless steels under anodic polarization. Unfortunately, there is no published data for the detailed performance of duplex stainless in oxalic acid, but reference 7 contains iso-corrosion curves for 316L and 904L. The limited information for 2205 and 2507 suggests that they are both satisfactory at 80°C up to the solubility limit, while 904L is a little better, presumably because of its higher nickel content[7].

4.6.2 Sulfonic Acids

The sulfonic acids are strong acids and are also reducing, but less so than hydrochloric acid. The more commonly encountered types include alkane sulfonic acids such as methyl sulfonic acid. The sulfonic acids have many uses in manufacturing phenols, dyes, electronic cleaning chemicals, drugs, soaps, detergents, etc. Methylsulfonic acid is also used as a catalyst for esterification, alkylation, in the production of peroxycarboxylic acids, and the reduction of amides. Despite this widespread industrial use, there is no published data on the performance of duplex stainless steels in sulfonic acids. There is some data for a limited range of austenitic stainless steels, nickel alloys, and refractory metals[11].

In industry, the maximum concentration of methylsulfonic acid is ~25% with a maximum temperature of 95°C[11]. Gaur and Srinivasan confined most of their testing to these conditions[11]. Table 4.4 shows the corrosion rate of some austenitic stainless steels and nickel alloys in 25% methylsulfonic acid at 95°C, with some additional data under the same conditions plus 1.5 g/L ferric chloride[11].

TABLE 4.4 Corrosion rate of some stainless steels and nickel alloys in 25% methyl sulfonic acid at 95°C[11].

ALLOY	UNS No.	CORROSION RATE (mm/y)	
		Pure Acid	Acid + 1.5 g/L $FeCl_3$
316L	S31603	2.27	0.17
904L	N08904	1.78	-
825	N08825	1.27	-
625	N06625	1.02	-
C-22	N06022	0.002	0.005
G-30	N06030	0.0005	0.002

Only two highly alloyed nickel alloys showed low corrosion rates in pure acid. Because these were both high chromium alloys, it would be interesting to obtain data for superduplex stainless steel and alloy 28 (UNS N08028: Fe/27Cr/32Ni/3Mo/1Cu), which would be more cost effective if their corrosion rates were low enough.

As with other acids in this chapter, the addition of oxidizing chemicals improves alloy performance, as shown in Table 4.4. The presence of chlorides increases the risk of pitting and crevice corrosion, particularly if oxidizers are also present in significant quantities. Gaur and Srinivasan pointed out that the alloys that corroded in sulfonic acid also showed a propensity to suffer crevice corrosion and/or SCC. Because of the paucity of corrosion data in the sulfonic acids, it is strongly recommended that corrosion tests be conducted before selecting alloys and these should include general corrosion, crevice corrosion, and SCC.

4.7 UREA

Urea is increasingly popular as a fertilizer because of its high nitrogen content compared with ammonium nitrate. It is produced by reacting ammonia and carbon dioxide together at high temperature, which creates ammonium carbamate as an intermediate product. The carbamate is then decomposed to give urea and water. Unconverted carbamate is removed and recycled into the reactor.

Urea is not very corrosive and is easily handled by low alloy stainless steels, but ammonium carbamate is corrosive and it can cause severe corrosion of urea grade 316L at elevated temperatures in the recycle lines and carbamate heat exchanger[12]. Notten describes the successful use of 2205 to replace 316L that was suffering severe corrosion, particularly where carbamate was condensing[12]. This was confirmed by Baumert, who also reported the successful use of 2205 to combat corrosion by ammonium carbamate, which was produced as a byproduct during the production of alkyl amines[13].

Notten describes the use of the Streicher test, ASTM A262 method B (boiling ferric sulfate and sulfuric acid) to prove resistance to carbamate[12]. The pass criteria include a maximum weight loss and a maximum penetration of intergranular corrosion. The criteria for 2205 and 25% Cr duplex are shown in Table 4.5, plus that for Safurex (UNS S32906), a special high chromium, low molybdenum grade developed for aggressive urea service. Table 4.5 also shows the author's unpublished data for Z100 superduplex in this test. It can be seen that superduplex more than meets the criteria for Safurex and should be much more readily available than this special grade.

TABLE 4.5 Requirements for duplex stainless steels in the Streicher test to resist ammonium carbamate attack[12].

ALLOY	MAX. WT. LOSS (g/m^2h)	MAX. DEPTH OF ATTACK (μm)
Requirement for 2205	1.6	100
Requirement for 25% Cr	0.9	100
Requirement for Safurex	0.7	100
Wrought Z100	0.20	12
Cast Z100	0.27	32
GTA Welded Z100	0.38	26

4.8 ALKALIS

The most common alkaline chemical is sodium hydroxide (caustic soda), although potassium hydroxide is sometimes used in some processes. There is much corrosion data for caustic soda and what there is for potassium hydroxide suggests it behaves similarly.

There is not a lot of corrosion data for duplex stainless steels in caustic soda, compared with, say, sulfuric acid. The general corrosion resistance of stainless steels in caustic soda is very good, particularly at low concentrations or lower temperatures.

Figure 4.24 shows the iso-corrosion curves (0.1 mm/y) for some common duplex stainless steels and 316L austenitic. The data for 316L is taken from Sedriks[14], while that for 2205 is taken from Turnbull et al.[15]. The superduplex data is a combination of the author's unpublished data and that of Turnbull et al.[15]. It can be

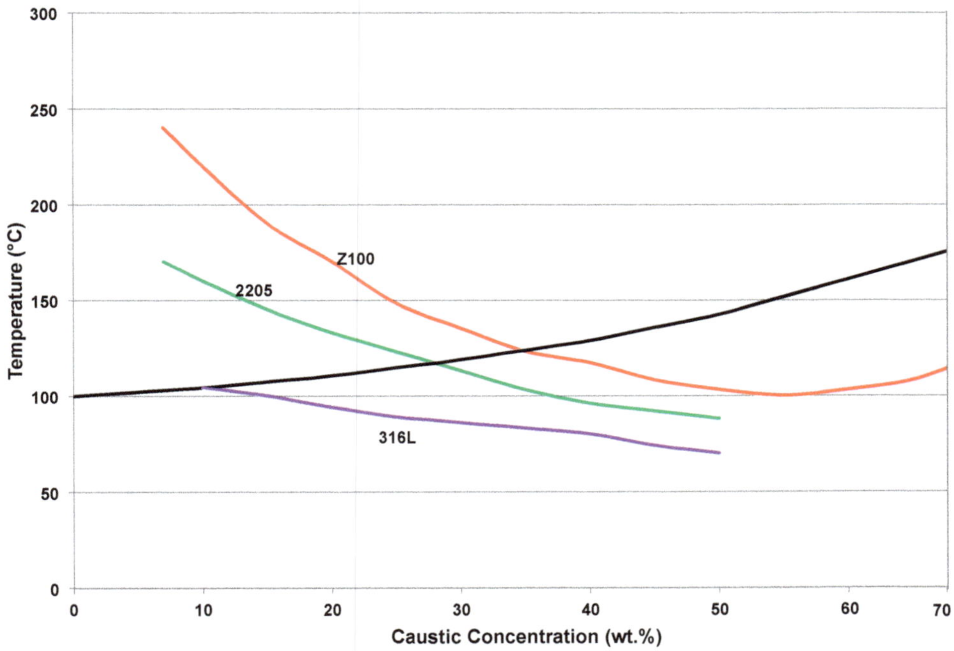

FIGURE 4.24 Iso-corrosion curves (0.1 mm/y) for some stainless steels in caustic soda.

seen that the corrosion resistance increases with increasing chromium content, particularly in dilute solutions. Above about 50 wt.% caustic, the corrosion resistance of superduplex stainless steel increases, and this is believed to be due to the very low water content in these solutions. Note that very few commercial operations store or use caustic solutions at concentrations greater than 50%. Figure 4.24 shows that 2205 has good resistance to caustic solutions at moderate temperatures up to around 20 wt.% and the petrochemical industry has made wide use of 2205 in caustic wash lines for piping, vessels, and heat exchangers, with superduplex being used at higher temperatures and concentrations (see Chapter 1).

The majority of the data on superduplex stainless steels in this section were generated using alloy Z100 (UNS S32760), but the limited data for 2507 (UNS S32750) suggests that it performs more or less the same as Z100, and because of their very similar compositions, there is no reason to think it would be significantly different.

One factor of note is that crossing the iso-corrosion line in caustic soda does not have the dramatic effects on corrosion rates that are seen, for instance, in sulfuric acid. Figure 4.25 shows the corrosion rate of Z100 superduplex in 75 wt.% sulfuric acid and 30 wt.% caustic soda. An increase of just 10°C in temperature results in a 300-fold increase in the corrosion rate in sulfuric acid, while a 40°C increase in temperature increases the corrosion rate by only a factor of 10 in 30 wt.% caustic soda. The reason for this is believed to be due to the nature of the passive film. In acid it is essentially a chromium oxide, while in caustic soda, the nickel content is also important[16].

There is little corrosion data for the lean duplex stainless steels in caustic soda, but because of their low nickel content, and the importance of nickel in controlling the corrosion rate, it is not safe to assume the corrosion rate will be similar to that of 316L. One data point for 2101 in 50 wt.% caustic soda suggests that a corrosion rate of 0.1 mm/y occurs at ~80°C, only a little below that for 2205. Corrosion testing is advised if the use of lean duplex stainless steels in caustic soda is contemplated.

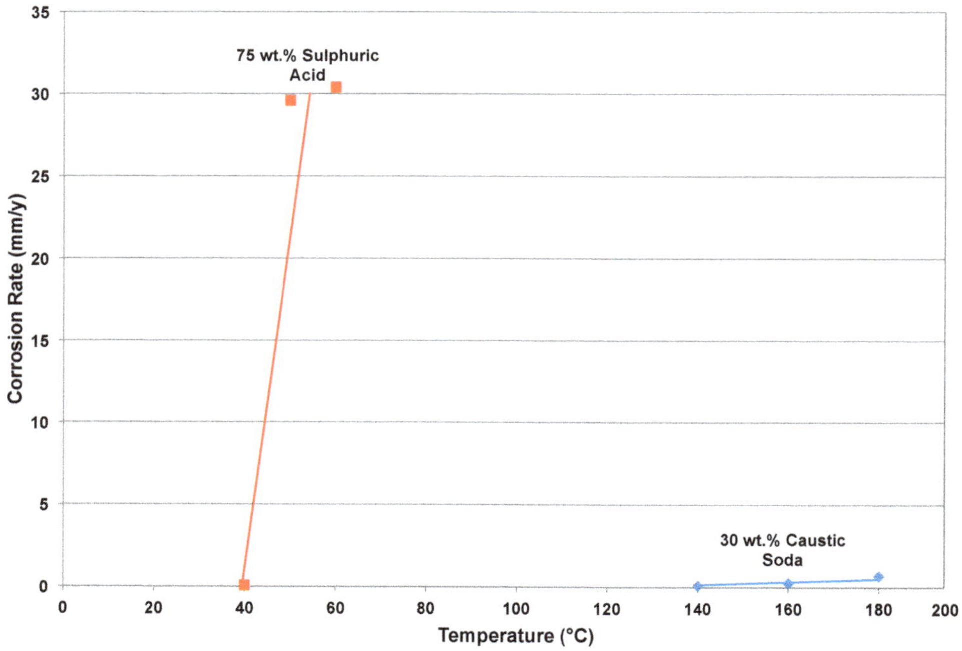

FIGURE 4.25 Comparison of the effect of temperature on corrosion in sulfuric acid and caustic soda.

The degree of aeration has a significant effect, as is shown in Table 4.6, where the corrosion rate of Z100 is shown in two caustic solutions in the presence of either pure nitrogen or pure oxygen. In 10% caustic at 200°C, the corrosion rate increases in the presence of oxygen, while in 50% caustic at 100°C, the corrosion rate decreases with oxygen. The reason for this is the shape of the polarization curve for stainless steels in caustic soda, which is similar to that in acid, as shown in Figure 4.5. In 10% caustic, oxygen moved the potential from the passive to the transpassive zone, thereby increasing the dissolution rate. In 50% caustic, the oxygen moved the potential from the active to the passive region, thereby reducing the corrosion rate. This also has implications for caustic SCC, as discussed in Chapter 6.

TABLE 4.6 The effect of oxygen on the corrosion of Z100 superduplex stainless steel in caustic soda.

CAUSTIC CONC. (wt.%)	TEMPERATURE (°C)	CORROSION RATE (mm/y)	
		Nitrogen	Oxygen
10	200	0.013	0.19
50	100	0.10	0.058

In solutions of caustic that are only mildly oxidizing or neutral, and at modest temperatures, the corrosion of duplex stainless steels is more or less uniform. However, in hot, oxidizing caustic soda the corrosion of duplex stainless steels begins as selective corrosion of the ferrite, particularly along phase boundaries (Figure 4.26).

Other oxidizers can also alter the corrosion rate of stainless steels in caustic soda. In the chloralkali process chlorate, which is a strong oxidizer, is often produced as a by-product of chlorine. Figure 4.27 shows the ef-

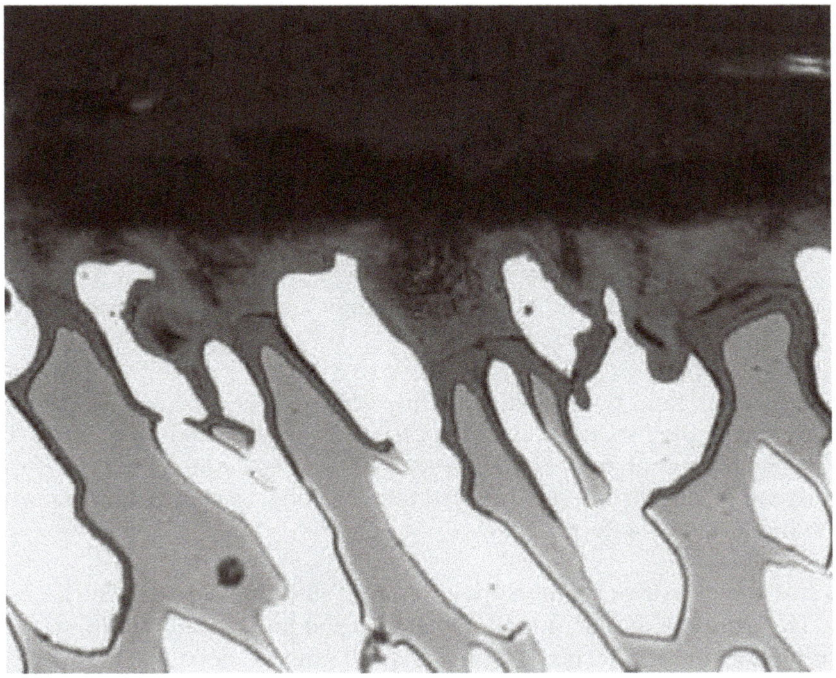

FIGURE 4.26 Selective corrosion of ferrite of Z100 superduplex in 20% NaOH at 160°C with 10 bar oxygen.

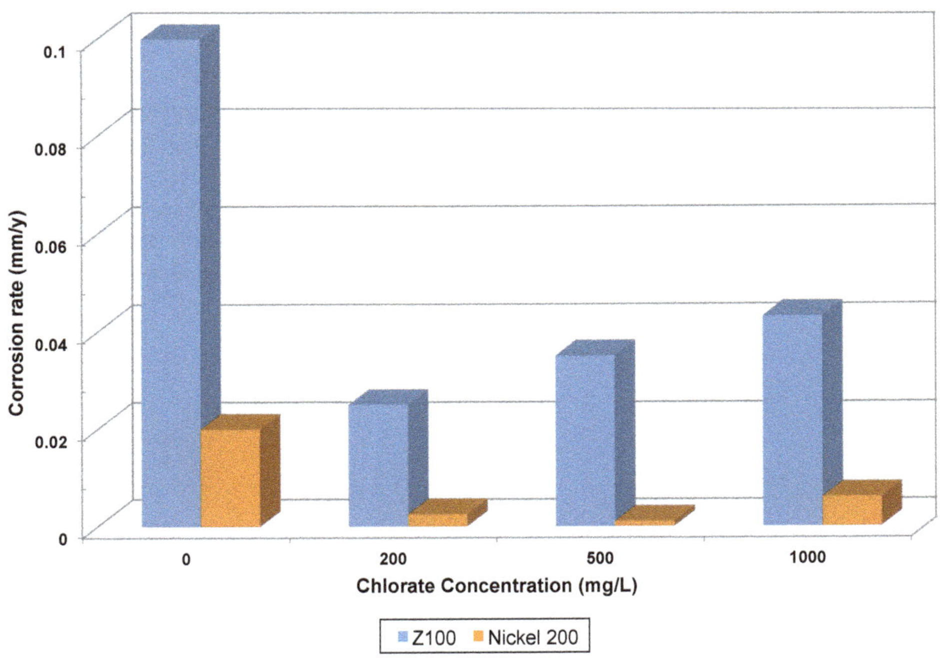

FIGURE 4.27 Effect of chlorate on the corrosion of Z100 superduplex and nickel 200 in 50 wt% caustic plus 7% NaCl at 100°C.

fect of chlorate concentration on the corrosion rates of nickel and Z100 superduplex in 50 wt.% caustic plus 7% sodium chloride at 100°C. The sharp reduction in corrosion rate with chlorate additions is because the potential has moved from the active to the passive region. The corrosion rate increases with further chlorate additions because the current density in the passive region increases slightly with potential (see Figure 4.5). Although the corrosion rate of nickel 200 is lower than that of superduplex, the results show that superduplex would be perfectly acceptable in such fluids, i.e., corrosion rate < 0.1 mm/y.

Chloride has no effect on the corrosion of duplex stainless steels up to 40 wt.% caustic, but in 50% caustic, the corrosion rate increased by a factor of three with 3% sodium chloride at 120°C[15]. Under these conditions, the corrosion rate in pure caustic exceeds 0.1 mm/y and the film is only semi-protective; this may explain the effect of chloride under these conditions.

The other common alkaline fluid is ammonium hydroxide and this is far less aggressive than caustic soda. Low alloy austenitic and duplex stainless steels can handle concentrated ammonium hydroxide up to the boiling point. If oxidizers and/or halides are present at elevated temperatures, then lower alloy stainless steels may go transpassive and/or pit, as discussed above. Under such conditions, corrosion testing is advised to determine the most cost effective alloy.

In the Kraft pulp and paper process, wood chips are heated with caustic soda, which produces a fluid known as "white liquor". This contains sulfide, which means that the hot caustic solution is reducing. During processing, the white liquor converts to black liquor. Typical compositions for simulated liquors are 150g/L NaOH + 50g/L Na_2S (white liquor), and 15g/L NaOH + 50g/L Na_2S (black liquor).

Carbon steel was used originally for Kraft digesters, but as fluids became more aggressive, corrosion and especially caustic SCC became a problem. This was solved with duplex stainless steels, initially 2205 and later with lean duplex 2304 (see Chapter 6).

The black liquor goes to flash tanks where the liquid is evaporated to leave the solids. These tanks were initially carbon steel, but increasingly aggressive solutions caused failures by high rates of corrosion and/or cracking. Clarke and Stead[17] tested several stainless steels in two commercial flash tanks with lower concentrations of caustic and sulfide than above, but with ~1.1 g/L NaCl and 3–6 g/L of sodium thiosulfate at pH 12.3. Samples were exposed for periods exceeding 100 days at ~120°C. In the more aggressive fluid samples, 304L and 316L suffered mild corrosion and some shallow pitting, while 2205 showed no corrosion or pitting. They concluded that 316L was marginal in this service and duplex stainless steel was the best choice; 2205 is now the standard alloy for flash tanks.

REFERENCES

1. R. Cottis and R. Newman, "Stress Corrosion Cracking Resistance of Duplex Stainless Steels," Offshore Technology Report OTH 94 440 (Bootle, United Kingdom: UK Health and Safety Executive, 1994).
2. S. Clarke and R. Francis, *The Selection of Materials for Mineral Processing Operations* (Houston, TX: NACE International, 2016).
3. R. Francis and D.R. McIntyre, "The Importance of Redox Potential on Corrosion Control," *MP*, 55, 5 (2016): p. 48.
4. R. Francis, "Zeron 100 Superduplex Stainless Steel in the Chemical and Process Industries," Rolled Alloys Technical Report TN 1387R3 (Temperance, MI: Rolled Alloys, 2008).
5. U.H. Kivisakk, "Significance of Activation of Test Specimens for the Determination of Iso-Corrosion Curves of Stainless Steels," *CORROSION*, 61, 6 (2005): p. 602.
6. R. Francis, "Zeron 100 Superduplex Stainless Steel in the Mining Industry," Rolled Alloys Technical Report TN 1346R4 (Temperance, MI: Rolled Alloys, 2008).
7. Avesta Sheffield, *Corrosion Handbook for Stainless Steels* (Stockholm, Sweden: Avesta Sheffield AB, 1994).
8. J.P. Audouard and P. Soulignac, "Duplex Stainless Steels for Phosphoric Acid Plants; Corrosion Properties and In-Service Experiences," Duplex Stainless Steels '91 (Les Ulis Cedex A, France: Les editions de physique, 1991): p. 1121.
9. R. Francis, "The Use of Zeron 100 Superduplex Stainless Steel in Phosphatic and Urea-Based Fertiliser Plants," Rolled Alloys Technical Report TN 1105R3 (Temperance, MI: Rolled Alloys, 2009).
10. U. Kivisakk and C. Nicolio, "Properties of Duplex Stainless Steels in Organic Acids," Duplex America 2000 (Zutphen, Netherlands: KCI Publishing, 2000): p. 379.
11. B. Gaur and H.S. Srinivasan, "Corrosion of Metals and Alloys in Methane Sulfonic Acid," *Brit. Corr. J.* 34, 1 (1999): p. 63.
12. G. Notten, "Application of Duplex Stainless Steel in the Chemical Process Industry," Vol. I, Duplex '97 (Zutphen, Netherlands: KCI Publishing, 1997): p. 9.
13. K.L. Baumert, "Resistance of Duplex Stainless Steels to Various Organic Acids," Duplex America 2000 (Zutphen, Netherlands: KCI Publishing, 2000): p. 61.
14. A.J. Sedriks, *Corrosion of Stainless Steels*, 2nd edition (New York, NY: J. Wiley and Sons, 1996).
15. A. Turnbull, A. Griffiths, and T. Reid, "Corrosion and Stress Corrosion Cracking of Duplex Stainless Steels in Caustic Solutions at Elevated Temperatures," *Stainless Steel World* (December 1999): p. 63.
16. S. Clarke, "Stress Corrosion Cracking of Stainless Steels in Oxygenated Caustic Environments," CORROSION 2000, paper no. 596 (Houston, TX: NACE International, 2000).
17. S. Clarke and N.J. Stead, "Corrosion Testing in Flash Tanks of Kraft Pulp Mills," *MP*, 38, 11 (1999): p. 62.

5 Pitting and Crevice Corrosion

Pitting occurs when there is a local breakdown of the passive film on a stainless steel and the local conditions are too aggressive for repassivation to occur. This is usually triggered by an aggressive ion, such as chloride, and may occur over a very wide pH range. However, it is unlikely to occur at very high pH, because propagation of a pit requires the development of a low local pH, and diffusion of high pH solution into the incipient pit prevents the reduction of pH to sufficiently low levels.

Crevice corrosion occurs where the metal is shielded from the bulk environment, such as at a flanged joint. This means that the environment in the crevice can change more easily than in a pit and crevice corrosion usually occurs more readily, e.g., at lower temperatures, than pitting. Because crevices are often present in engineering components, it is usual to consider crevice corrosion as the limiting factor. However, there are some applications where pitting is the dominant mechanism. Figure 5.1 shows a 2507 superduplex stainless steel bismuth leach tank that was handling an acid chloride solution at 80°C (100 g/L chloride at pH=0.8) and the tank failed by pitting after just 48 hours operation. The fluid also contained 9.5 g/L ferric ions, which made the solution very oxidizing and ensured that the superduplex stainless steel was in the transpassive zone. Superduplex was clearly not a suitable alloy for this application.

The schematic polarization curve for stainless steels, shown in Figure 4.5, is relevant for localized corrosion. Pitting occurs when the passive film breaks down locally and metal dissolution occurs before the film can re-form. This is more likely to occur under oxidizing conditions when the metal is in the transpassive region. Crevice corrosion occurs when the crevice is tight enough and deep enough that oxygen is consumed faster than it can diffuse in from the bulk solution. Chromium chloride is then produced in the crevice, and hydrolysis causes the pH to decrease. When the pH decreases to less than the depassivation pH, general dissolution of the metal in the crevice follows. The metal in the crevice is then in the active region.

FIGURE 5.1 Pitting of parent 2507 superduplex stainless steel and weld metal in a bismuth leach tank. [Photo courtesy of Steve Clarke]

5.1 PITTING IN SOME COMMON SOLUTIONS

Pitting usually occurs in the presence of halides, and the most common of these is chloride. Hence, there are a lot of test data in simple chloride solutions, usually around neutral pH under mildly oxidizing conditions.

A common way to determine the limit of use is the critical pitting temperature (CPT), which is the temperature where pitting initiates in any particular fluid. This may be determined by exposure tests at various temperatures, but it is more commonly determined by an electrochemical test at constant potential, where the temperature is slowly increased until pitting initiates.

Figure 5.2 shows the CPT for some stainless steels as a function of chloride concentration at a potential of +300 mV SCE, which is quite oxidizing. The curves show that the CPT increases with alloy content, and the CPT depends on both the chloride content and the temperature, particularly at lower chloride concentrations.

Figure 5.3 shows the CPT at +600 mV SCE as a function of chloride content for some higher alloy duplex stainless steels. This is very oxidizing, such as might occur in the presence of ~1 mg/L chlorine/hypochlorite, used either as a disinfectant or to control fouling. At the high chloride concentrations in Figure 5.3, the CPT is almost independent of chloride concentration, decreasing by only approximately 10°C over two orders of magnitude of chloride concentration. This is a common feature with stainless steels at higher potentials.

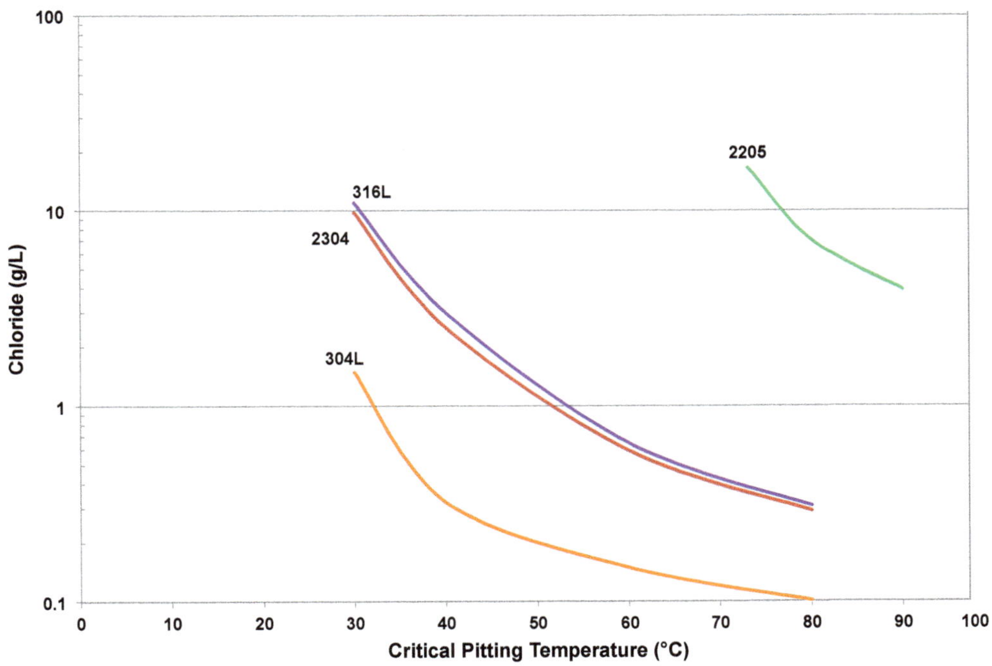

FIGURE 5.2 Effect of chloride on the CPT of some stainless steels at +300 mV SCE and pH 7.

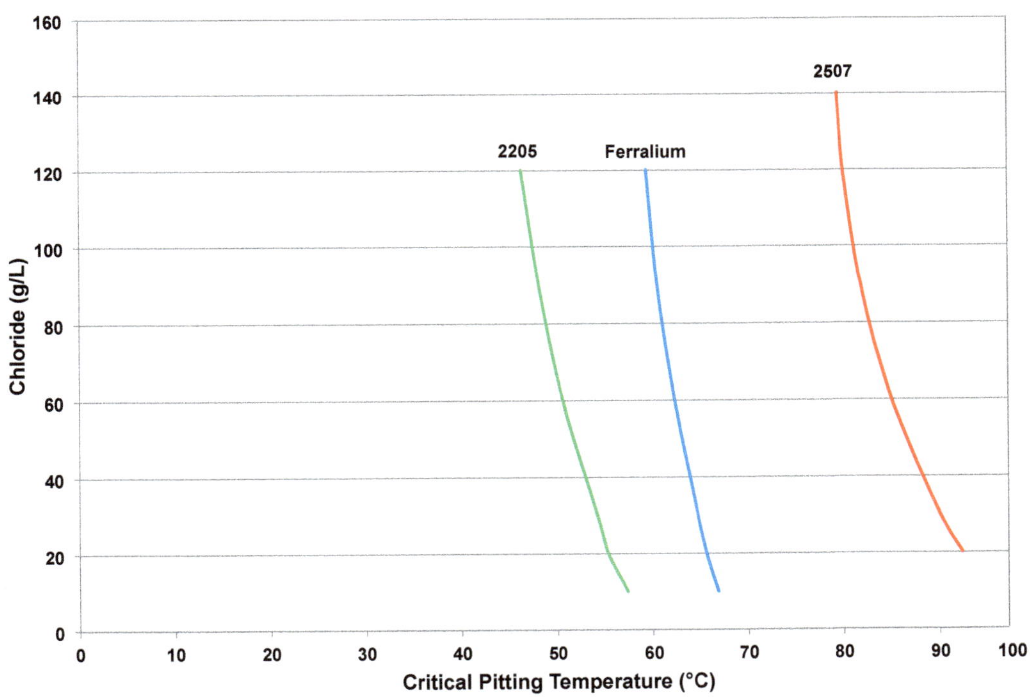

FIGURE 5.3 Effect of chloride on the CPT of some stainless steels at +600 mV SCE and pH 7.

Figure 5.4 shows the effect of pH on the CPT of some duplex stainless steels at +600 mV SCE in 3% sodium chloride solution. The CPT is almost independent of pH in the range from 1–7.

When the redox potential is low, due to the presence of reducing species, duplex stainless steels may often be used at much higher temperatures than under oxidizing conditions[1]. In the oil and gas industry, the produced fluids often contain carbon dioxide and hydrogen sulfide, in addition to oil and gas. The water also present can have high chloride concentrations, and the presence of CO_2 means that the pH is low, usually in the range from 3–5. Although H_2S does not cause pitting of stainless steels on its own, it reduces the severity of conditions at which chloride can cause pitting. However, H_2S is reducing, so the redox potential is low, while temperatures often exceed 100°C. The usual concern in the presence of H_2S is SCC (see Chapter 6), but pitting may also occur.

Francis and Byrne reported pitting tests on three duplex stainless steels in a sour brine at 121°C[2]. The alloys were cold worked (110 ksi) because they were intended for downhole tubulars. The water contained 46,000 mg/L chloride with 93 bar CO_2 and varying H_2S contents, giving a pH of 4.3 at temperature and pressure. The results in Table 5.1 show that all three alloys resisted pitting up to 0.25 bar H_2S, but at 0.375 bar H_2S there was pitting on all the alloys. However, the pitting was noticeably less severe as the PREN of the alloys increased, where PREN is given by:

$$\text{PREN} = \%\,\text{Cr} + 3.3(\%\,\text{Mo} + 0.5\text{x}\%\,\text{W}) + 16\text{x}\%\,\text{N} \text{ (see Chapter 2)}$$

TABLE 5.1 Results of pitting tests on some cold worked duplex stainless steels in a sour brine containing 46,000 mg/L chloride and 93 bar CO_2 at 121°C and pH 4.3[2].

ALLOY	H_2S PARTIAL PRESSURE (bar)		
	0.125	0.25	0.375
2205	No Pits	No Pits	Deep Pitting (>0.5 mm) Both Samples
Ferralium (low copper)	No Pits	No Pits	Deep Pitting (>0.5 mm)/ 3 Shallow Pits
Z100	No Pits	No Pits	No Pits/ 3 Shallow Pits

There are a number of standard tests that are used to rank stainless steels by their resistance to pitting. These should not be used as an indicator of service performance in any other fluid. The first of these is the ASTM G48 test in 6% ferric chloride[3], which is very aggressive because of the high chloride concentration, the low pH (1.0–1.5), and the high redox potential because of the ferric ions. The most common way to rank alloys is to measure the CPT as described in ASTM G48, method E[3].

Figure 5.5 shows a superduplex sample where pitting occurred after a ferric chloride test. Figure 5.6 shows the CPT for a range of stainless steels, and it can be seen that alloys of a similar PREN value have similar CPT values. The problem with this test is that the CPT of low alloy stainless steels (304L, 316L, and lean duplex) is less than ambient temperature, and this makes determination very difficult. To get around this, various people have proposed modified solutions for low-alloy stainless steel (PREN<26). The suggested solutions are usually more dilute ferric chloride with other additions, such as sodium nitrate, to make it less aggressive, as for example in reference 4.

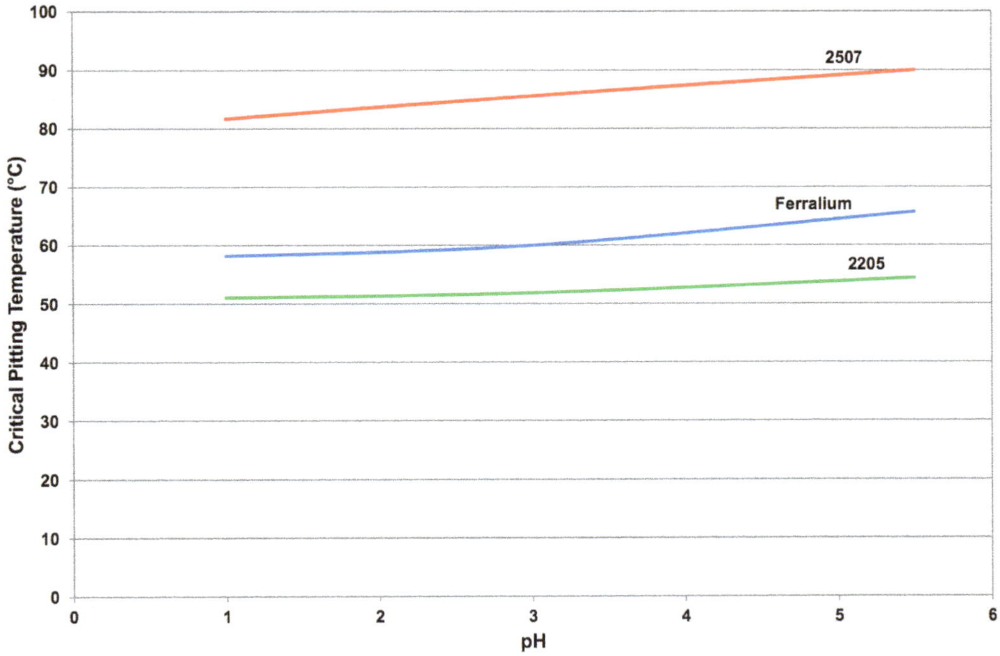

FIGURE 5.4 Effect of pH on the CPT of some stainless steels at +600 mV SCE in 3% sodium chloride.

FIGURE 5.5 Pitting of superduplex stainless steel in ferric chloride solution (ASTM G48).

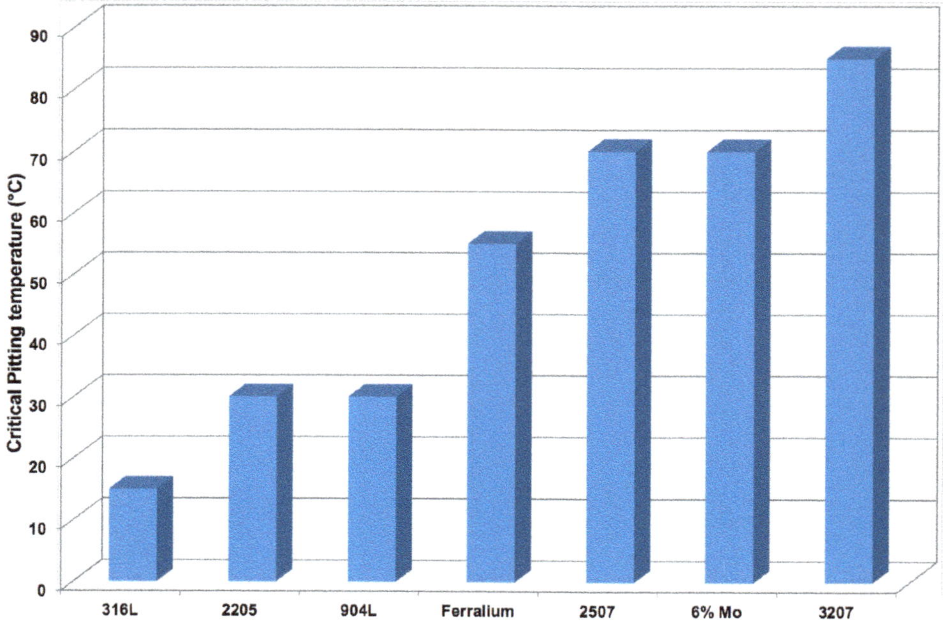

FIGURE 5.6 CPT of some stainless steels in ferric chloride to ASTM G48.

The regular ASTM G48 test has been adopted by some specifications and standards, such as NORSOK, the Norwegian oil and gas standard[5]. The latter requires superduplex stainless steel to pass a G48 test at 50°C with no pitting and a weight loss less than 4 g/m^2. This is a quality control test to demonstrate that there are no third phases present and there is no chromium-depleted layer on the surface (the weight loss criterion). NORSOK used to include a test at 25°C for 2205, but this was dropped. However, service failures of poor quality material suggest that it should be restored[6, 7].

It has been claimed that the ASTM G48 test is not very discriminating and there is now an electrochemical CPT test in ASTM G150[8]. This involves polarization at +700 mV SCE in 1M sodium chloride, and the temperature is then increased until pitting occurs, as defined by a current density of 100 μA/cm^2 or more.

Figure 5.7 shows some CPT's for a range of stainless steels in the ASTM G150 test, and at first sight it appears to show some distinction between grades. This is true for the duplex stainless steels as a family, but the austenitic alloys of similar PREN to duplex alloys, such as 904L and 2205, are not really any better in service. They look better in this test because of the fail criterion of 100 μA/cm^2. Oldfield and Sutton[9] showed that localized corrosion initiates at a current density of 10 μA/cm^2, and further increases in current density denote an increase in the corrosion propagation rate. Oldfield[10] showed that alloys with higher nickel contents propagate more slowly than those with low nickel contents, hence the higher CPT at 100 μA/cm^2 for the higher nickel austenitic alloys, compared with duplex alloys. The test does have the advantage that it can be used on all grades of duplex stainless steel.

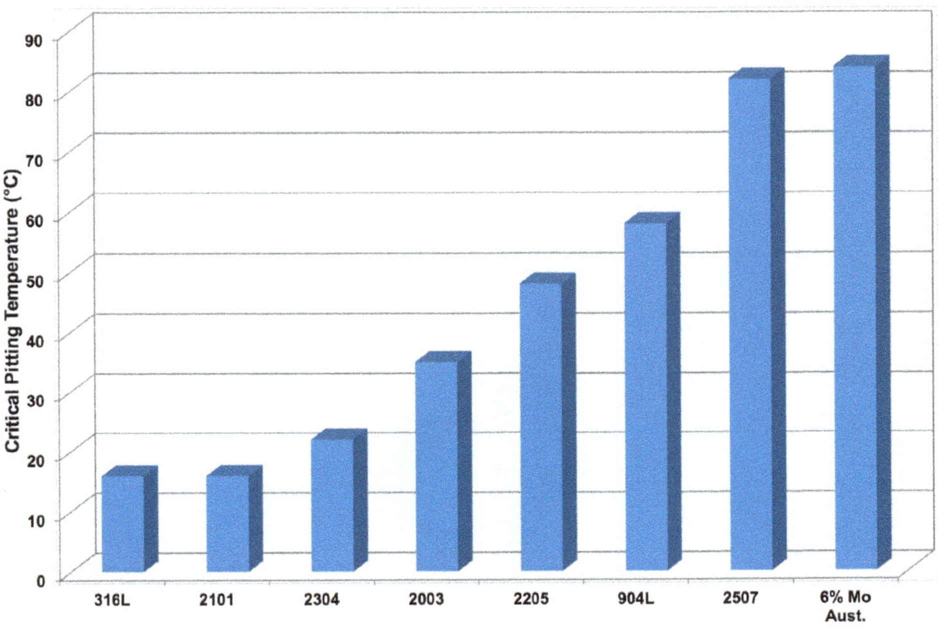

FIGURE 5.7 CPT of some stainless steels in 1M NaCl to ASTM G150.

5.2 CREVICE CORROSION

Crevices can be created by a wide range of components in industrial service, and include flanged joints, screwed couplings, and 'O' ring seals. Oldfield and Sutton showed that the susceptibility to crevice corrosion is controlled by the crevice geometry, in particular the tightness and the depth[8]. Hence, screwed couplings are a severe crevice, while 'O' rings are less severe, because although they are tight, they are also often shallow. The ASME flange has a spiral groove machined on the flat face and forms a series of tight and shallow crevices when assembled. This means concentrated solution in the tight crevices can be diluted by solution from the wider ones. Hence, this need not be a particularly severe crevice, although the choice of gasket also affects this, as discussed below.

Testing crevice corrosion resistance is difficult because potential crevice formers vary so widely in tightness and depth. One attempt to standardize this has been the castellated crevice washer, or INCO washer, as described in ASTM G78[11]. (INCO was formerly known as the International Nickel Company.) Figure 5.8 shows a typical washer assembly on a flat coupon. Originally, it was tightened to a preset torque, with a typical value being 7 Nm. However, bolting is not uniform and the use of nylon coated spring washers to obtain a uniform load gives better reproducibility from laboratory to laboratory[12]. This technique works well on flat plates, but is not suitable for curved surfaces. The author has carried out crevice tests on pipes with specially machined curved crevice washers that worked well on pipes with a diameter of ≥ 50 mm. For small diameter cylinders, it has been common to use 'O' rings, but these are rather narrow crevices (~1 mm). The author has found that square section silicone rubber rings (≥5 mm wide) create a reasonably tight and deep crevice, which gives a lower critical crevice temperature (CCT) than 'O' rings. For example, a crevice was created with a

FIGURE 5.8 Picture of a castellated washer assembly on a flat coupon (ASTM G78).

7 mm bore ring, 5 mm wide, on a 10 mm diameter cylinder. The effects of different crevice formers and tightness on crevice corrosion initiation in laboratory tests have been discussed by Oldfield and by Carranza et al.[10, 13]. What is missing is a good correlation between different laboratory crevice formers and common crevices found in service (flanges, screwed couplings, etc.).

From an engineering point of view, the most relevant information is the temperature and chloride limits of different grades of stainless steel. The author conducted a series of tests using the castellated crevice washer at a torque of 7 Nm in sodium chloride solutions at near neutral pH. The solutions were open to the air, but were not agitated so they were mildly oxidizing.

The results are shown in Figure 5.9 for the lower alloy stainless steels. The curves are more or less parallel except for that of 2101 lean duplex. The reason for this may be that all the other alloys have 4–10% nickel and 2–3% molybdenum, while 2101 has only 1% nickel and 0.3% molybdenum. It is quite possible that nickel and/or molybdenum has an effect on the temperature response to crevice corrosion initiation. The curves follow the trend of increasing chloride limit with increasing PREN. Although the slope for 2101 is somewhat different, its crevice corrosion resistance is similar to that of 316L.

Fresh waters often have additions of chlorine, either for disinfection and/or to prevent microbial fouling. Mameng and Pettersson reported crevice corrosion test data for some stainless steels in low chloride waters with chlorine contents from 0.2–1.0 mg/L after 30 days exposure[14]. The results in Figure 5.10 show that 2205 was more resistant than 316L, which was more resistant than 304L at chlorine concentrations of 0.5–1.0 mg/L. No corrosion was seen with 2205 even under the most aggressive conditions tested. It was not

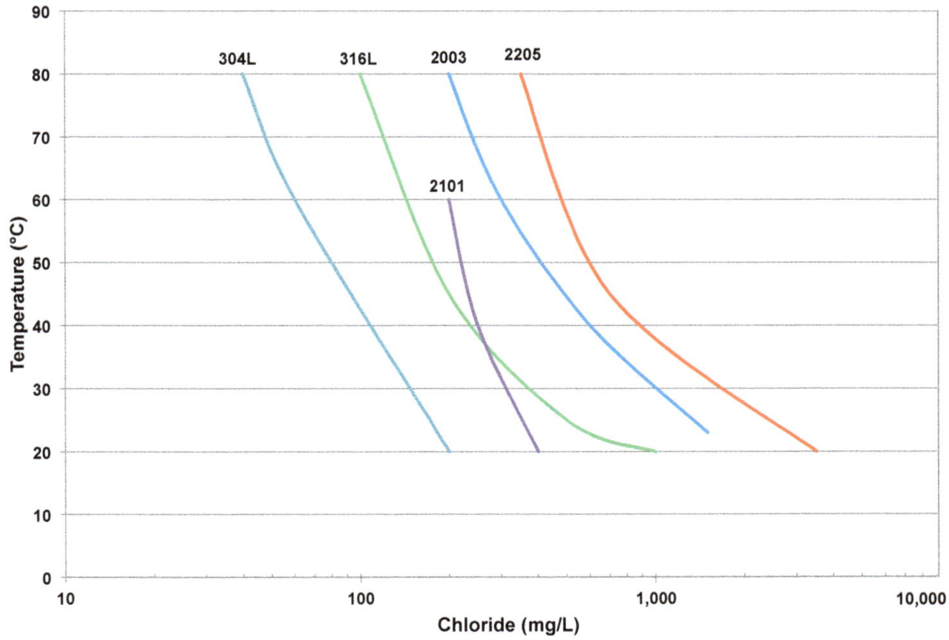

FIGURE 5.9 Temperature for the onset of crevice corrosion of some stainless steels in sodium chloride solutions at pH~7.

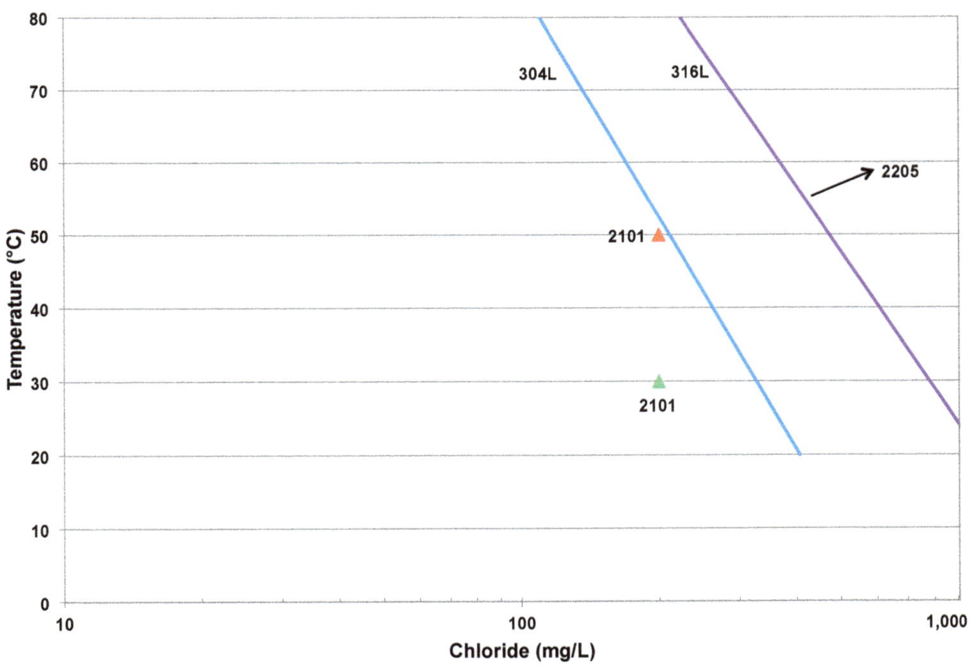

FIGURE 5.10 Effect of chloride on the safe operating temperature to resist crevice corrosion in chlorinated water[14].

possible to draw a line for 2101, but one pass and one fail point are shown in Figure 5.10 (green and red). It is surprising that 2101 was not as resistant as 304L, but this may be because insufficient data was collected with 2101. It is noticeable that the data for 304L and 316L show higher limits than the same alloys in Figure 5.9. However, the torque used in the work of Mameng and Petterssson was only 1.58 Nm, while the author's work had a torque of 7 Nm, creating tighter crevices, similar to that in a flanged joint.

The potentials of stainless steels in chlorinated waters are high, typically in the range +550 to +750 mV SCE. Francis et al. reported the CCT for two duplex stainless steels in waters of 3,000–100,000 mg/L chloride[15]. The results in Figure 5.11 show that the CCT increases as the potential decreases, but much more for Z100 superduplex than for 2205. The CCT shows little dependence on chloride content in this range, decreasing by only 10–20°C.

As discussed above, the oil and gas industry can produce reducing fluids with high chlorides and H_2S. Francis and Byrne tested samples of three cold-worked duplex stainless steels, plus a solution annealed superduplex[2]. The water contained 46,000 mg/L chloride with 93 bar CO_2 and varying H_2S contents, giving a pH of 4.3 at temperature and pressure. The crevices were created with a castellated washer as described in ASTM G78[11]. The results in Table 5.2 showed no attack on any alloy up to 0.25 bar H_2S, but decreasing attack with 0.375 bar H_2S as the PREN increased. These alloys were cold worked (110 ksi minimum proof strength) and the solution-annealed superduplex alloy (80 ksi minimum proof strength) showed no crevice corrosion even with 0.375 bar H_2S.

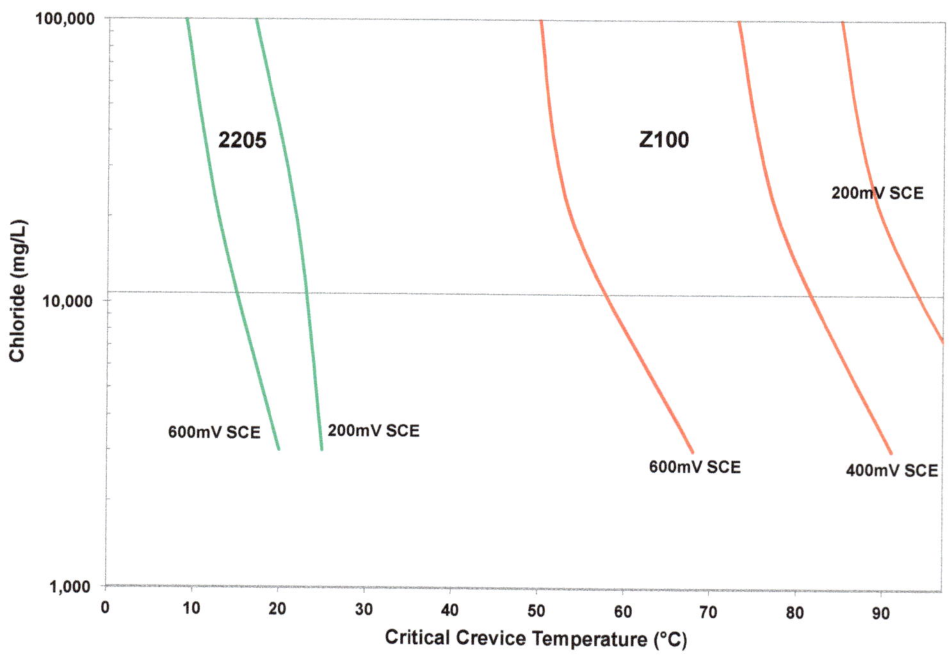

FIGURE 5.11 Effect of chloride on the CCT of some duplex stainless steels at various potentials[15].

TABLE 5.2 Results of crevice tests on some cold worked duplex stainless steels in a sour brine containing 46,000 mg/L chloride and 93 bar CO_2 at 121°C and pH 4.3[2].

ALLOY	H_2S PARTIAL PRESSURE (bar)		
	0.125	0.25	0.375
2205	No Attack	No Attack	6 segments attacked
Ferralium (low copper)	No Attack	No Attack	4 segments attacked
Z100	No Attack	No Attack	2 segments attacked
Z100 (Solution Annealed)	No Attack	No Attack	No Attack

Both ASTM G48 and G150, which were described for assessing the CPT in the previous section, can also be used for comparing the crevice corrosion resistance of materials. However, while the results of one set of data are internally consistent, it is difficult to compare the results with those of other workers because of the variety of crevice formers used and, in particular, the vast range of crevice tightness used. Hence, comparing results from different tests should be done with a great deal of caution. These tests are only suitable for ranking alloys and tests more closely mimicking service conditions are best suited to select materials.

5.3 CREVICE CORROSION IN SEAWATER

Seawater is one of the most widely used cooling mediums and it is important that any alloys used in such a system have good resistance to crevice corrosion. Figure 5.12 shows crevice corrosion of a 316L stainless steel pipe under the rubber seal of a high pressure coupling in a seawater reverse osmosis (SWRO) desalination plant. Note that 316L is not a suitable alloy for use in aerated seawater as is shown by the heavy crevice corrosion.

The two factors that most influence the CCT for stainless steels are potential and alloy content, in particular chromium, molybdenum, and nitrogen. In seawater, stainless steels can adopt a wide range of potentials, as shown in Figure 5.13[16]. In natural seawater, a biofilm forms over a period of a few days to a few weeks, which depolarizes the cathodic reaction (the reduction of dissolved oxygen), elevating the potential to around +300 mV SCE. If chlorine/hypochlorite is added to the seawater to control fouling, the biofilm does not form, but chlorine/hypochlorite is a powerful oxidizer and the potential rises to around +600 mV SCE. If the seawater is heated to 25 to 30°C above the normal ambient temperature, the biofilm does not form and the potential is then much lower, around +150 mV SCE. If the seawater is partially deaerated, then the potential falls further, and with 200 ppb oxygen, it is around −100 mV SCE. When the seawater is fully deaerated the potential is around –450 mV SCE.

In SWRO desalination plants, there is a low pressure seawater section, where the seawater is filtered and chemical additions are made, followed by a high pressure section, where the water enters the membrane cells[17]. Figure 5.14 shows a schematic layout of a typical SWRO plant. The redox potential must not exceed 350 mV $Ag/AgCl_{SAT}$ (where the subscript "SAT" indicates saturated silver chloride) in the high-pressure sec-

FIGURE 5.12 Crevice corrosion of a 316L pipe under the rubber seal of a high pressure coupling in a SWRO desalination plant.

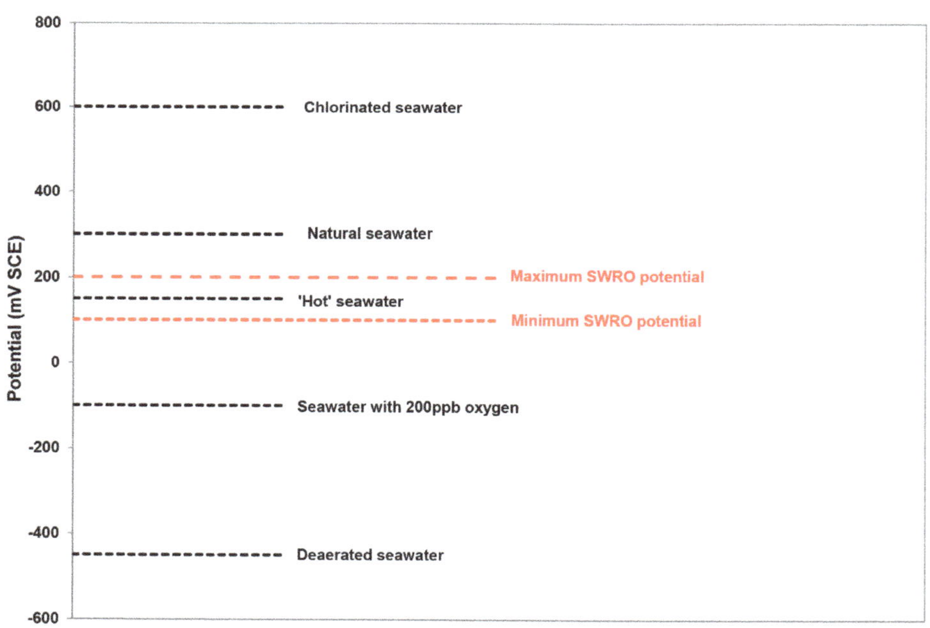

FIGURE 5.13 Potentials of stainless steels in different types of seawater[15, 16].

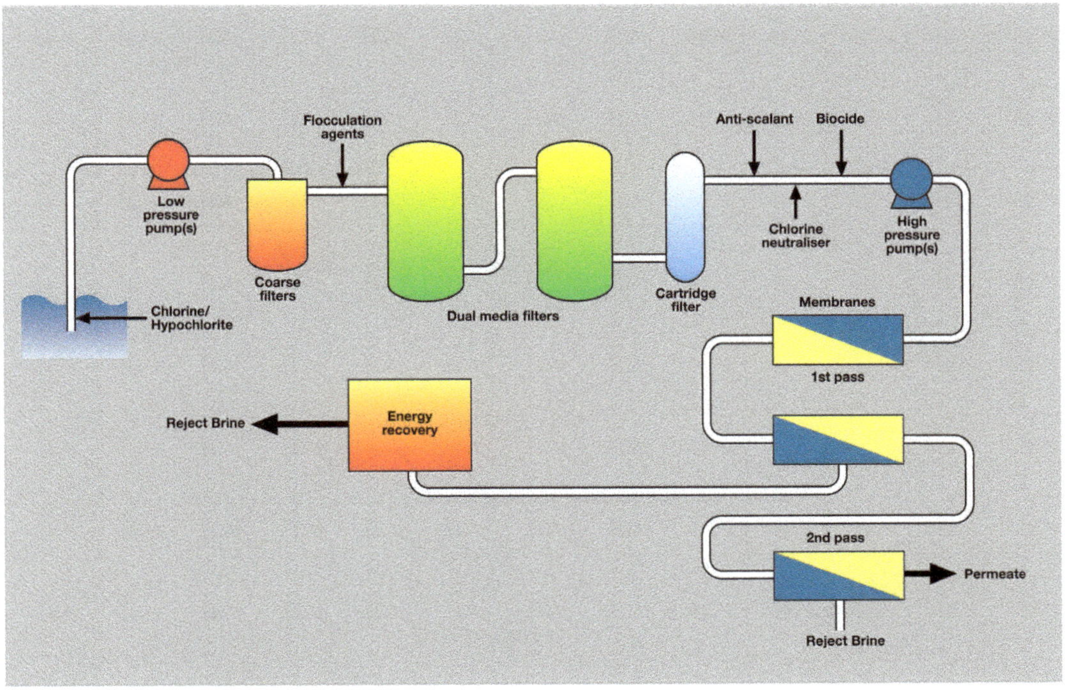

FIGURE 5.14 Schematic diagram of a typical SWRO plant.

tion to preserve the membranes, so the water is often dosed with sodium metabisulfite to reduce the redox potential to a safe value. Francis et al. showed that the potential of stainless steels in the high-pressure section of an SWRO plant is in the range +100 to +200 mV SCE, with values commonly nearer the upper limit to reduce the cost of dosing chemicals[15]. These upper and lower potentials are shown in red in Figure 5.13.

Kovach and Redmond reviewed a large quantity of crevice corrosion test data for stainless steels in warm seawater and concluded that to resist crevice corrosion an alloy must have a CCT in the ASTM G48 ferric chloride test ≥ 35°C[18]. They then plotted the CCT in the G48 test against PREN and found separate correlations for ferritic, duplex, and austenitic stainless steels, as shown in Figure 5.15. From their data, a ferritic alloy must have a PREN ≥ 35, a duplex alloy ≥ 40, and an austenitic alloy ≥ 43, to resist crevice corrosion in warm seawater. The superduplex and 6% Mo austenitic alloys meet these criteria, while the lower alloys do not.

Francis et al. carried out electrochemical tests at constant potential in synthetic seawater on a range of stainless steels[15]. The results in Figure 5.16 show that in chlorinated and natural seawater alloys 2205 and 904L only resisted crevice corrosion up to around 22°C. Alloy 316L suffered attack in all the tests at ambient temperature. Only the superduplex and 6% Mo alloys resisted crevice corrosion at temperatures normally found in seawater cooling systems and SWRO desalination plants.

The inflexion in the lines for 2205 and 904L at 200 mV SCE shows why these alloys have suffered crevice corrosion in many SWRO plants, where temperatures are usually in the range from 25–40°C[15, 19]. Where these alloys have worked the stainless steel potential has been significantly lower than +200 mV SCE.

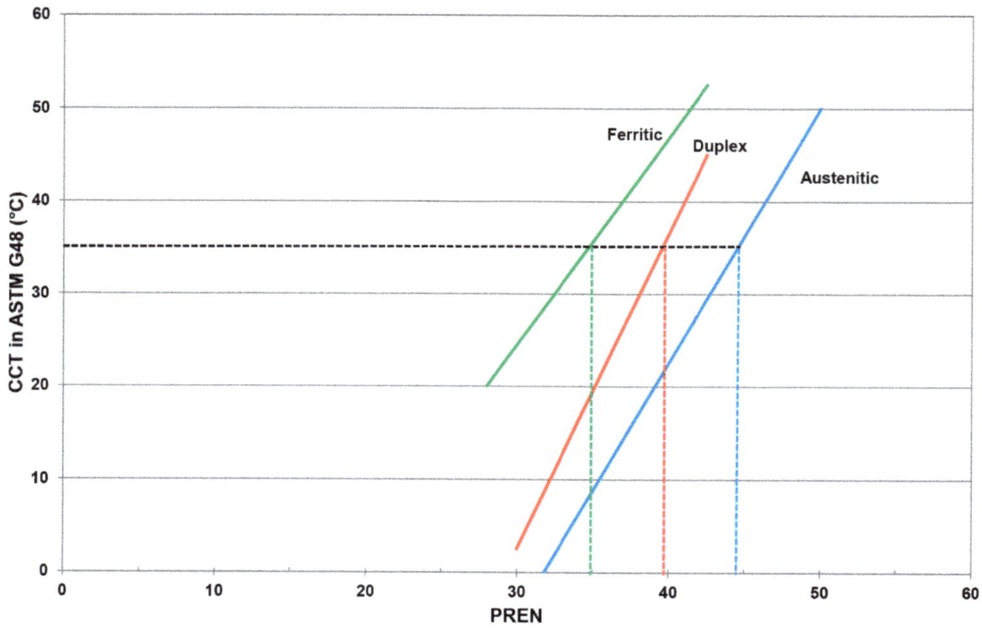

FIGURE 5.15 Effect of PREN on the CCT in ASTM G48 ferric chloride test[18].

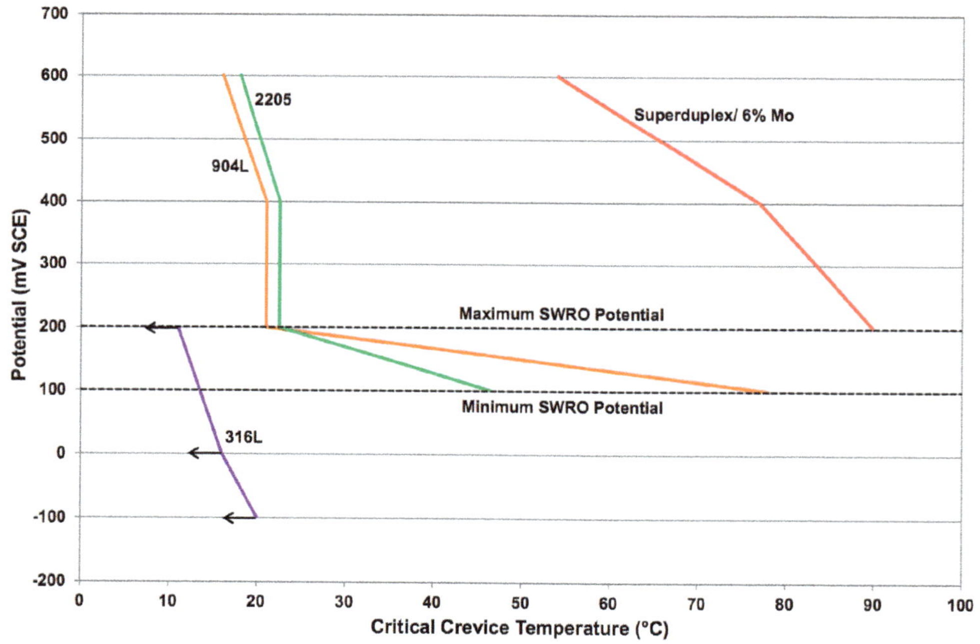

FIGURE 5.16 CCT for some stainless steels in synthetic seawater as a function of potential[15].

In long term exposure tests in natural and chlorinated seawater, the electrochemical results have been confirmed. Figure 5.17 shows the results from several laboratories in natural seawater[20-22], where only the superduplex and 6% Mo alloys showed no attack. The results in Figure 5.18 for chlorinated seawater show no attack for superduplex and 6% Mo austenitic at 16°C, but the 6% Mo alloy showed a little attack at 40°C, while there was none on the superduplex. Figure 5.19 shows pitting along the weld of a 2205 seawater filter vessel, handling chlorinated seawater at ~30°C, which also had crevice corrosion at the flanged joints, clearly showing the unsuitability of this alloy in aerated warm seawater.

Zeuthen and Kain conducted some crevice corrosion tests on high alloy stainless steels using two types of crevice formers in seawater at 20–25°C[23]. With the conventional crevice washer (ASTM G78) they found no crevice corrosion of the several 6% Mo austenitic or superduplex alloys. Using tube samples joined by vinyl tubing with plastic clamps, they created a tighter, deeper crevice. The attack on all the 6% Mo alloys, Ferralium, and 2507 was from 0.01–0.13 mm deep after 60 days, while there was no attack on Z100 and DP3W superduplex. This does not represent a difference in the performance of these alloys in seawater, but is probably connected with varying tightness of the crevices. Francis et al.[15] showed that the surface roughness of the outside of tubes and pipes affected the risk of crevice corrosion. Using a smooth machined finish on the outside of the tube did not increase the CCT, but it did reduce the scatter, confining the results to close to the maximum seen without machining (Figure 5.20). These tests were undertaken to investigate crevice corrosion under the rubber boot of high-pressure Victaulic-type couplings in SWRO plants and were conducted at a potential of +200 mV SCE.

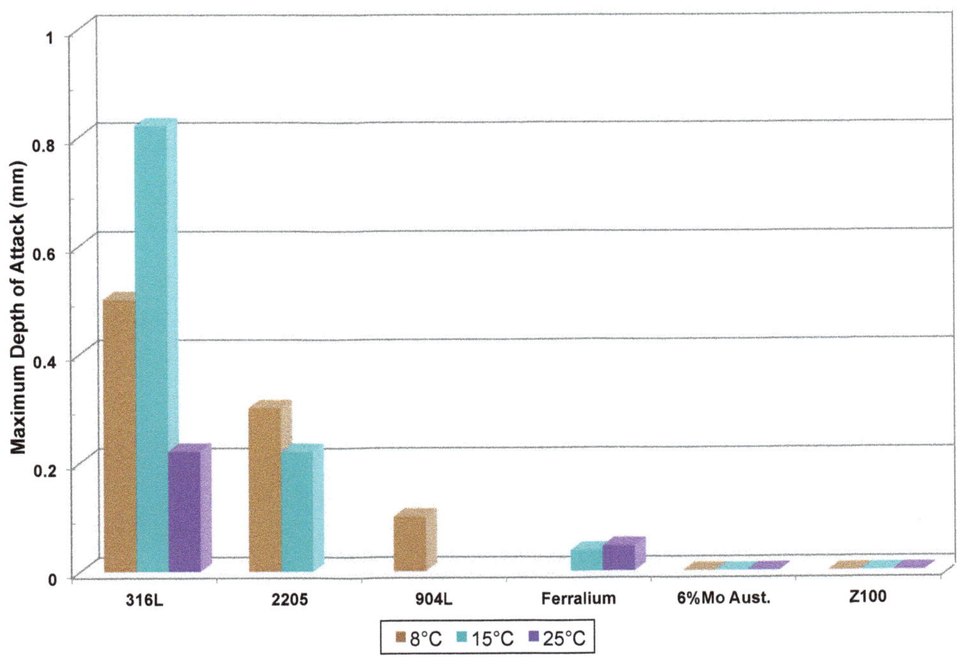

FIGURE 5.17 Maximum depth of crevice corrosion after exposure for 60 days in natural seawater[20-22].

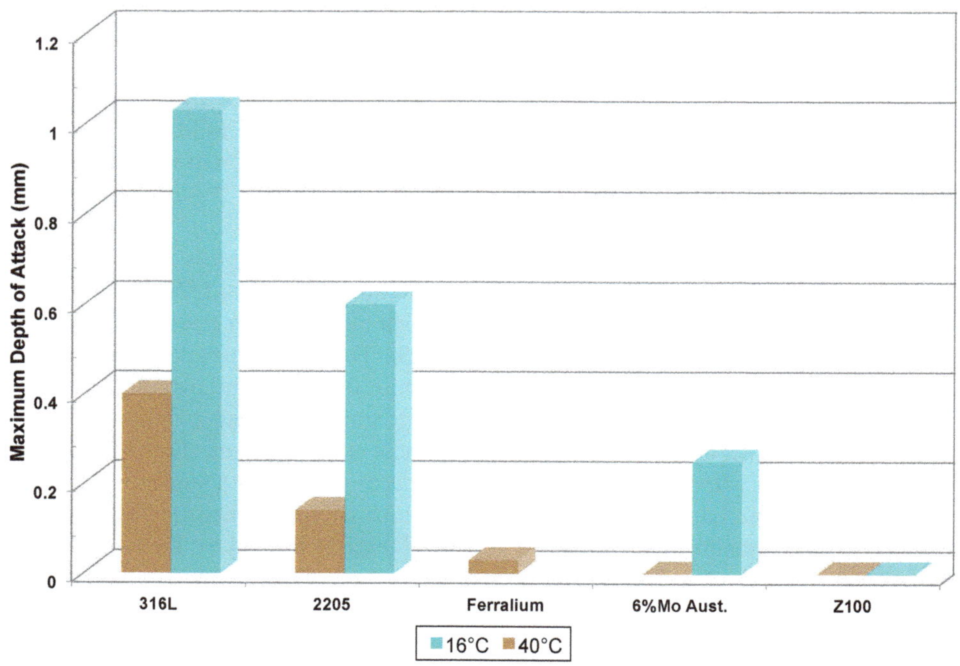

FIGURE 5.18 Maximum depth of crevice corrosion after exposure for 60 days in chlorinated (1mg/L) seawater[21]

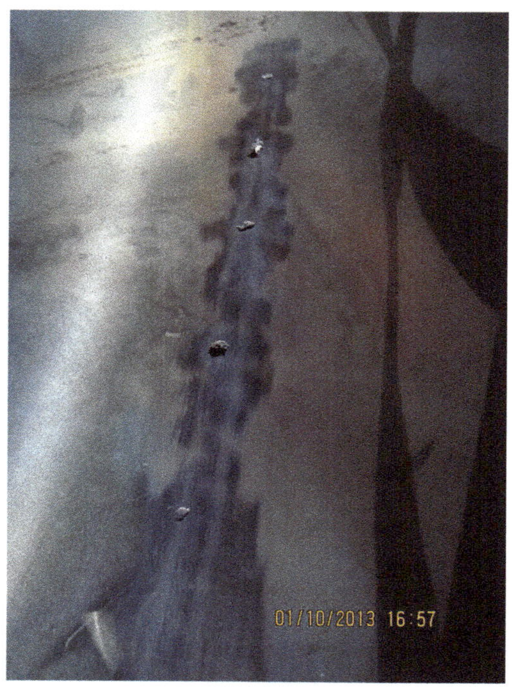

FIGURE 5.19 Pitting of weld on a 2205 seawater filter vessel carrying chlorinated seawater at ~30°C.

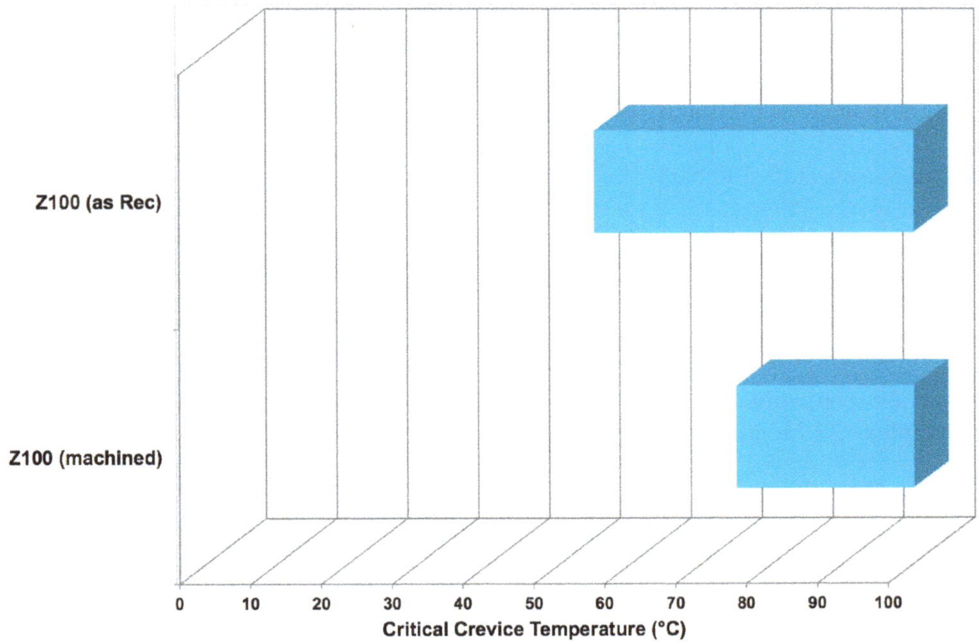

FIGURE 5.20 Effect of surface finish on the critical crevice temperature in seawater at +200 mV SCE[15].

Francis and Byrne summarized over 20 years' service experience with superduplex stainless steel in seawater, and combined laboratory data with field results to show the seawater temperature limit at different chlorination levels[24], as shown in Table 5.3. The results clearly show the tolerance of superduplex for chlorine/hypochlorite in colder water.

TABLE 5.3 Maximum operating temperatures for superduplex stainless steel as a function of chlorine content[24].

CHLORINE (mg/L)	TEMPERATURE (°C)
200	10
5	20
1	30
0.7	40

Flanged joints are a common crevice in service and the gasket material can significantly affect the severity of the crevice and the risk for corrosion initiation. Francis and Byrne reviewed the published data and came to the following conclusions[25]:

A. Low Pressure Systems (up to 10 bar).

1. Avoid the use of PTFE or graphite loaded gaskets.

2. Use gaskets made of synthetic rubber, rubber bonded aramid, or synthetic fiber.

B. High Pressure Systems (10–100 bar)

1. Avoid PTFE coated gaskets.

2. Graphite-containing gaskets are acceptable provided the graphite is sealed from the seawater and is never wetted.

3. Only metals compatible with high alloy stainless steel should be used for spiral wound gaskets. (Superduplex, 6% Mo austenitic, Ni-Cr-Mo alloys, where Mo >7%, titanium).

The authors cite service failures with PTFE and graphite-loaded gaskets in low-pressure service (chlorinated seawater) from the North Sea and the Middle East, and failures in high-pressure service of spiral wound gaskets coated with PTFE or made from low alloy materials. Figure 5.21 shows corrosion of a superduplex stainless steel flange fitted with a graphite-loaded gasket and used in a chlorinated seawater system.

As the dissolved oxygen decreases, the susceptibility to crevice corrosion should decrease and Rogne and Johnsen demonstrated this in brines with varying CO_2 and oxygen contents for both 6% Mo austenitic and 2205 duplex[26]. They were simulating the injection of deaerated, produced water mixed with partially deaerated seawater into oil-producing strata. Their results in Figure 5.22 show that 2205 can resist crevice corrosion to modest temperatures in brines with low oxygen content.

Francis et al. published the results of autoclave crevice tests on Z100 superduplex stainless steel in a strong brine (100,000 mg/L chloride) with 1 bar CO_2 and varying oxygen contents[27]. The results in Table 5.4 show that superduplex can be used up to high temperatures in low oxygen brines. The service history with superduplex injection pumps handling mixed waters has been good to date, but the author knows of no fields where temperatures have approached those in Table 5.4 where crevice corrosion of cast superduplex occurred.

TABLE 5.4 Results of crevice and pitting corrosion tests of Z100 in a simulated mixed injection water[27]

OXYGEN (ppb)	TEMPERATURE (°C)	RESULT (Pass/Fail)		
		Wrought	Weld*	Cast
50	120	P	P	P
	140	P	P	P/F
200	100	P	P	P
	120	P	P	F
410	80	P	P	P/F
	100	P	P	F

*pitting test

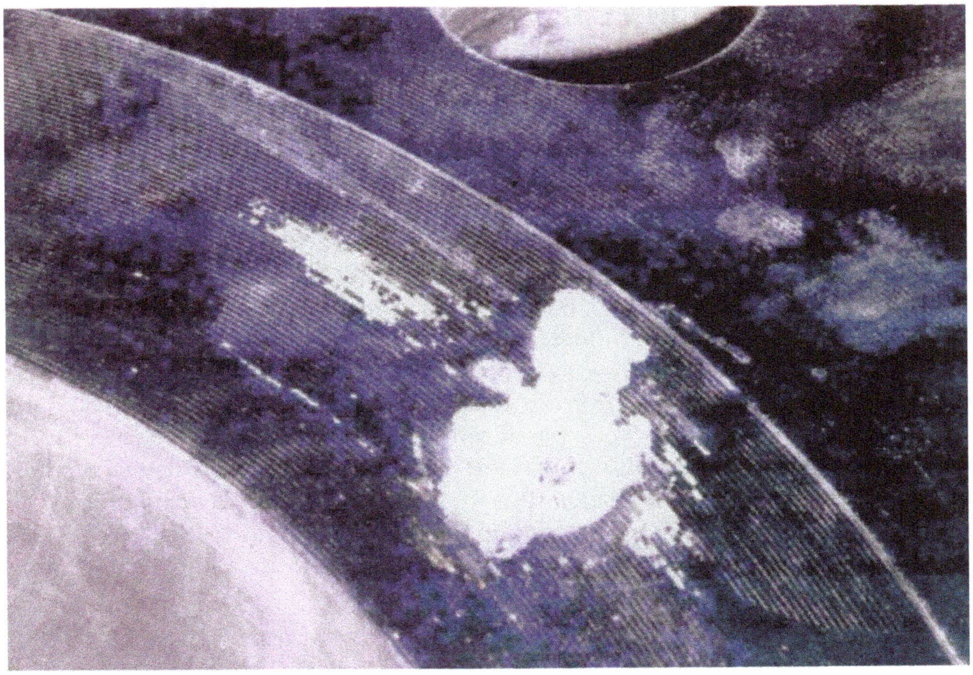

FIGURE 5.21 Crevice corrosion of a superduplex flange fitted with a graphite-loaded gasket in a chlorinated seawater system at ~30°C[24]. [Photo courtesy of Rolled Alloys]

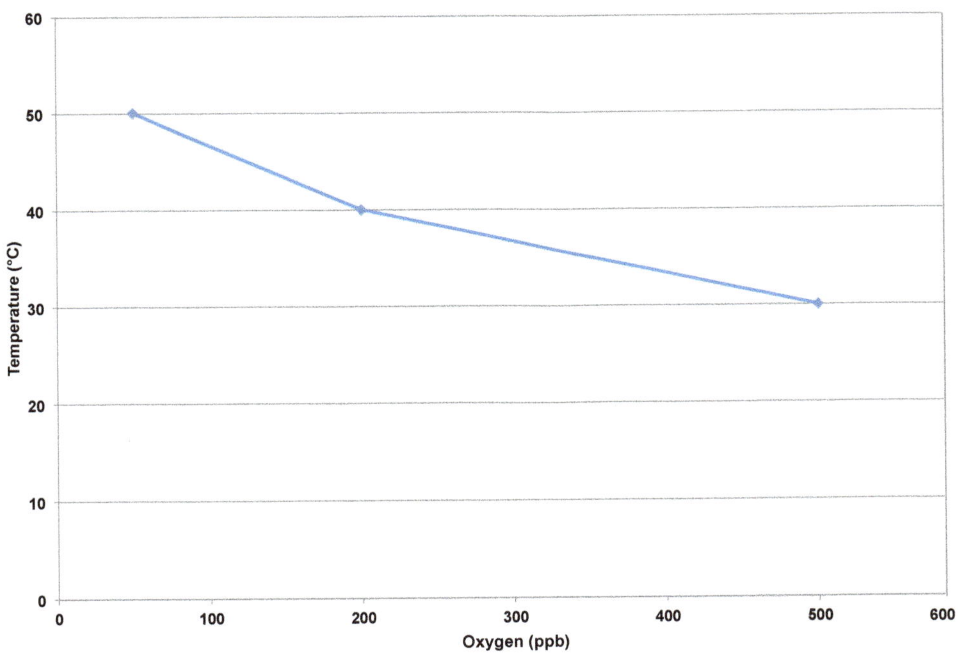

FIGURE 5.22 CCT for 2205 in a brine with 2 bar CO_2 as a function of dissolved oxygen content[26].

5.4 CREVICE CORROSION IN FLUE GAS DESULFURIZATION

Another common chloride-rich environment that is very corrosive is that in the wet limestone scrubbers of flue gas desulfurization (FGD) plants. Figure 5.23 is a schematic diagram of a typical absorber tower. The hot flue gas contains SO_x and NO_x gases plus some fly ash. The acid gases are responsible for acid rain if not removed, and they react with the slurry of crushed limestone ($CaCO_3$) mixed with local water. The slurry is agitated and aerated to encourage the formation of calcium sulfate, which may be separated to use in plasterboard. The recirculating slurry has a high chloride content, which varies with the coal being burnt, and it also has a low pH, again a function of the coal source. In addition, there are also fluorides and ferric ions present. Because of the presence of SO_x, there are also partially oxidized sulfur species present, such as sulfite (SO_3^{2-}), thiosulfate ($S_2O_3^{2-}$), and dithionate ($S_2O_6^{2-}$).

There are two chief coal sources, lignite (or brown coal) and anthracite, and Francis et al. derived synthetic solutions to reproduce these two slurry types, based on the composition of the slurries in some European power stations[28]. The compositions are shown in Table 5.5 and it is clear that lignite slurries are typically lower in chloride and higher in fluoride than anthracite slurries. The partially oxidized sulfur species were added as manganese dithionate, because this is easy to make in the laboratory[28]. The other factor of importance is the pH and lignite slurries tend to run in the range of 4.5–6, with a value of 5.0–5.5 being common.

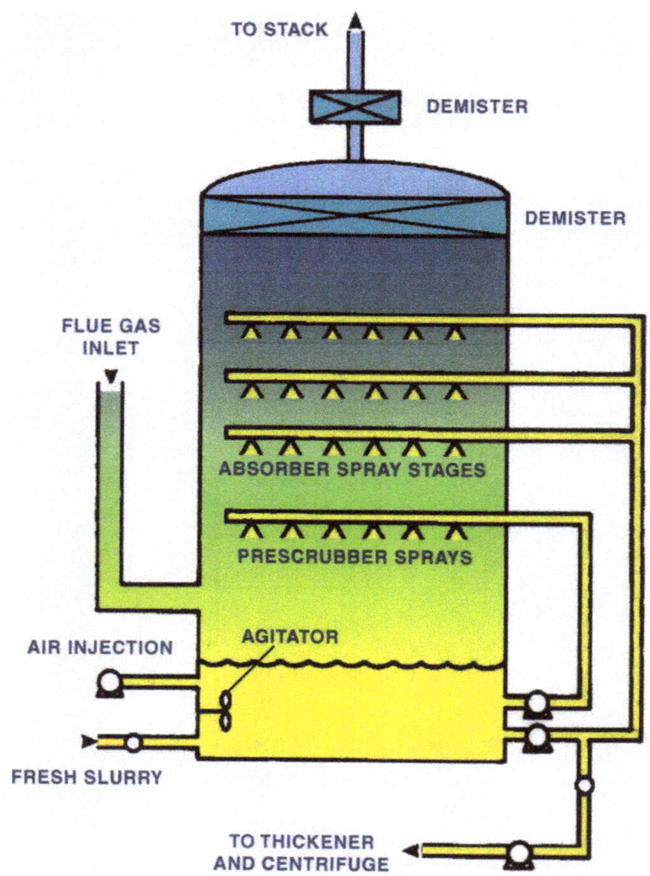

FIGURE 5.23 Schematic diagram of the absorber tower in a wet limestone FGD system.

Anthracite slurries run at a pH of 4–5.5, and 4 is typical for stations making plasterboard grade calcium sulfate, while a slightly higher pH of 4.5 or more is common for other stations. The operating temperature of FGD slurries is around 60–70°C for lignite slurries, and 45–50°C for anthracite slurries.

TABLE 5.5 Composition of synthetic FGD slurry for two types of coal[28].

CHEMICAL SPECIES	ADDED AS	CONCENTRATION	
		LIGNITE	ANTHRACITE
$CaSO_4$	Anhydrous	100 g/L	100 g/L
Chloride	$CaCl_2$:NaCl (19:1 as Cl)	15 g/L	40 g/L
Fluoride	NaF	200 mg/L	50 mg/L
Bromide	NaBr	500 mg/L	500 mg/L
Dithionate	MnS_2O_6	200 mg/L	200 mg/L
Fe^{3+}	$Fe_2(SO_4)_3$	10 mg/L	10 mg/L
Al^{3+}	$Al_2(SO_4)_3$	30 mg/L	30 mg/L

Francis et al. monitored the potentials of some higher alloy stainless steels in these slurries and showed that they stabilized around a value of +300 mV SCE, much more positive than is found in simple solutions of chloride and sulfate[28]. They polarized a series of stainless steels to this potential in both slurries and slowly increased the temperature until crevice corrosion initiated. The results in Table 5.6 show that the anthracite slurry was more aggressive at the same pH of 4 and that the 25% Cr cast alloy (similar to Ferralium but with no copper and a PREN ~35) was the only alloy that had a CCT below the normal operating temperature in both slurries.

TABLE 5.6 Results of electrochemical CCT tests in simulated FGD slurries at pH 4[28].

FORM	ALLOY	CCT (°C)	
		Lignite	Anthracite
Wrought	Z100	>83	>80
	6% Mo Aust.	66.7	64.7
Welded*	Z100	63.4	58.5
Cast	Z100	62.6	57.8
	25% Cr Duplex	39.6	42.0

* Critical Pitting Temperature

In further work, the author carried out tests on Z100 superduplex in the simulated anthracite slurry as a function of pH[29]. The results in Figure 5.24 show that the wrought alloy was not sensitive to pH over the range of 3–5, while the cast alloy was. Tests in other fluids also show the cast alloy to be slightly inferior to the wrought alloy in oxidizing chloride solutions when the conditions are particularly severe[29].

Figure 5.25 shows the CCT as a function of chloride for two cast duplex stainless steels at pH4. The 25% Cr alloy was similar to Ferralium, but with no copper and a PREN ~35, while the other alloy was cast Z100 (ASTM A995, grade 6A). It is clear that PREN has a big effect on the CCT, particularly at higher chloride concentrations.

In further tests in both lignite and anthracite slurries at pH 4, the author showed that the CCT of wrought superduplex, 6% Mo austenitic and C-276 exceeded the normal operating temperatures of these slurries, while that of 2205 did not (Table 5.7).

TABLE 5.7 Electrochemical CCT data for some wrought stainless steels in synthetic FGD slurry at pH 4.

COAL TYPE	CHLORIDE (mg/L)	pH	CCT (°C)			
			2205	Z100	6% Mo	C-276
Anthracite*	40,000	4	47.8 42.3 56.6 42.4	84.5 80.1 61.3	>84.5 >81.5 >84.5	>84.5 >81.5 >84.5
Lignite *	15,000	4	68.0 79.8 57.1	>83 >83 >84.5	>83 >83 >84.5	>83 >83 >84.5

* See Table 5.5

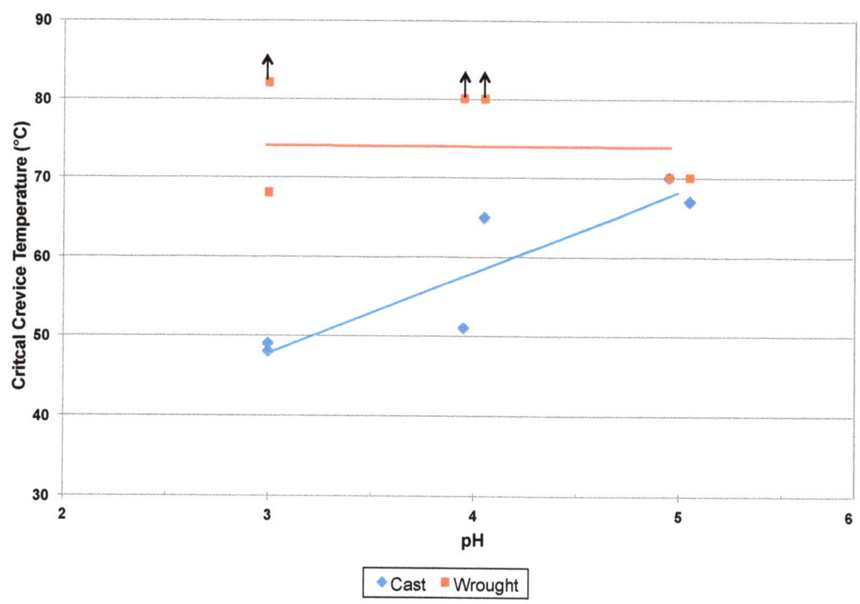

FIGURE 5.24 Effect of pH on the CCT in a simulated FGD slurry with 40,000 mg/L chloride for cast and wrought Z100 superduplex[29].

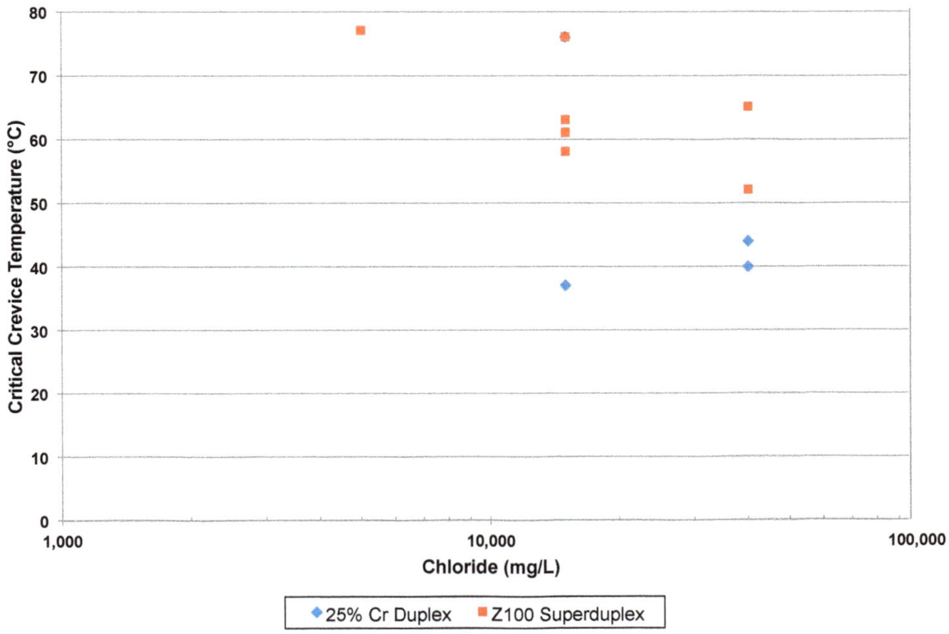

FIGURE 5.25 Effect of chloride on the CCT of some cast duplex stainless steels in a simulated FGD slurry at pH 4.

Hagen et al. tested a range of cast duplex stainless steels in synthetic FGD slurries over a range of temperatures, pH values, and chloride and fluoride concentrations[30]. They found that a cast alloy similar to cast Z100 (with tungsten and copper) gave the best resistance over the full range of conditions up to 70°C.

A number of authors have reviewed materials for FGD environments and have concluded that superduplex or 6% Mo austenitic alloys are required in the absorber slurry area[31, 32]. Phull et al. exposed a wide range of stainless steels and nickel alloys in some FGD systems in the United States, and concluded that in most cases, 2205 duplex would be adequate[33]. However, the stations used for exposure tests had pH values in the range of 5.2–5.6 and operated at the lower end of the temperature range (around 50°C). On the basis of this conclusion, a number of U.S. FGD systems used 2205 for the absorbers and subsequently experienced severe corrosion in the form of pitting and crevice corrosion[34].

Tossey et al. reported further details of the failures and the repair procedures to get the 2205 absorbers working without leaking[35, 36]. They also carried out field tests of 10 alloys in FGD slurries in two U.S. power plants[37]. They concluded that only higher alloy materials were satisfactory, provided they had a good surface finish. Richard reported the use of Ferralium in the FGD absorbers at the Gibson power plants in the USA[38]. Apart from a little corrosion at the flue gas inlet, the Ferralium was satisfactory after 20+ years. The absorbers in this plant are a little unusual in that they are horizontal rather than vertical, as shown in Figure 5.23.

In addition to the absorber vessels in FGD systems, superduplex stainless steel has been successfully used in one UK power station for the gas distribution plates[39]. These are flat sheets with numerous drilled holes to thoroughly mix the rising flue gas with the falling limestone slurry. In the same station, cast superduplex has been used for the slurry recirculation pumps and agitator stools. In another UK plant, superduplex was used for the slurry return lines to replace fiberglass reinforced plastic (FRP) that failed by erosion corrosion, and for the slurry centrifuges that separate the gypsum from the slurry (Figure 5.26). When inspected after 10 years, the centrifuges were in excellent condition[39].

FIGURE 5.26 Superduplex stainless centrifuge basket from a UK FGD system. [Photo courtesy of Rolled Alloys]

REFERENCES

1. R. Francis and D.R. McIntyre, "The Importance of Redox Potential on Corrosion Control," *MP*, 55, 5 (2016): p. 48.
2. R. Francis and G. Byrne, "The Corrosion of Duplex Stainless Steels in Sour Service," CORROSION/94, paper no. 64 (Houston, TX: NACE International, 1994).
3. ASTM G48, "Standard Test Methods for Pitting and Crevice Corrosion Resistance of Stainless Steels and Related Alloys by Use of Ferric Chloride Solution" (West Conshohocken, PA: ASTM International).
4. S. Ekman, M. Johansson, J.Y. Jonsson, and E. Johansson, "Inhibited Ferric Chloride Corrosion Test of Lean Duplex Stainless Steel Welds," CORROSION 2014, paper no. 4076 (Houston, TX: NACE International, 2014).
5. NORSOK, M-630, "Material Data Sheets for Piping, Edition 5" (Oslo, Norway: Standards Norway, 2010).
6. G. Byrne, R. Francis, G. Warburton, and M. Maligas, "The Selection of Superduplex Stainless Steel for Oilfield Applications," CORROSION 2004, paper no. 123 (Houston, TX: NACE International, 2004).
7. R. Francis, G. Warburton, G. Byrne, and J. Wilson, "QA/QC Tools to Ensure the Quality of Duplex Fittings and Other Components," Duplex Stainless Steel 2010 (Zutphen, Netherlands: KCI Publishing, 2010).
8. ASTM G150, *Standard Test Method for Electrochemical Critical Pitting Temperature Testing of Stainless Steels* (West Conshohocken, PA: ASTM International).

9. J.W. Oldfield and W.H. Sutton, "Crevice Corrosion of Stainless Steels, Parts I and II," *Brit. Corr. J.*, 13 (1978): pp. 13 and 104.
10. J.W. Oldfield, "Crevice Corrosion of Stainless Steels in Seawater," *Bull. Electrochem.*, 3, 6 (1987): p. 597.
11. ASTM G78, "Standard Guide for Crevice Corrosion Testing of Iron-Base and Nickel-Base Stainless Steels in Seawater and Other Chloride-Containing Environments" (West Conshohocken, PA: ASTM International).
12. T. Rogne, S. Huizinga, E. Rislund, B. Espelid, G. Ferrari, A. Mollica, D. Feron, L. Wegrelius, U. Kivisakk, and O. Strandmyr, "Crevice Corrosion Testing of Stainless Steels in Natural and Synthetic Seawater—Results from a Round Robin Test Involving 19 Laboratories," Stainless Steel America 2004 (Zutphen, Netherlands: KCI Publishing, 2004): p. 17.
13. R.M. Carranza, C.M. Giordano, M.A. Rodriguez, and R.B. Rebak, "Electrochemical Methods for Repassivation Potential Measurements," CORROSION 2009, paper no. 09427 (Houston, TX: NACE International, 2009).
14. S. Mamang and R. Pettersson, "Localized Corrosion of Stainless Steels Depending on Chlorine Dosage in Chlorinated Water," EUROCORR 2011, European Federation of Corrosion (London, United Kingdom: Institute of Materials, Mining and Minerals, 2011).
15. R. Francis, G. Warburton, G. Byrne, and J. Wilson, "Electrochemical Potential and the Corrosion Resistance of Stainless Steels in SWRO Applications," International Desalination Conference (Topsfield, MA: International Desalination Association, 2009).
16. R. Francis, *Galvanic Corrosion: A Practical Guide for Engineers, 2nd edition* (Houston, TX: NACE International, 2017): p. 10.
17. R. Francis, *The Selection of Materials for Seawater Cooling Systems: A Practical Guide for Engineers* (Houston, TX: NACE International, 2006): p. 113.
18. C.W. Kovach and J.D. Redmond, "Correlations Between the Critical Crevice Temperature, PRE Number and Long Term Crevice Corrosion Data for Stainless Steels," CORROSION/93, paper no. 267 (Houston, TX: NACE International, 1993).
19. J.O. Olsson and M.M. Snis, "Don't Repeat Mistakes! An SWRO Plant Case Study," IDA World Congress, paper no. SP050-036 (Topsfield, MA: International Desalination Association, 2005).
20. E.B. Shone, R.E. Malpas, and P. Gallagher, "Stainless Steels as Replacement Materials for Copper Alloys in Sea Water Handling Systems," *Trans. Inst. Mar. Eng.*, 100 (1988): p. 193.
21. R. Francis, "Effects of Chlorine on Corrosion of High Alloy Stainless Steel in Seawater," Stainless Steel '87 (London, United Kingdom: Institute of Materials, Mining and Minerals, 1987): p. 192.
22. R.M. Kain, "Crevice Corrosion Resistance of Duplex Stainless Steels in Chloride-Containing Waters," Duplex Stainless Steels '97 (Zutphen, Netherlands: KCI Publishing, 1997): p. 627.
23. A.W. Zeuthen and R.M. Kain, "Crevice Corrosion Testing of Austenitic, Super Austenitic, Super Ferritic and Superduplex Stainless Steel Alloys in Seawater," in ASTM STP 1300, *Corrosion Testing in Natural Waters: Second Volume* (West Conshohocken, PA: ASTM International, 1996).
24. R. Francis and G. Byrne, "Experiences with Superduplex Stainless Steel in Seawater," *Stainless Steel World*, Vol. 16 (Zutphen, Netherlands: KCI Publishing, June 2004): p. 53.
25. R. Francis and G. Byrne, "Gasket Selection for Stainless Steels in Seawater," *MP*, 46, 10 (2007): p. 50.
26. T. Rogne and R. Johnsen, "The Effects of CO_2 and Oxygen on the Corrosion of S31254 and S31803 in Brine Solutions," CORROSION/96, paper no. 295 (Houston, TX: NACE International, 1996).
27. R. Francis, G. Byrne, and G. Warburton, "The Corrosion of Superduplex Stainless Steel in Different Types of Seawater," CORROSION 2011, paper no. 11351 (Houston, TX: NACE International, 2011).
28. R. Francis, G. Byrne, G. Warburton, and S. Hebdon, "Applications and Experiences with Superduplex Stainless Steel in Wet FGD Scrubber Systems," CORROSION/98, paper no. 479 (Houston, TX: NACE International, 1998).

29. R. Francis and S.A. Hebdon, "A Comparison of the Corrosion Resistance of Cast and Wrought Superduplex Stainless Steel," 58th T&O Conference of the Steel Founders Society of America (Crystal Lake, IL: Steel Founders Society of America, 2004).
30. M. Hagen, W. Schoffler, and W. Bollinger, "Corrosion Resistance of Cast DSS in the FGD Process containing Fluoride," Duplex Stainless Steels '97 Conference (Zutphen, Netherlands: KCI Publishing, 1997): p. 923.
31. J.P. Audouard, M. Verneau, and J. Grocki, "Corrosion Performance and Field Experience with Superduplex and Super Austenitic Stainless Steels in FGD Systems," CORROSION 2000, paper no. 828 (Houston, TX: NACE International, 2000).
32. L. Shoemaker, J. Crum, D. Maitra, R. Muro, and M. Neighbor, "Recent Experience with Stainless Steels in FGD Air Pollution Control Service," CORROSION 2011, paper no. 167 (Houston, TX: NACE International, 2011).
33. B.S. Phull, W.L. Mathay, and R.W. Ross, "Corrosion Resistance of Duplex and 4–6% Mo-Containing Stainless Steels in FGD Scrubber Environments," CORROSION 2000, paper no. 578 (Houston, TX: NACE International, 2000).
34. B. Tossey, "Corrosion in Wet Flue Gas Desulfurization Absorbers," *Power Eng.* 115, 1 (2011): p. 20.
35. B.M. Tossey, B. Padgett, and J. Shingledecker, "Technical Root Cause Analysis of Localized Corrosion in Wet Flue Gas Desulfurization Slurry at Coal Fired Power Stations," CORROSION 2014, paper no. 4442 (Houston, TX: NACE International, 2014).
36. B.M. Tossey, A. Agrawall, and J. Shingledecker, "Effect of Postweld Clean Up Method on the Localized Corrosion Resistance of Duplex Stainless Steels," CORROSION 2014, paper no. 4384 (Houston, TX: NACE International, 2014).
37. B.M. Tossey, B. Padgett, and J. Shingledecker, "Field Testing of Ten Corrosion Resistant Alloys in Wet Flue Gas Desulfurization Absorber Slurries," CORROSION 2014, paper no. 4441 (Houston, TX: NACE International, 2014).
38. R.L. Richard, "20+ Years of Successful FGD Experience with Superduplex at the Gibson Generating Station," Duplex Stainless Steel 2007 (Zutphen, Netherlands: KCI Publishing, 2007).
39. R. Francis, "The Use of Zeron 100 in Flue Gas Desulfurization Plant," Rolled Alloys Technical Report TN780R6 (Temperance, MI: Rolled Alloys, 2008).

6 Stress Corrosion Cracking

6.1 INTRODUCTION

Stress corrosion cracking (SCC) requires both a stress and an aggressive environment to initiate and propagate. With different alloys, the environments that can cause SCC vary, but for duplex stainless steels they are chlorides, sulfides, caustic solutions, and hydrogen. The reason for selecting a duplex stainless steel is often because of its good resistance to SCC, particularly when compared with 300 series austenitic alloys. The SCC resistance of duplex stainless steels in all these environments—as well as ways to avoid SCC—are discussed in detail by McIntyre and Francis[1]. This chapter will provide the data required to help select the most appropriate grade of duplex stainless steel, but for more detail consult reference 1. The four environments causing SCC of duplex stainless steels are discussed below, as are the problems with standard SCC test methods and the importance of selecting the most appropriate one.

6.2 CHLORIDE SCC

Figure 6.1 shows a typical chloride SCC crack in duplex stainless steel. It propagates mostly through the ferrite or along ferrite/austenite phase boundaries, although transgranular cracking through the austenite can occur[2]. It has far fewer branches than is typically seen in an austenitic stainless steel, like 316L.

Useful data can be obtained from domain diagrams, where chloride concentration is plotted against the threshold temperature for SCC. Figure 6.2 shows data compiled from manufacturers' data sheets and references 2 and 3. This data was compiled from samples that were stressed under constant load. Pettersson and Johansson[2] also show data obtained with U-bends, with much higher threshold temperatures for austenitic alloys, while those for duplex alloys were similar to the constant load results. These higher threshold temperatures exceeded those observed in service for the austenitic alloys. This highlights the importance of the stressing method when testing for SCC susceptibility.

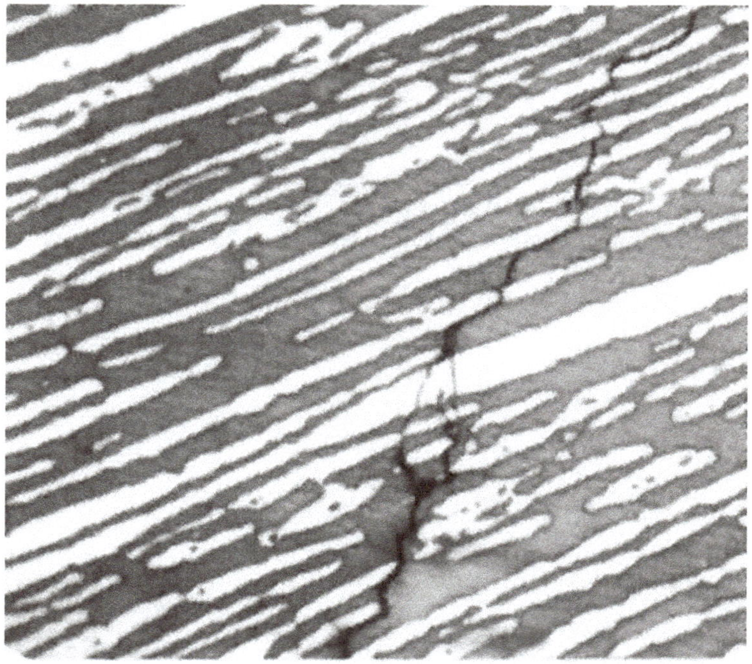

FIGURE 6.1 Typical chloride SCC crack in duplex stainless steel.

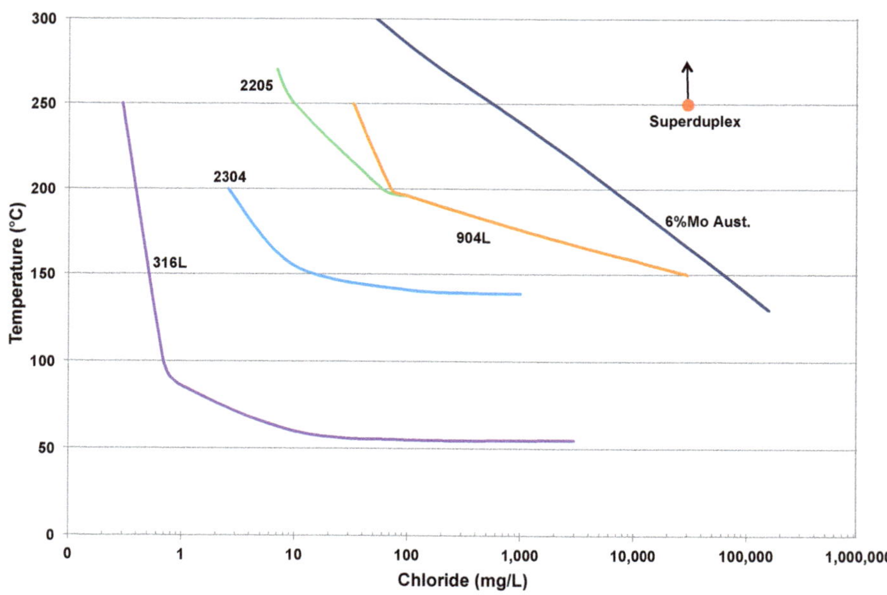

FIGURE 6.2 Threshold temperature/chloride regions for some common stainless steels. (Samples stressed under constant load.)

Cottis and Newman have criticized the results of testing with plain, smooth tensile specimens as being unrealistic compared to service experience, where crevices are often present[4]. They present data showing that SCC occurs under less severe conditions when crevices are present. Gerlock and Fritz showed that cracking of 6% Mo austenitic stainless steel occurred under the ceramic washers on loading bolts for U-bends in aerated seawater at 150 and 170°C, unlike those in pure sodium chloride solutions[3]. However, there was no cracking of the U-bends. This is because the environment in the crevice is very acidic and the crevice corrosion acts as a stress raiser, increasing the local stress.

One other point is that these domain diagrams are only valid under mildly oxidizing conditions around neutral pH. Other pH values or redox potentials can change the SCC susceptibility substantially, as discussed below.

As discussed in Chapter 2, duplex stainless steels have a phase balance around 50/50 austenite/ferrite and third phases may precipitate if the heat treatment or welding is not done correctly. Cottis and Newman[4] and Sedriks[5] show data on the effect of phase balance on SCC resistance, and it remained high across the range 30 to 60% ferrite.

The presence of third phases, such as sigma, has been observed to increase susceptibility to chloride SCC[4, 5]. Figure 6.3 shows the effect of aging time at 700°C and 800°C (producing sigma phase) on the stress/time-to-failure curve in boiling 35% magnesium chloride. The curves show that there is a reduction in resistance after five hours at 700°C, but just one hour at 800°C produces a bigger decrease in time to failure. Extended heat treatment produces even more rapid SCC failure.

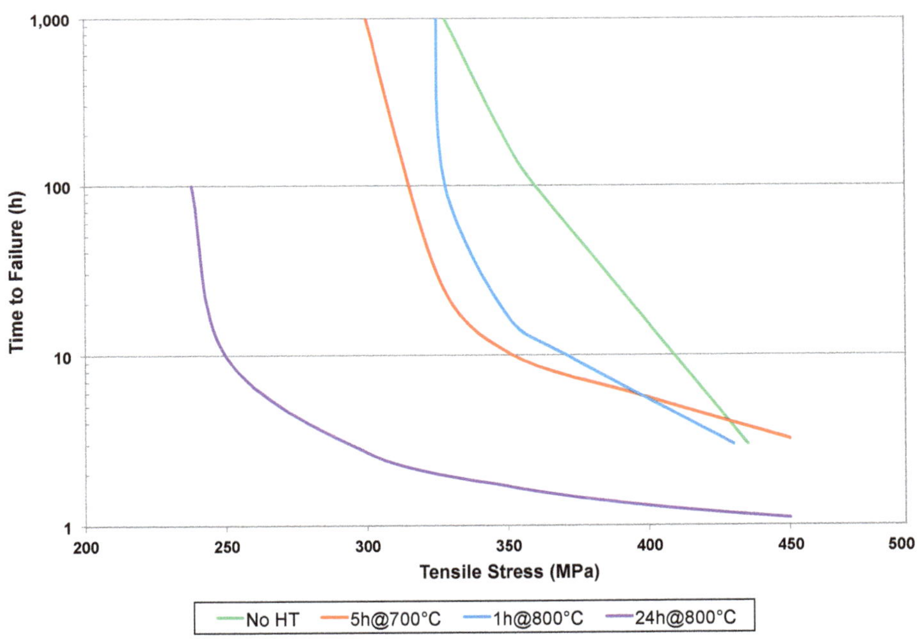

FIGURE 6.3 Effect of heat treatment (sigma) of 2205 on the SCC stress and time to failure in boiling 35% magnesium chloride[4].

Sometimes alloys are cold worked to increase their mechanical strength. In the process industries, the author is only aware of lightly cold worked (5–10%) Z100 superduplex bar being used for fasteners, with equivalent strength to ASTM A193 Gr B7 steel. Cottis and Newman state that increasing levels of cold work increase the susceptibility of duplex stainless steels to SCC, but point out that this is well documented for sulfide SCC (see next section), but is not well researched for chloride SCC[4]. The author has never seen any chloride SCC of cold worked superduplex bolting, but these fasteners are rarely loaded beyond 60–70% of the 0.2% proof stress, so this is not surprising.

Laitinen and Hanninen compared the SCC resistance of 2205 and 2507 components produced by both hot isostatic pressing (HIP) and conventional forging[6]. They tested 4-point bent beams over a range of stresses in 50% calcium chloride at 100°C and pH 6.5–7.0. They found that the HIP 2205 performed better than the conventionally forged material, but the HIP superduplex was inferior to conventional forged material. When the HIP superduplex material was forged and solution annealed after HIP, it performed the same as a forging.

With some duplex product forms, such as hot rolled plate, the properties can be different in the longitudinal and transverse directions. This is because the laths of austenite become orientated in the primary rolling direction. This means that the susceptibility to SCC can be different depending on orientation, with duplex alloys being more susceptible when stressed in the transverse direction[7].

It has been demonstrated that there is a threshold stress for SCC of many alloys, including duplex stainless steels[1]. However, in practical plant design it is common to assume that some components will end up with local stresses (residual plus applied stress) close to the 0.2% proof stress. Hence, it is the SCC resistance at that stress that is of chief interest.

The data presented above have not really considered pH and potential as variables, but Cottis and Newman point out the importance of both these variables and the fact that a change in either can shift an alloy from a noncracking to a cracking domain[4]. Figure 6.4 shows a schematic polarization curve for a stainless steel in an aggressive solution, which includes aggressive chloride-containing solutions. Under neutral or slightly oxidizing conditions, the alloy will be in the passive region and SCC is not possible. However, if conditions, in the bulk solution or locally, move the potential to the transpassive or active boundary regions, then SCC is possible. Such a change in potential might be caused by a variety of mechanisms. Some metal species, such as copper and iron, are oxidizing when in solution and will move the potential towards the transpassive region. The actual potential change will depend on the species and its concentration. Species such as nitrites and sulfites are reducing and will move the potential towards the active region. Coupling a duplex stainless steel to a less noble alloy, such as steel or a copper alloy, may move the potential into the active region, if the anode area is sufficiently large. A couple with a more-noble alloy is unlikely, but graphite is noble to most stainless steels and might produce a potential change to the transpassive region if the area was sufficiently large.

A change of pH can also cause similar changes, in that the potential of both the transpassive and active regions change with increasing acid content.

Spaehn showed the effect of potential on the time to failure of 304 and 2205 in 35% magnesium chloride at 125°C, as shown in Figure 6.5[8]. The potential regions where these two alloys are susceptible to SCC are not so different, but in a less aggressive solution it is likely that the potential ranges of susceptibility would be very different, because of their differing resistance to chloride.

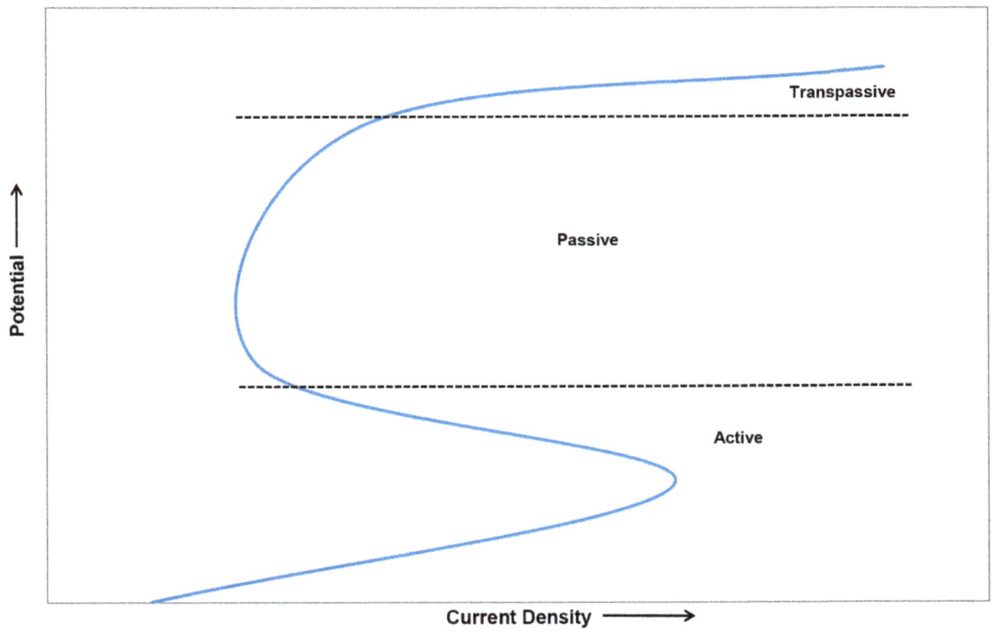

FIGURE 6.4 Schematic polarization curve for a stainless steel in an aggressive solution.

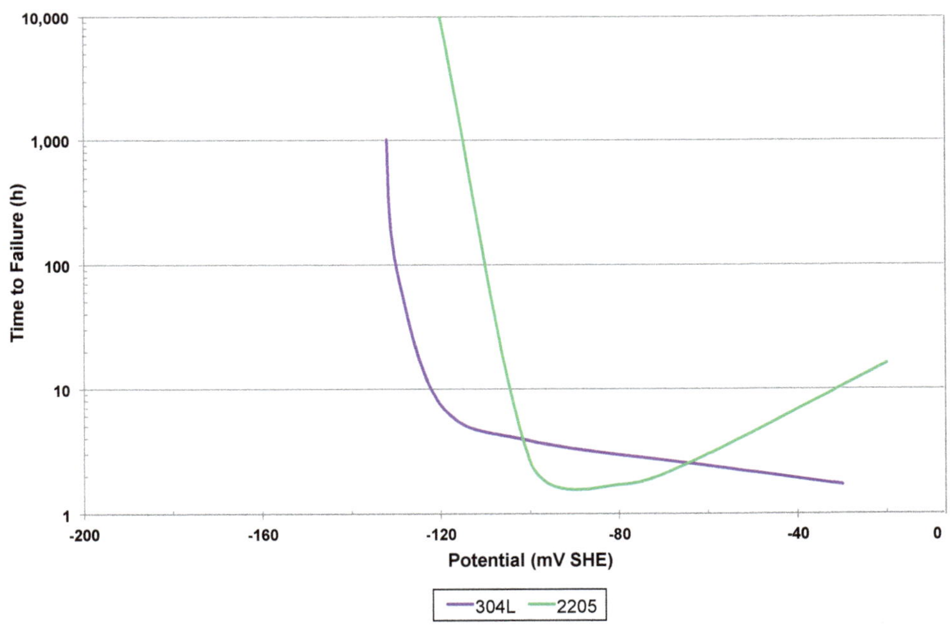

FIGURE 6.5 Effect of potential on the time to failure in 35% magnesium chloride at 125°C on samples stressed to their 0.2% proof stress[8].

Gras and Ternon tested U-bends of several stainless steels in water containing 100 mg/L chloride at both 150°C and 200°C, and 500 mg/L chloride at 250°C, at neutral pH[9]. The autoclaves were over-pressured with 3 bar of oxygen at 80°C, so the environment was extremely oxidizing. The results, in Table 6.1, show the superior resistance to SCC of 2205 compared with 316L at 150°C and 100 mg/L chloride. At 200°C, the higher alloyed austenitic alloys were superior to 2205 and Ferralium. However, no superduplex alloy was included in these tests and it is expected that they would be at least as resistant to SCC as the 6% Mo alloy.

TABLE 6.1 Results from U-bend tests in highly oxidizing low chloride waters at high temperatures[9].

TYPE	ALLOY	RESULT (Crack/ No Crack)					
		Chloride = 100 mg/L				Chloride = 500 mg/L	
		150°C	Time (h)	200°C	Time (h)	250°C	Time (h)
Austenitic	316L	C	1,000	-	-	C	500
	904L	NC	5,000	NC	2,000	C	500
	6%Mo	NC	5,000	NC	2,000	C	500
Duplex	2304	C	4,000	C	1,000	C	500
	2205	NC	5,000	C	1,000	C	500
	Ferralium	NC	5,000	C	1,000	C	500

Atkinson and Saithala and their co-workers reported a series of tests on a range of stainless steels in low chloride waters (15–1,000 mg/L) at 130°C[10, 11]. They used slow strain rate (SSR) testing and investigated the effect of potential on the resistance to cracking. Their initial work investigated the effect of potential on the strain to failure, but the reduction of area (ROA) proved a more useful indicator of SCC susceptibility. Figure 6.6 shows the ROA versus potential for 304L austenitic and Z100 superduplex stainless steels. Both alloys show a clear threshold potential at which cracking initiates. Figure 6.7 shows the results of further testing, in which they identified three zones of potential, which did not change significantly from 15–1,000 mg/L chloride[10] for Z100 superduplex stainless steel. Under neutral and slightly oxidizing conditions (0–400 mV Ag/AgCl, there was no cracking and the alloy remained passive. At strongly oxidizing potentials (≥ 600 mV Ag/AgCl), cracking initiated, and between the two potentials was a transition region where cracking initiated and became more severe as the potential moved electropositively.

Figure 6.8 shows the pitting and cracking potentials as a function of PREN in water with 15–30 mg/L chloride at 130°C[11]. It can be seen that the cracking potential is more negative than the pitting potential and the difference increases with PREN.

The results described above were in very highly oxygenated water[10] or in water with ~8 mg/L oxygen[10, 11]. Saithala and Atkinson also tested Z100 superduplex with <20 ppb oxygen and 1,000 mg/L chloride at 130°C, and commented that this increased the time to failure[12]. The most noticeable difference was the much greater reduction in area under low oxygen conditions, but they did not report the numerical values. Clearly, SCC is much harder to initiate at moderate potentials when there is no alternative reaction to the reduction of dissolved oxygen.

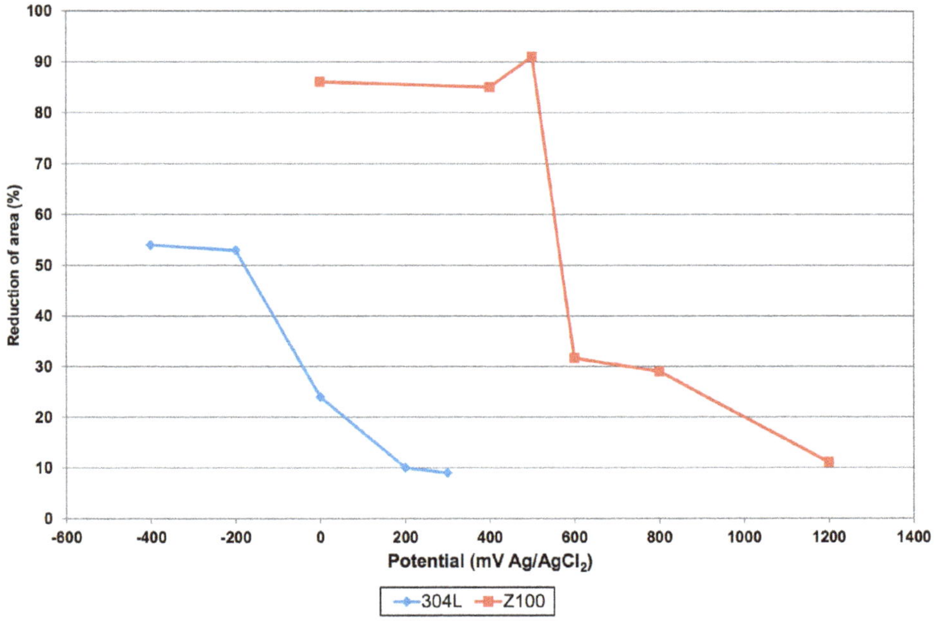

FIGURE 6.6 Effect of potential on the reduction of area in SSR tests in aerated 15–30 mg/L chloride solution at 130°C[10].

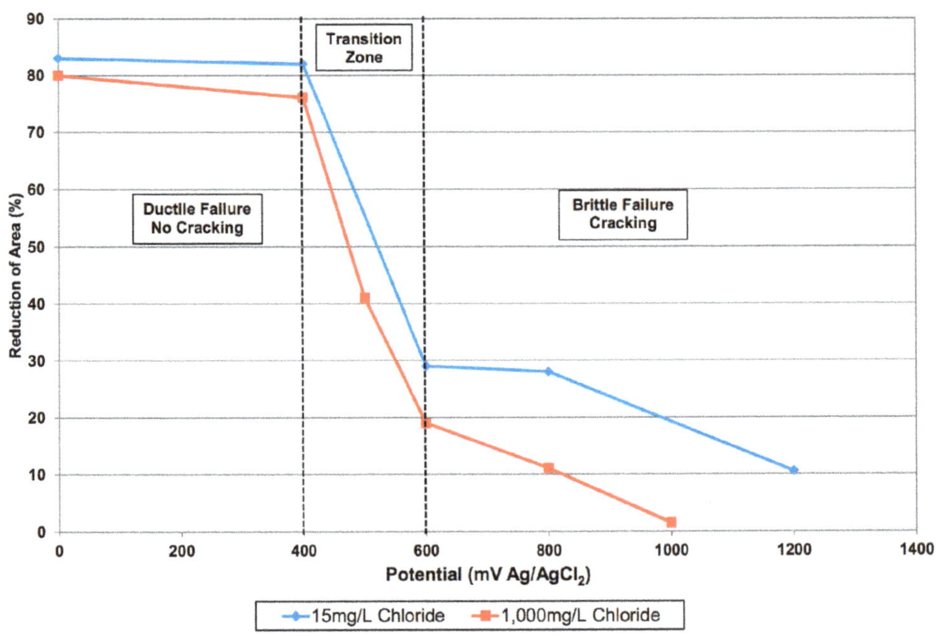

FIGURE 6.7 The three different ductility regions for Z100 superduplex stainless steel in aerated solution at 130°C[10].

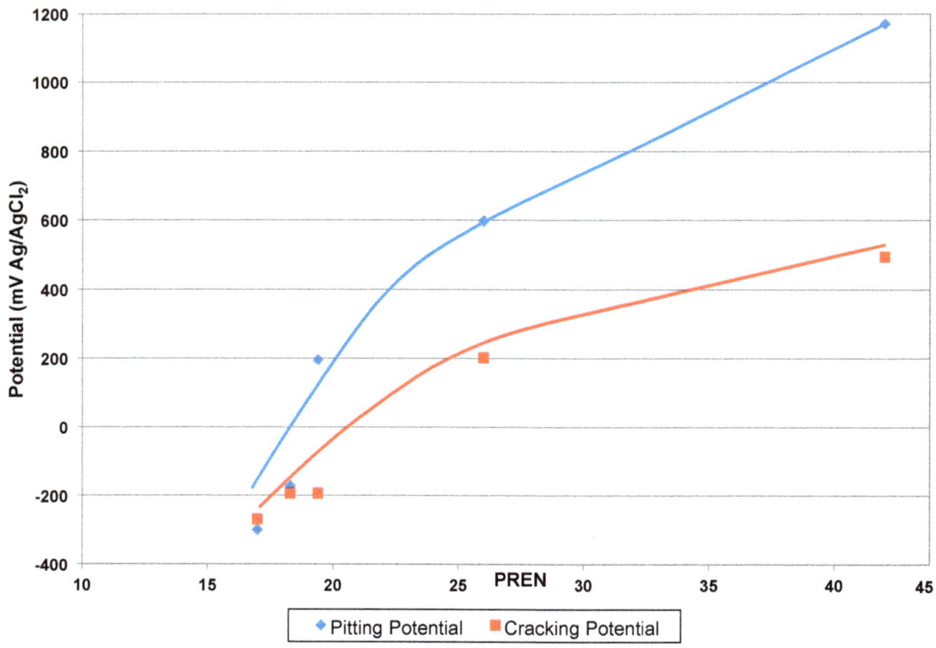

FIGURE 6.8 The relationship between pitting potential and cracking potential with PREN in aerated water with 15–30 mg/L chloride at 130°C[12].

Notten describes heat exchanger failures by SCC of 316L in a urea plant and a naphtha cracking plant[13]—both failures occurred on the shell side, which was carrying steam at 140°C and 180°C, respectively. The failures were due to traces of chloride contaminating the steam. In both cases, 2205 duplex proved a satisfactory replacement.

Shell reported an SCC failure of 2205 duplex at 140°C in methane gas containing a relatively low chloride brine. However, the water content was very low, and where there was a big reduction in pressure, evaporation occurred concentrating the brine[14]. Huizinga et al. reported that they were able to simulate the failure in deaerated laboratory tests using brines of 74% concentration and greater[14]. Superduplex cracked at brine concentrations exceeding 84%. They also commented that if the initial water content had been higher, the concentration after the pressure drop could never have increased the chloride concentration sufficiently to cause cracking. However, Turnbull et al. also tried to duplicate the failure and got no cracking of 2205 and Z100 superduplex in concentrated, fully deaerated brines at 140°C[15]. They postulated that there was a small amount of oxygen contamination in both the lab tests and the field conditions, and the reduction of this provided the cathodic reaction to enable cracking to occur. This demonstrates the effect that even small amounts of oxygen may have on SCC.

Mahajanam et al. reported SSR tests on 2205 and 2507 in a strong brine saturated with CO_2 at 80°C[16]. They tested material aged for different times to produce varying quantities of sigma phase, to simulate poor heat treatment during production. They found that the ROA only decreased significantly for sigma contents >7%. This is a high volume fraction of sigma and shows that a large degradation is required before the cathodic reaction in CO_2 can support cracking.

Cottis and Newman commented that simple laboratory tests can show immunity to SCC, but the application of a small anodic current can often result in SCC[4]. Aerts and Notten reported an SCC failure of 2205 in a PVC slurry with 20–50 mg/L chloride at pH 3 and 105°C[17]. The failure was adjacent to a weld and the failure was investigated using SSR testing in 25% calcium chloride brine at pH 3 and 107°C. There was no SCC in this solution, but the application of an anodic current caused increasing lowering of the ROA as the current was increased. They postulate that SCC occurred in the HAZ of the weld, because that was the site of the highest stresses and the galvanic current from the over-alloyed weld bead would have provided enough current to stimulate SCC.

The effect of pH on SCC in boiling 40% calcium chloride solution for several stainless steels can be found in several Sandvik data sheets. (Note that all Sandvik data sheets can be obtained directly from Sandvik at http://smt.sandvik.com/en/.) The results at pH 6.5 and 1.5 are shown in Figure 6.9 and it can be seen that

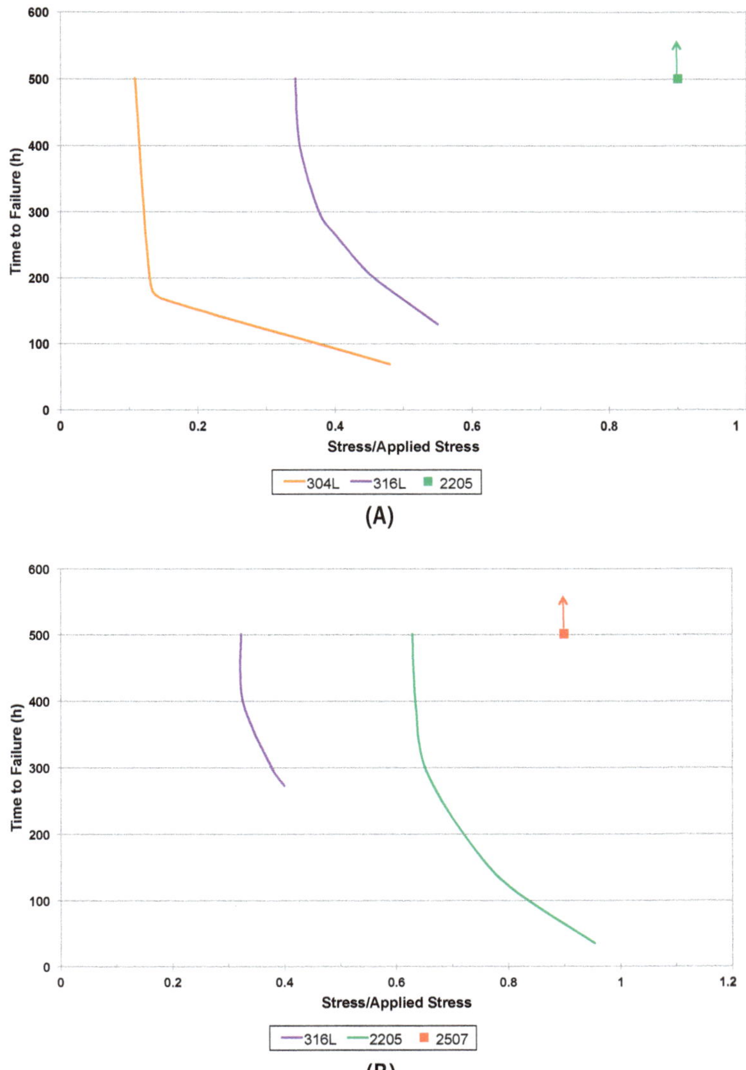

FIGURE 6.9 Threshold stress versus time to failure in 40% calcium chloride at 100°C and different pH values (taken from Sandvik data sheets). A = pH 6.5; B = pH 1.5

the reduction in pH results in significant SCC with 2205 at the lower pH. Although not shown in Figure 6.9, data at pH 3.5 showed that 2205 was immune to SCC in this solution. Under these conditions, 316L is not very resistant at pH 6.5, and a reduction to pH 1.5 only makes things slightly worse. Superduplex 2507 was not affected by the change in pH in this solution and remained resistant to SCC at pH 1.5.

Seawater is frequently used as a coolant and there is always the risk of seawater dripping onto hot process pipes, either bare or insulated, and causing SCC. This first came to prominence with the failure of a 2205 vessel on the Gyda platform in the North Sea[18]. The failure occurred in welds under insulation, which was of the rockwool type and absorbed water easily. The vessel temperature was 90–100°C and it was determined that the chlorides had concentrated under the insulation and the welds had ≥70% ferrite. TWI (formerly The Welding Institute) in the UK produced a report for the UK Health and Safety Executive describing a number of external SCC service failures and highlighting inconsistent operating practices[19]. This prompted a lot of research to determine the risk of cracking of bare pipe, because problems with insulated pipe could largely be solved by the correct choice of insulation and putting in drainage channels on lower surfaces.

Unpublished work by Newman at the University of Manchester Institute of Science and Technology (UMIST) showed that no cracking occurred when 3.5% sodium chloride solution was dripped onto hot, stressed 2205 and Z100 superduplex at 150°C, but cracking occurred readily when 1.3 g/L magnesium was added to the solution as magnesium chloride (the concentration in seawater).

Work at the National Physical Laboratory (NPL) in the UK showed that the evaporating solution always contains significant quantities of sodium and never gets hotter than ~107°C, because that is its boiling point[20, 21]. They recommended a modified drip test onto stressed samples, using improved techniques to determine the temperature where cracking occurred.

Turnbull and Hinds[20] carried out drop evaporation tests using aerated seawater with both 2205 and Z100 superduplex and found that cracking could occur at temperatures above 70°C when stressed to 100% of the 0.2% proof stress, and above 100°C when stressed to 70% of the 0.2% proof stress. Ozturk et al.[22] reported that duplex alloy 2003 resisted SCC in the drip test up to the same temperature as 2205 duplex. The Norwegian oil and gas standard NORSOK[23] permits the use of 2205 up to 100°C and superduplex up to 110°C without additional protective measures to combat external chloride SCC.

One important factor to remember is the criticality of the drip rate. If it is too fast, a salt crust never forms. If it is too slow, the bottom of the crust dries out and the metal is no longer wetted. After the Gyda failure, the author contacted a number of offshore operators who were known to have bare duplex or superduplex process pipes operating above 120°C. They all reported significant salt crusts, but when these were removed, there was no corrosion beneath them, demonstrating the criticality of the drip rate.

If duplex or superduplex is to be used above the NORSOK guideline temperatures, the most common coating used is thermally sprayed aluminum. It is not necessary to spray all the piping, only those areas considered at risk from seawater wetting. Such a policy was carried out on the Phillips Embla platform in the North Sea in the mid 1990s and it has been very successful. Since the introduction of the NORSOK guidelines, there have not been any major offshore failures due to external SCC of duplex process lines.

6.3 SULFIDE SCC

It is well known that stainless steels can suffer stress corrosion cracking in the presence of hydrogen sulfide (H_2S). These environments are essentially deaerated with pH values from 3 to ~7 and are principally found in oil and gas production and refineries. Because H_2S is very toxic, any leak could be very dangerous, so there are standards controlling the limits of use for metals in these applications. NACE MR0175/ISO 15156[24] governs the use in upstream oil and gas production, while NACE MR0103[25] applies to refineries. MR0103 has tended to take its limits of use from NACE MR0175/ISO 15156 and so these tend to be very conservative. The reason for this is that environments in refineries generally have higher H_2S than in upstream applications, but the chlorides are lower, the pH is higher, and the pressures and stresses are lower, making the environment more benign. This is described in more detail in the following paragraphs.

Figure 6.10 shows a typical sulfide SCC crack in duplex stainless steel. The crack goes through both austenite and ferrite and tends to be broader and more branched than a chloride SCC crack.

In sour (H_2S containing) environments, it is not the concentration of H_2S in solution that is specified, but rather the partial pressure of H_2S in the gas phase. NACE MR0175/ISO 15156 permits solution annealed duplex to be used up to 10 kPa H_2S for alloys with a 30<PREN<40, while alloys with a PREN >40 may be used up to 20 kPa H_2S. The alloys must have a ferrite content in the range of 35–65% without deliberate cold work or aging heat treatment. These definitions mean that all lean duplex alloys are excluded except 2003 and cast Z21 LCD. Both of these alloys have a PREN >30 and have passed tests at 90°C in 120,000 mg/L chloride with 10 kPa H_2S at pH 3.5. The inclusion of lean duplex, such as 2101, would require adequate test data to enable the creation of a new category for duplex alloys with PREN <30.

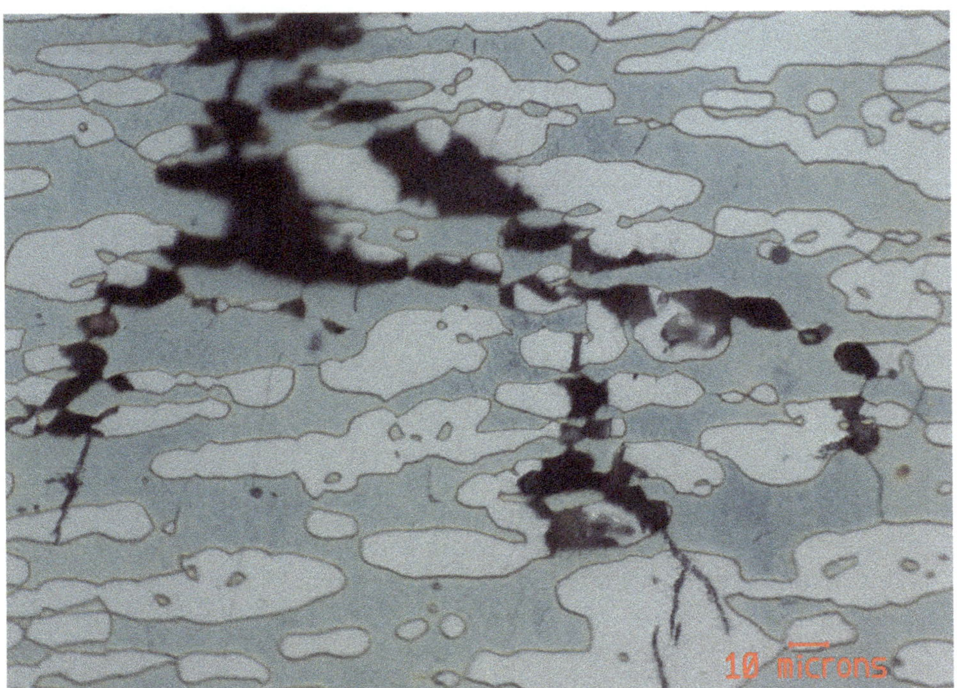

FIGURE 6.10 Typical appearance of a sulfide SCC crack in duplex stainless steel. [Photo courtesy of INTERTEK]

Further tests on alloys 2003 and cast Z21 LCD were described by O'Loughlin at higher pH values and with much higher H_2S partial pressures[26]. The conditions were 120,000 mg/L chloride with 1 or 1.5 bar H_2S at pH 4.5 and at 90°C, and 1.5 bar H_2S at pH 4. The 2003 alloy passed both tests at pH 4.5, but cracked at pH 4. The cast Z21 LDX cracked in all three tests, although the cracking with 1 bar H_2S at pH 4.5 was slight, suggesting the H_2S limit is only slightly lower.

As with chloride SCC, the method of applying the stress affects the threshold H_2S partial pressure for cracking. It has been demonstrated by numerous workers that SSR testing gives much lower threshold H_2S partial pressures for duplex stainless steels than constant load or constant strain tests, for example references 27 and 28.

Figures 6.11 and 6.12 show the effect of temperature on the threshold H_2S partial pressure for cracking of solution annealed and cold-worked materials[29]. Some of these data were from SSR testing, while some were from bent beams or C-rings. The absolute H_2S partial pressures cannot be regarded as the limits of use, but the curves show that solution annealed and lightly cold worked material can show a maximum susceptibility to sulfide SCC at 70–100°C. Heavily cold-worked material (0.2% proof stress >950 MPa) shows no such susceptibility and the H_2S threshold decreases slowly with increasing temperature.

As with chloride SCC, there is a clear trend of increasing resistance to sulfide SCC with PREN[30]. Table 6.2 shows the results from C-ring tests on three duplex stainless steels at three H_2S partial pressures. The results clearly show the superior performance of the high PREN, Z100 superduplex alloy[31].

TABLE 6.2 The effect of PREN on the susceptibility to sulfide SCC of some duplex stainless steels[31]. (All samples stressed to actual 0.2% proof stress in a solution containing 46,000 mg/L chloride with a pH = 4.3 at 80°C)

ALLOY	CONDITION*	PREN+	RESULT (Cracking/No Cracking)		
			12.5 kPa H_2S	25 kPa H_2S	37.5 kPa H_2S
2205	CW	34.1	NC/C	C/C	C/C
Ferralium	CW	38.1	NC/NC	C/C	C/C
Z100	CW	41.7	NC/NC	NC/NC	NC/NC
Z100	SA	41.2	NC/NC	NC/NC	NC/NC

* SA = Solution Annealed; CW = Cold Worked (110 ksi)
+ PREN = %Cr + 3.3(%Mo + 0.5x%W) + 16x%N

Nisbet et al.[32] examined the effect of phase balance on sulfide SCC of Z100 superduplex stainless steel, and found that the resistance to SCC increased as the austenite content increased. The biggest increase was from 15–24%, but the resistance increased further with increasing austenite content. Cottis and Newman[4] commented that the optimum resistance to sulfide SCC occurs with 40–50% ferrite.

The effect of cold work in increasing the susceptibility to sulfide SCC was discussed above. Figure 6.13 shows the effect of increasing levels of cold work on the susceptibility of 2205 to SCC in NACE TM0177 solution[33]. There have been numerous service failures by sulfide SCC of cold worked (>10%) 2205 duplex stainless steel.

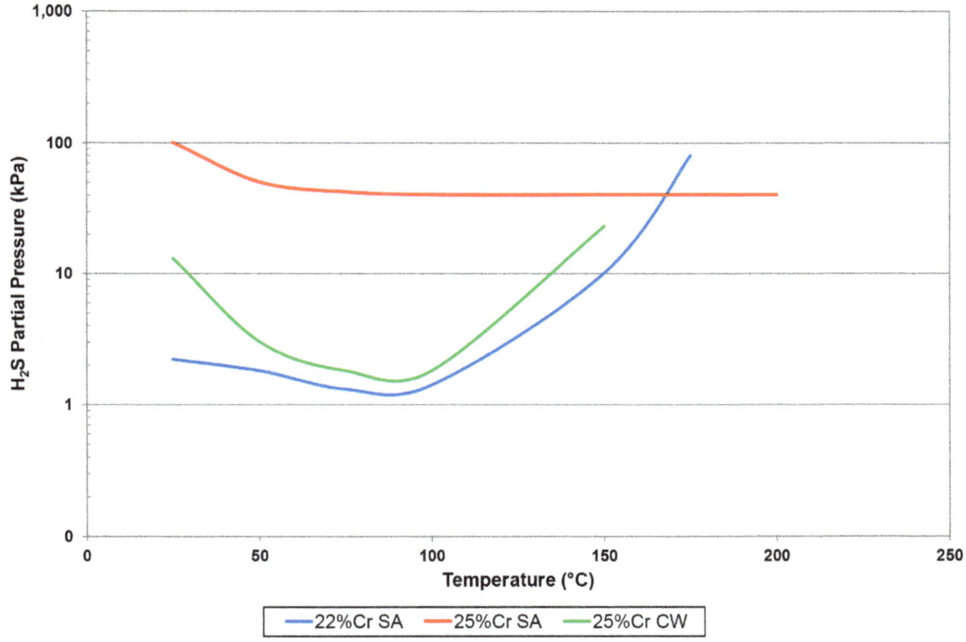

FIGURE 6.11 Effect of temperature on the threshold partial pressure of H_2S for solution annealed and lightly cold-worked duplex stainless steels[29].

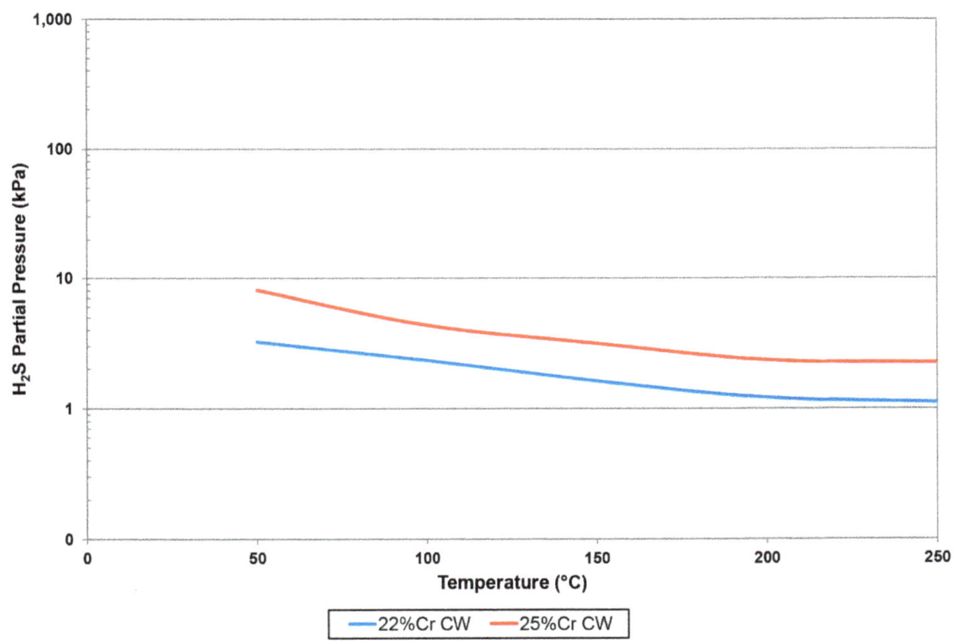

FIGURE 6.12 Effect of temperature on the threshold partial pressure of H_2S for heavily cold-worked duplex stainless steels[29].

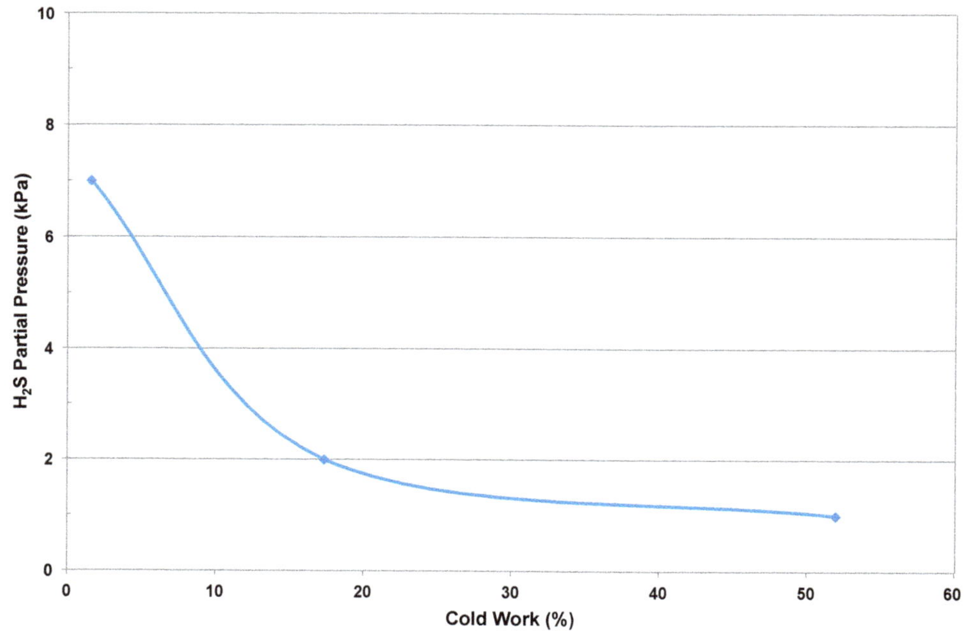

FIGURE 6.13 Threshold H_2S partial pressure as a function of cold work for 2205 in NACE TM0177 solution[33].

As mentioned in the section on chloride SCC, duplex alloys made by the HIP route can be attractive for complex shapes; HIP 2205 is approved by NACE MR0175/ISO 15156 up to the same H_2S limits as conventional material. However, superduplex HIP material has never been balloted for inclusion in the standard. An early test showed cracking of HIP superduplex under the same conditions specified as safe for superduplex in NACE MR0175/ISO 15156 (20 kPa H_2S, 120,000 mg/L chloride at pH 3.5). However, several North Sea projects have use Z100 superduplex made by HIP and it has passed the sulfide SCC tests under simulated field conditions, involving higher pH values or lower H_2S partial pressures. Some of these components have been in service since 1995.

As with chloride SCC, the resistance of poorly heat-treated material that has third phases can be reduced. The supply to the European oil and gas industry of a large quantity of poorly heat-treated fittings of Z100 superduplex–which also contained significant quantities of sigma phase–sparked a lot of research to determine the limits of use under various conditions. Rorvik et al.[34] described the problem and showed that up to 5% sigma did not affect the sulfide SCC resistance in sour, produced fluids.

Byrne et al. showed that nitrides can reduce both the toughness and corrosion resistance of duplex stainless steels[35]. Byrne et al. compared a nitride-free Z100 superduplex alloy with another commercial heat with a high nitride content[35]. Both were tested in a brine containing 120,000 mg/L chloride and 20 kPa H_2S at pH 3.5 and 90°C–the conditions that were used to get Z100 into NACE MR0175/ISO 15156. While the normal material showed no cracking, there was sulfide cracking of the material with nitrides, as shown in Figure 6.14. Nitrides can be seen both within the ferrite (darker phase) and along the ferrite subgrain boundaries.

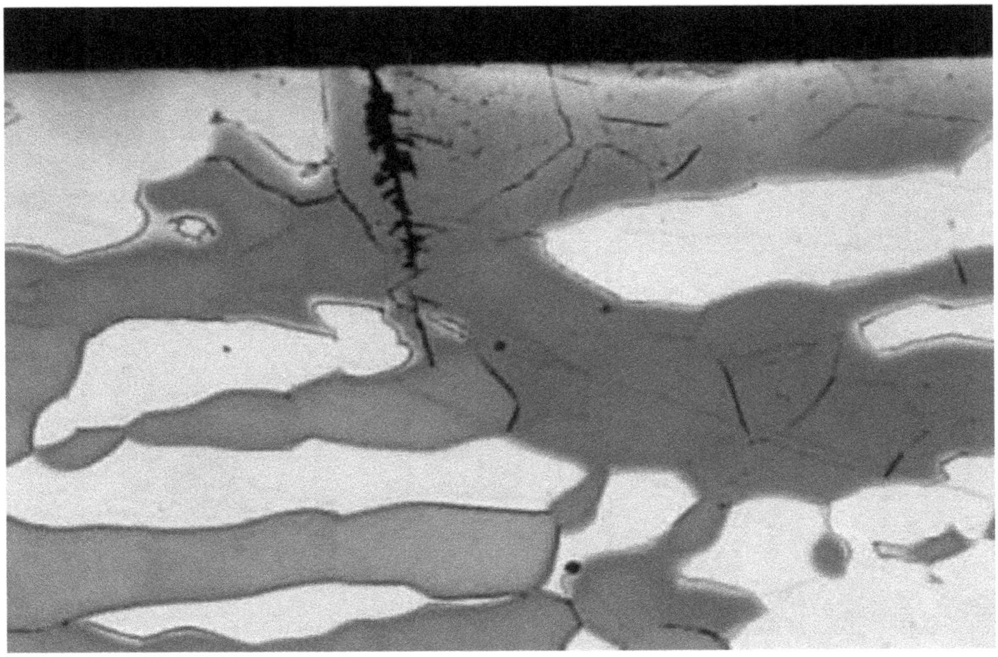

FIGURE 6.14 Sulfide stress corrosion cracks in Z100 superduplex containing nitrides[35], under conditions permitted by NACE MR0175/ISO 15156.

It has been accepted for some time that the resistance of duplex stainless steels to sulfide SCC is not just a function of the H_2S partial pressure, but also the pH and chloride concentration (e.g., references 36 and 37). Although alloy 2003 has passed a test with 10 kPa H_2S and 120,000 mg/L chloride at pH 3.5, there is no body of data showing where the cracking limits might be. Even for 2205 in the solution-annealed condition, there is a limited amount of reliable data. The curves in Figure 6.15 at pH 3.5 and 4.5 have been constructed from data in references 37, 38, and 39, and Cassagne[37] commented that more data is needed. Holmes et al. describe a SCC failure of 2205 at 80°C with 35 kPa H_2S with 100,000 mg/L chloride at pH 4, so more work is clearly needed to refine the limits of use for 2205[40]. However, even these results suggest that 2205 can be used at higher H_2S levels than allowed by NACE MR0175/ISO 15156 when the pH is higher or the chlorides are lower.

The data in Figure 6.16, for Z100 superduplex, was collected from a number of reliable sources by Francis et al.[41] and the passes and fails show a clear trend. Combining this with data at other pH values, Francis et al. produced the series of curves shown in Figure 6.17[41]. This data clearly shows the much greater H_2S levels that are acceptable at higher pH values or lower chloride concentrations. This data has been used to justify the use of superduplex rather than nickel alloys in a number of instances, and the service experience has shown that superduplex has performed well.

Woollin and Maligas[42] reported the failure by sulfide SCC of superduplex stainless steel in a brine containing 20,000 mg/L chloride and 200 kPa H_2S at pH 3.5. This point lies above the boundary curve in Figure 6.17 for pH 3.5, so the failure is not surprising.

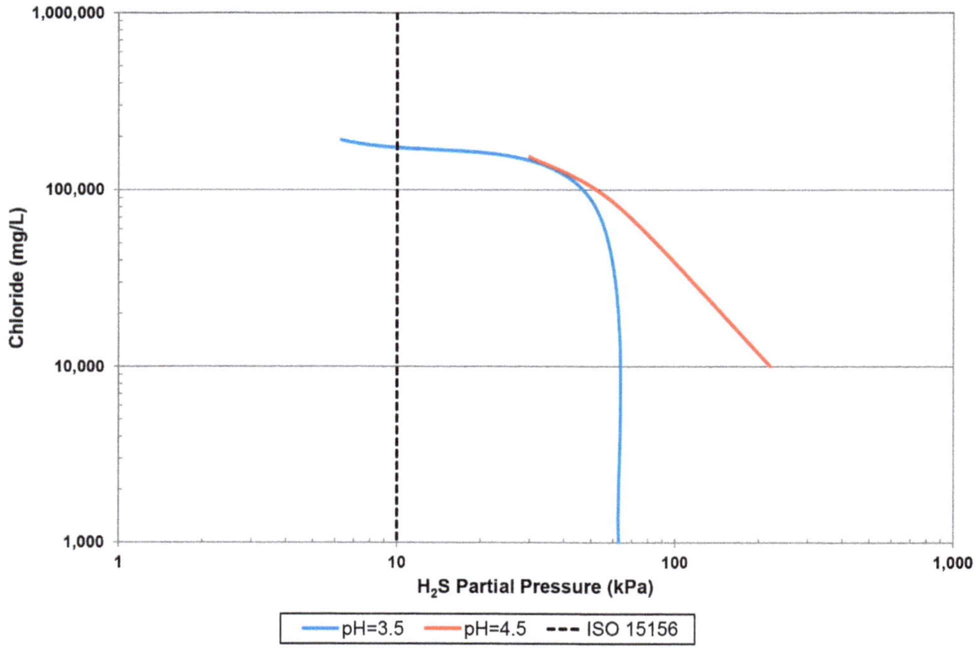

FIGURE 6.15 Effect of pH and chloride on the sulfide SCC resistance of 2205[37-39].

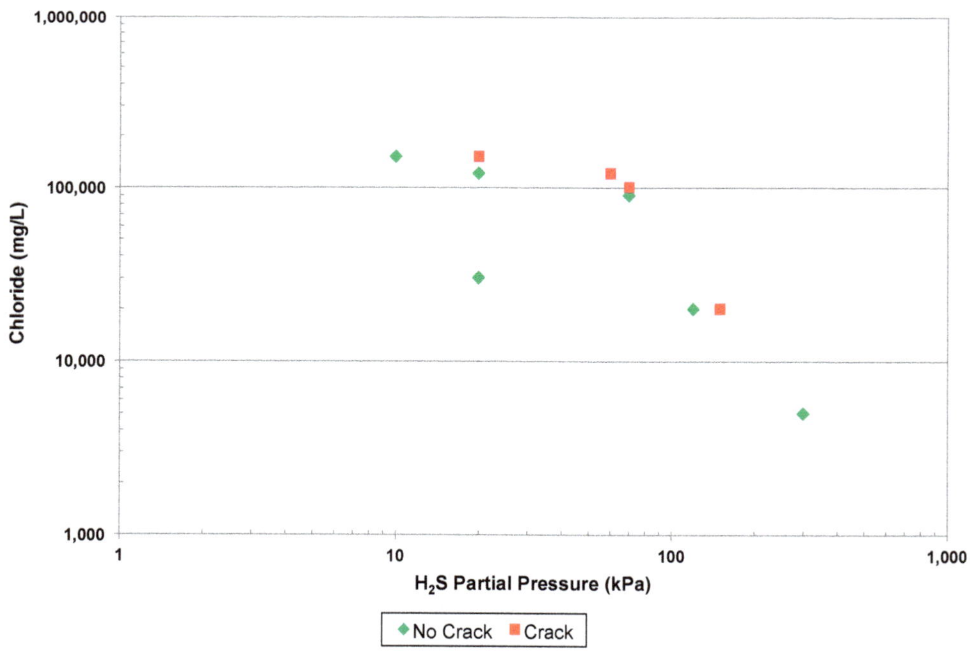

FIGURE 6.16 Results of sulfide SCC tests on Z100 superduplex at pH 3.5 from a number of sources[41].

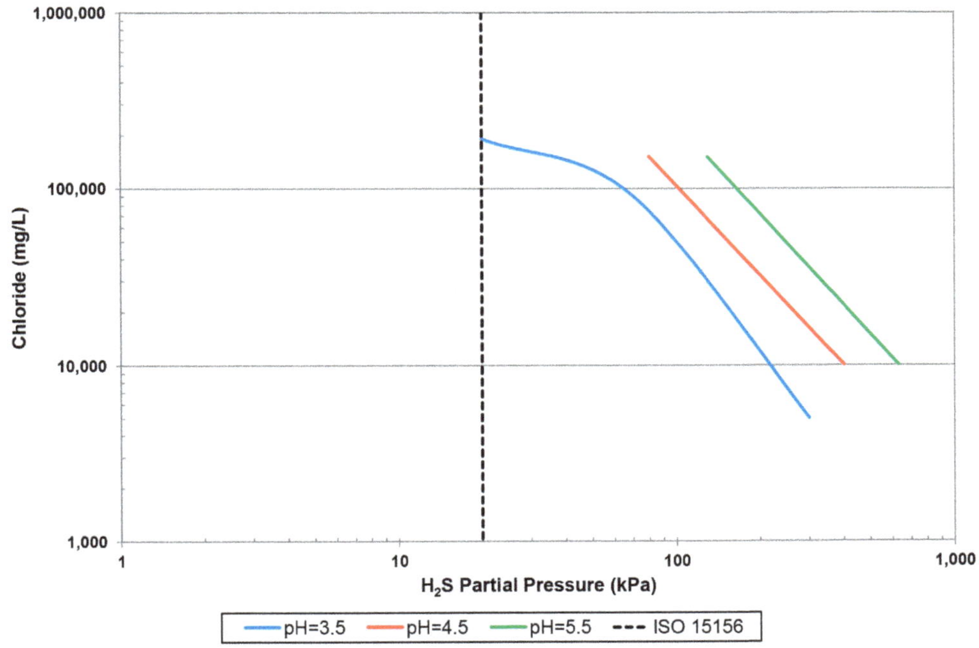

FIGURE 6.17 Effect of pH and chloride on the sulfide SCC resistance of Z100 superduplex[41].

Griffiths and Turnbull examined the cracking resistance of 2205 in sour brines when coupled to carbon steel[43] by slow strain rate testing (SSRT). They found that there was a steady increase of hydrogen uptake by the duplex stainless steel as the galvanic current increased (see also Section 6.5).

The extensive data for duplex stainless steels means that 2205 has become a workhorse corrosion resistant alloy (CRA) for the oil and gas industry for both sweet and mildly sour applications. Superduplex has also been widely used where higher strength or greater corrosion/cracking resistance is required (see Table 1.1 in Chapter 1).

6.4 CAUSTIC SCC

It is well known that austenitic stainless steels can suffer stress corrosion cracking in caustic soda at elevated temperatures[5]. Sedriks criticized some of the published data for 316L austenitic and suggested a "safe" curve at lower temperatures[5]. This is shown in Figure 6.18. Sedriks also commented that SCC seems to be more severe under oxidizing conditions. Increasing environmental restrictions on discharges in some countries means that caustic/sulfide-containing solutions must be treated to remove the sulfide. One common way of doing this is to heat the solution to around 160°C or higher with about 1 MPa oxygen pressure. Clarke reported the caustic SCC of 300 series stainless steels in Canadian pulp mills in 10–14% caustic soda at temperatures from 160–200°C[44]. Given the limits for 316L in Figure 6.18, this is hardly surprising.

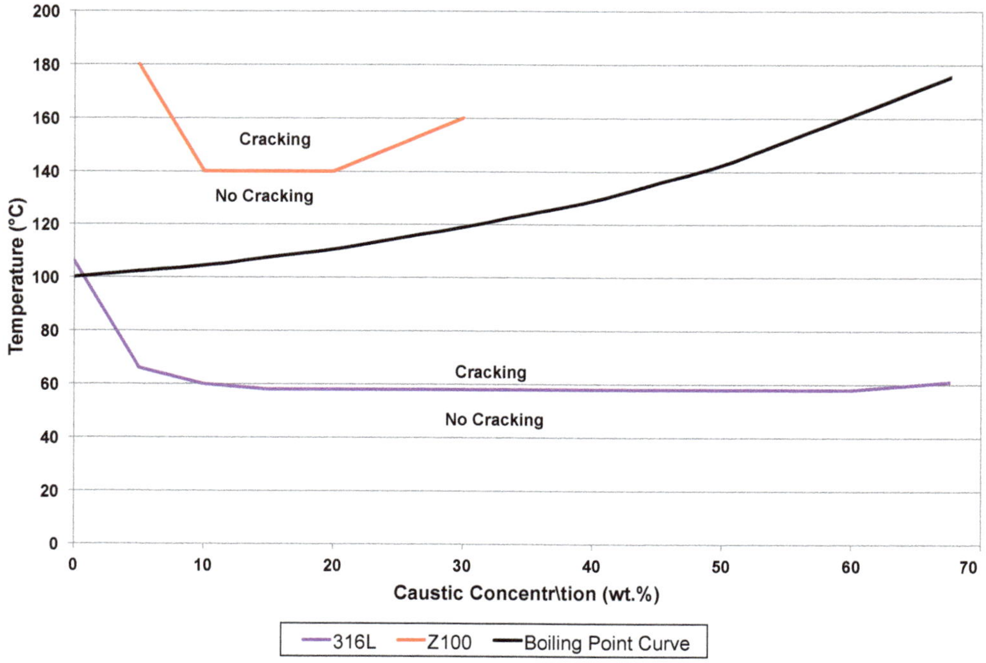

FIGURE 6.18 Caustic SCC limits for some stainless steels under oxidizing conditions.

Clarke carried out laboratory tests of 2205 in 15% sodium hydroxide plus 10% sodium sulfate at various temperatures[45]. He found SCC at 160°C, but not at 140°C. Unfortunately, there is little published data for SCC of 2205 in caustic soda. Following the failures reported above, the author carried out a series of autoclave tests of Z100 superduplex stainless steel in pure sodium hydroxide over a range of temperatures, with 1 MPa oxygen pressure. The large number of tests, recording both cracking and no cracking, enabled the limits of use curve in Figure 6.18 to be constructed. The pass/fail limits in 15% NaOH were the same as reported by Clarke[45] for 2205, but until more work on 2205 is carried out, it is not known what differences there might be in the resistance of the two alloys to caustic SCC. Figure 6.18, shows that, under oxidizing conditions, Z100 superduplex is only susceptible to SCC over a narrow range of caustic concentrations and at elevated temperatures (≥160°C). Welded C-rings were included in these tests and the Z100 welds were no more susceptible to caustic SCC than parent metal.

Oxygen is not the only species that can create the oxidizing conditions for caustic SCC of duplex. Some Z100 piping was supplied to an alumina plant that used caustic to refine bauxite. The piping was carrying ~7% NaOH plus bauxite at around 230°C, and SCC was found after ~2 years of service. The conditions were not thought to be oxidizing, but an examination of the bauxite composition being processed showed that it contained large quantities of vanadium (~0.5 g/L), a powerful oxidizer in its higher valency states. The schematic polarization curve in Figure 6.4 also applies to duplex alloys in caustic soda, and any species that moves the potential up to the transpassive transition may cause SCC.

The corrosion of duplex stainless steels in oxidizing caustic becomes intergranular, as shown in Figure 6.19. However, when cracking occurs it appears as shown in Figure 6.20, with a single main crack and few, if any branches. The cracking is mostly at ferrite/austenite boundaries or in the ferrite.

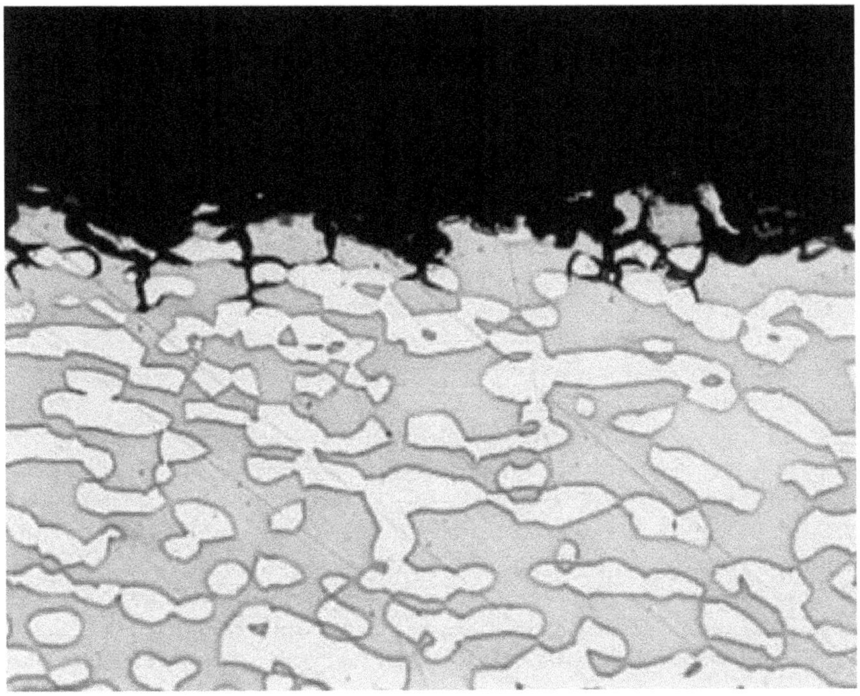

FIGURE 6.19 Intergranular corrosion of Z100 in 20% caustic at 180°C with 10 bar oxygen. [Photo courtesy of Rolled Alloys]

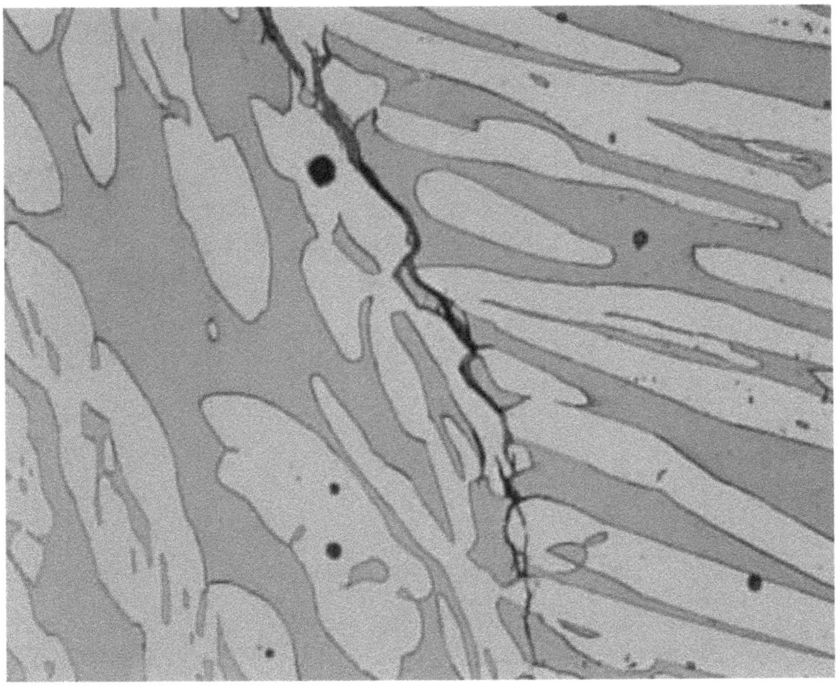

FIGURE 6.20 Caustic SCC of cast Z100 in 20% caustic soda at 160°C with 10 bar oxygen. [Photo courtesy of Rolled Alloys]

The author also investigated the effect of the oxygen partial pressure on caustic SCC of Z100 and found the relationship shown in Figure 6.21. Clearly cracking only occurs when the oxygen partial pressure exceeds ~5 bar (500 kPa).

Turnbull et al. carried out corrosion and SCC tests on both 2205 and Z100 in dilute caustic solutions (7–50 wt.%) at temperatures from 120–200°C[46]. These were under neutral/slightly oxidizing conditions, and no SCC was found with either alloy. Turnbull et al. also reported that chloride had no significant effect on corrosion or SCC of duplex alloys in caustic solutions. This is because chloride works by reducing the pH in crevices and shielded areas by hydrolysis, and the pH of caustic soda is too high for the pH to be reduced below the depassivation pH, when fresh caustic is diffusing into the crevice.

Rondelli et al. carried out tests on a number of stainless steels in 200 g/L sodium hydroxide plus 10 g/L sodium chloride at 200°C and 250°C under both aerated and deaerated conditions[47]. Note that the corrosion rate of all the duplex alloys would exceed 0.1 mm/y under these conditions with 2507 having a lower corrosion rate than 2205 (see Chapter 4). Table 6.3 shows the results for the U-bend tests, while Table 6.4 shows results for SSR tests. No cracking occurred with the duplex alloys at 200°C, but there was some cracking at 250°C. Although the lower alloyed materials appear to be more resistant to SCC, this is probably because of the higher corrosion rate of these alloys under these conditions, making it more difficult to initiate cracks, as is discussed by Cottis and Newman for chloride SCC[4]. It is interesting that Rondelli et al. found no cracking of 2507 under aerated conditions at 200°C, while the data for Z100 showed cracking at temperatures of 160°C and higher. It may be that Rondelli's oxygen pressure was in the boundary region of Figure 6.21 (1–5 bar)

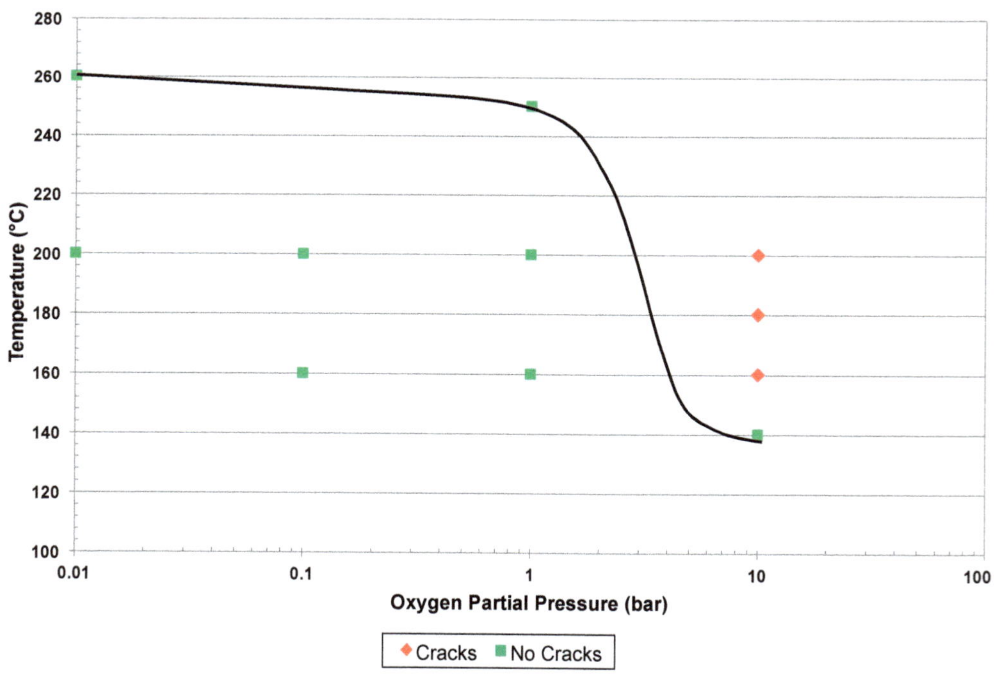

FIGURE 6.21 Effect of oxygen partial pressure on SCC of Z100 in 10% caustic soda.

TABLE 6.3 Results of U-bend tests of some stainless steels in 200 g/L NaOH plus 10 g/L NaCl[46].

ALLOY	RESULTS			
	AERATED		DEAERATED	
	200°C	250°C	200°C	250°C
316L	No Cracks	Cracks	No Cracks	Cracks
904L	No Cracks	Cracks	Cracks	Cracks
2304	No Cracks	No Cracks	No Cracks	No Cracks
2205	No Cracks	Cracks	No Cracks	Cracks
2507	No Cracks	Cracks	No Cracks	No Cracks

TABLE 6.4 Results of SSR tests of some stainless steels in 200 g/L NaOH plus 10 g/L NaCl at 200°C[46].

STRAIN RATE (s^{-1})	ALLOY	REDUCTION IN AREA (Test/Inert)	EXAMINATION
5.0×10^{-6}	316L	0.83	SCC
	904L	0.79	SCC
	2304	0.99	No SCC*
	2205	1.00	No SCC*
	2507	1.01	No SCC*
9.2×10^{-7}	2304	0.87	No SCC*
	2205	0.96	No SCC*
	2507	0.67	SCC

* Very shallow micro-cracks.

Leinonen et al. carried out U-bend tests on 2205 in 50% NaOH at 140 and 190°C[46], where the corrosion rate would have been greater than 1 mm/y[48]. Caustic SCC occurred under both conditions, but was more severe at the higher temperature.

Notten et al. reported the failure by SCC of alloy C-276 (UNS N10276) clamps on a 2205 heating coil in a fluid consisting of 20% organics, 8% sodium chloride, 4% caustic soda with the balance water, at 160°C[49]. This is not an oxidizing environment and laboratory tests showed severe cracking of C-276, while duplex and superduplex alloys suffered no significant attack. Horn et al. also reported the excellent resistance of duplex and superduplex alloys to caustic SCC under neutral/mildly oxidizing conditions[50].

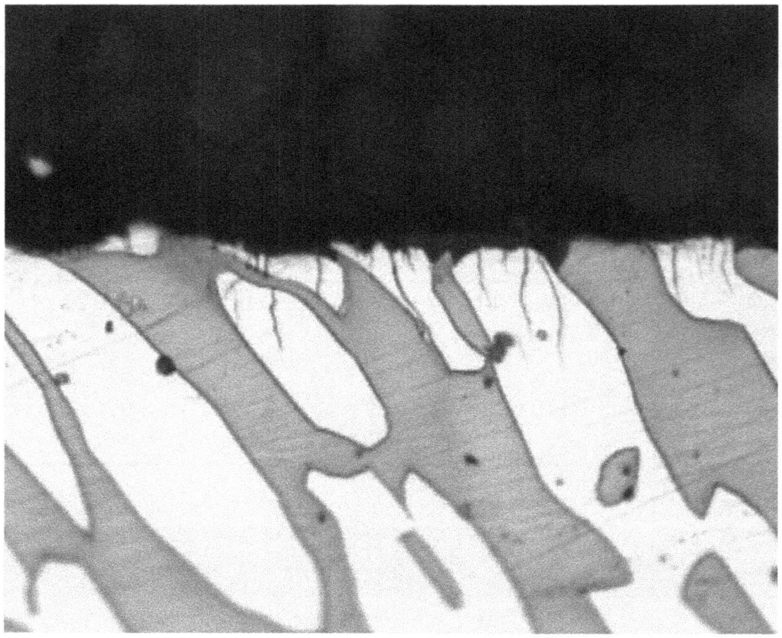

FIGURE 6.22 Appearance of caustic cracking of Z100 coupled to steel, after exposure to 10% NaOH at 240°C with 10 bar nitrogen. [Photo courtesy of Rolled Alloys]

So far, all the cracking has been observed on alloys operating around the passive/transpassive transition. However, there is no reason that SCC may not occur when the potential is around the passive/active transition. The author tested Z100 in hot caustic soda sparged continuously with nitrogen, but the potential remained in the passive region. However, it is possible that the potential could be moved into the active region if the duplex alloy were connected to a less noble alloy, like carbon steel. The author tested Z100 C-rings coupled to an equal area of carbon steel in 10% sodium hydroxide at 240°C with 10 bar nitrogen. Cracking occurred, as shown in Figure 6.22, and this is clearly very different to that in Figure 6.20, because it occurs entirely in the austenite. The cracking is very shallow compared to that in Figure 6.20, but it may become more severe if the area of carbon steel were larger, so as to depress the potential further.

6.5 PULP AND PAPER

In the Kraft pulping process, wood chips are heated under pressure in a solution of sodium sulfide and caustic soda, which is known as *white liquor*. This solution is strongly reducing. During processing, the white liquor dissolves the lignin in the wood and is converted to *black liquor*. Typical compositions for simulated liquors are 150 g/L NaOH + 50 g/L Na_2S (white liquor), and 15 g/L NaOH + 50 g/L Na_2S (black liquor). The black liquor is burned in the recovery boiler, and the molten residue "smelt" runs out from smelt spouts into a smelt-dissolving tank, where it is dissolved in water. This is called "raw green liquor." This is then filtered and clarified to remove impurities and entrained solids to make clarified green liquor.

Carbon steel was used originally for Kraft digesters, but as liquors became more aggressive and changes were made to cooking practice, corrosion and especially caustic SCC, became a problem. This was solved initially by weld overlays of 312, 309, or even alloy 625. Anodic protection was used successfully to prevent caustic cracking, and it also reduced corrosion rates.

It has been common for many years to use 2205 duplex for these vessels and for piping, but there have been a few SCC failures and substantial research has been conducted to determine the risk of SCC under Kraft conditions. The 2205 digesters have suffered far fewer failures than carbon steel or stainless steel lined carbon steel[51, 52]. Because molybdenum does not contribute much to corrosion resistance in hot caustic liquors (see Chapter 4), the lean duplex stainless steels have become more common, particularly 2304, although 2101 is also being used[52]. These alloys offer superior corrosion and stress corrosion resistance compared with 304L and 316L, as well as greater strength. Figure 6.23 shows the first 2205 batch digester for a Canadian mill after fabrication.

Within the pulp and paper industry, there are many other applications where duplex stainless steels, and particularly the lean grades, have been used successfully instead of 300 series austenitic grades (see Chapter 1). Suction rolls, which remove white water from the pulp and are very large, from 0.3 to 1.8 m diameter and from 2.5 to >10 m long. Because they are rotating, they are subject to cyclic stresses and there have been failures of austenitic stainless steels by corrosion fatigue[51]. The duplex stainless steels have superior resistance to corrosion fatigue and 2205 has been widely used for suction rolls with great success. This is discussed in more detail in Chapter 7.

Duplex stainless steels have been used for a wide range of other applications in pulp and paper plants including mechanical pulping, recycled paper mills, oxygen delignification (see Figure 1.4), black liquor evaporators (see Chapter 4) and paper making machinery. The lean duplex stainless steels are very popular because of their lower cost compared to 316L, and the additional savings possible by designing to their

FIGURE 6.23 Picture of 2205 batch digester for a Canadian pulp and paper mill.

higher strength. Black, white, and green liquor storage tanks are commonly made of lean duplex stainless steels. Case studies of all these applications and the advantages of duplex stainless steels are reviewed by Wegrelius[51, 52].

Despite the generally excellent experiences with duplex stainless steels in Kraft mills, there have been one or two SCC failures. Singh et al. used SSR testing to look at the resistance of 2304 and 2205 to SCC in black, white, and mixed liquors[53] (where "mixed liquor" is an intermediate between white and black liquor, i.e., partly processed). No cracking was found in black liquors taken from several mills or in mixed liquors at 170°C, while there was some SCC in simulated white liquor. The rate of crack propagation was approximately two orders of magnitude less than that of austenitic stainless steels, showing the greater resistance of duplex alloys to caustic SCC with sulfides. Battacharya et al. examined the performance of 2205 and 2304 welds in simulated white liquor at 170 and 200°C[54, 55]. They found that the welds were no more susceptible than parent metal to caustic SCC at 170°C, except those made by submerged arc welding (SAW). At 200°C, the susceptibility to SCC increased, but not particularly in the weld except for the SAW welds. They also found that the use of over-alloyed filler metals gave improved resistance to SCC, so that 2304 welded with 2209 filler and 2205 welded with 2509 filler, gave the best performance.

Leinonen et al. carried out similar tests to those of Singh et al., described above, using U-bends exposed to simulated white and black liquors at 190°C[56, 57]. They found SCC of 304L in both solutions, but 2304 and 2205 only cracked in the corrosion product layers, with no cracking in the metal.

Chasse et al. tested 2101 and 2205 duplex alloys in a simulated white liquor at 170°C with various chloride contents[58]. They found that chlorides favored dissolution of the austenite, particularly in the 2–10 g/L range for 2205 and the 0.5–10 g/L range for 2101. They found that SCC was more severe when increased austenite dissolution was occurring, and SCC was less severe for 2101 compared with 2205. This may have been because of the higher corrosion rate of 2101 in this solution, which reduces the risk of crack initiation, as discussed above.

In further work, Chasse and Singh[59] demonstrated that 2101 and 2205 duplex cut from hot rolled plate was more susceptible to cracking in the transverse than the longitudinal direction, although the threshold stresses were very high (>0.2% proof stress). They tested in a synthetic white liquor with various concentrations of chloride, and found that the difference between cracking in the two directions was most marked over narrow chloride concentration ranges, which were different for the two grades of duplex.

The overall conclusions of the research were that SCC of duplex stainless steels in Kraft liquors requires very high stresses (≥0.2% proof stress) and a poor microstructure.

6.6 HYDROGEN INDUCED STRESS CRACKING (HISC)

Duplex stainless steels can also suffer from cracking due to hydrogen induced stress cracking (HISC), also known as hydrogen embrittlement (HE). This occurs on stressed components when the cathodic reaction on the duplex stainless steel is hydrogen evolution and the hydrogen can diffuse into the steel. This most commonly occurs with duplex stainless steels that are used subsea and are exposed to cathodic protection. However, it could also occur with duplex alloys used underground and exposed to cathodic protection (CP).

During the 1980s, there were numerous failures of subsea duplex bolting in the North Sea that were exposed to CP. Figure 6.24 shows some 25% Cr duplex bolts that failed on a subsea wellhead[60]. This failure was due to an unsuitable heat treatment and it is discussed in more detail below.

It is not necessary for CP to get HISC of duplex stainless steels. If a duplex alloy is connected to a significant area of carbon steel in an environment where hydrogen evolution is the cathodic reaction, then HISC can also occur. Figure 6.25 shows a 25% Cr duplex cold-worked tubular from the Marathon East Brae field in the North Sea[61]. The cracking did not occur in service (at 120°C)–it occurred when the tubulars suffered

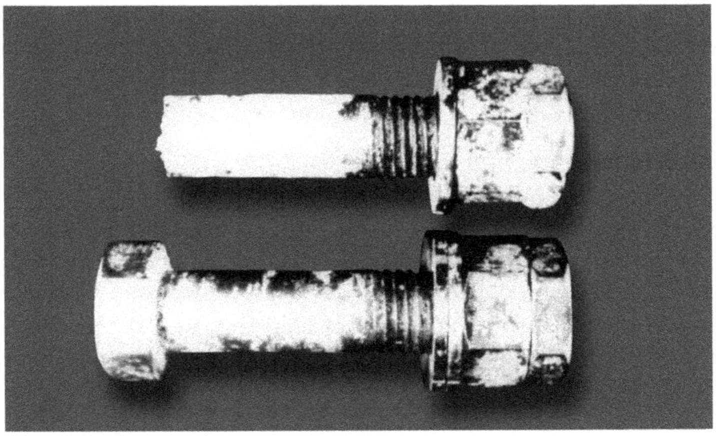

FIGURE 6.24 25% Cr duplex stainless steel bolt (UNS S32550) that failed by HISC[60].

FIGURE 6.25 25% Cr cold-worked tubular that failed by HISC due to coupling with carbon steel[61].

local mechanical overload when being removed during a routine workover. The tubulars were connected to a large area of carbon steel in the form of the well casing.

Cottis and Newman pointed out that HISC of duplex stainless steels is more likely at low pH values, where hydrogen evolution is easier and the passive film is less robust, so that hydrogen can enter the steel more easily[4].

In the late 1990s there was an HISC failure of a superduplex stainless steel hub forging on the Foinaven oilfield[62]. This was followed by a couple of failures of 2205 on subsea equipment and a further failure of superduplex[63]. All of these failures were associated with high local stresses. Because of the safety implications, a lot of research was undertaken to understand the phenomenon better and define how to avoid failures in future.

6.6.1 Effect of Potential, Temperature, and Pressure

Figure 6.26 shows the effect of cathodic potential on the elongation ratio from slow strain rate tests[60]. The results show that embrittlement gradually becomes more severe at potentials of -800 mV SCE and below, and the structure is very brittle at potentials of -900 mV SCE and below. Figure 6.27 shows the threshold strain for crack initiation and propagation from constant load tests as a function of potential[64]. The results show that high strains are required at potentials down to -800 mV, but at -900 mV and below, cracks initiate at relatively low strains, but require much higher strains (>10%) to propagate.

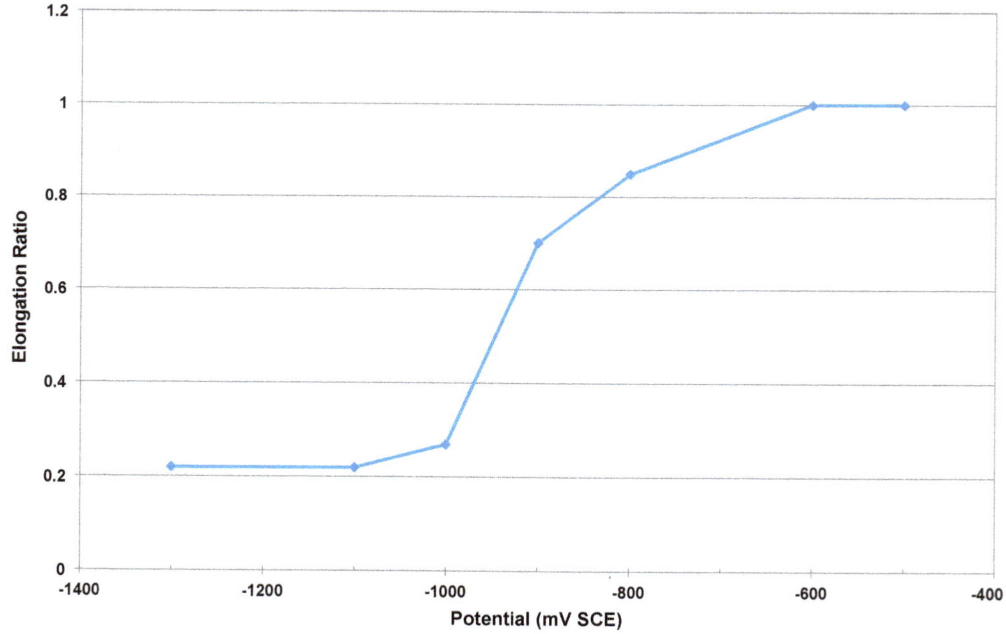

FIGURE 6.26 Effect of cathodic potential on the elongation ratio for Z100 superduplex in 3% sodium chloride solution[60].

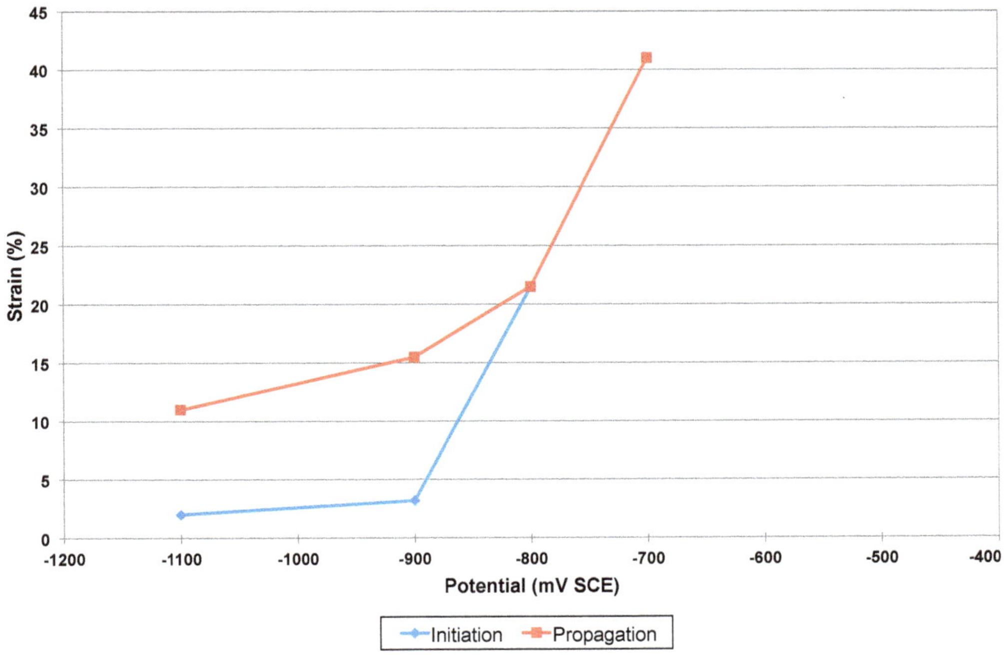

FIGURE 6.27 Effect of cathodic potential on the strain for crack initiation and propagation[64].

Griffiths and Turnbull showed that when 2205 duplex stainless steel is connected to carbon steel in a low-pH (3.5) brine, there is a limiting current density (~100 μA/cm^2), above which there is no increase in the rate of hydrogen entry[43]. This is because hydrogen atoms are combining to form molecules faster than atomic hydrogen can diffuse into the metal. However, if there is a chemical present that can poison the recombination reaction, such as H_2S, then the hydrogen content continues increasing with current density.

Chasse et al. investigated the effect of coupling 2205 to carbon steel in a synthetic white liquor (see section 6.5) using slow strain rate testing[65]. They only found a significant decrease in the reduction of area at potentials of −1100 mV SCE and below, compared with an open circuit potential of −700 mV. This means that a large cathodic current density would be needed to achieve such potentials and increase the risk of embrittlement.

It has been commented that duplex stainless steel used subsea may be at depths with high external pressure, which might affect hydrogen embrittlement. Bahrami et al. tested duplex under cathodic protection at both 1 bar and 100 bar external pressure and found no difference in the threshold stress for crack initiation[66].

When duplex stainless steels are carrying hot process fluids while under CP in seawater, the high temperature might affect hydrogen embrittlement susceptibility. Cottis and Newman commented that hydrogen embrittlement susceptibility should be at a maximum at around 80°C, with a lower susceptibility around room temperature[4]. However, Bahrami and Woollin found no difference in the threshold stress for crack initiation at 20°C and 80°C[66].

6.6.2 Effect of Structure

HISC occurs in the ferrite and large laths of austenite and can block the cracks, provided the stress is below that required to mechanically fracture the austenite[4]. Figure 6.28 shows a HISC crack from a notch, being blocked by austenite[60]. Hence, the phase balance can affect the susceptibility to HISC, with higher ferrite contents increasing the susceptibility. Murata et al. experimented with modified 25% Cr duplex compositions to give a range of austenite contents, and determined the threshold stress for cracking (Figure 6.29)[67]. The results show a decrease in the threshold stress above ~55% ferrite.

Woollin and Gregori showed that many duplex stainless steel product forms have a threshold stress for crack initiation around the actual 0.2% proof stress, while the threshold stress for crack propagation is around 130% of the proof stress[63]. Because of the necessity for dynamic straining, Woollin and Murphy defined the threshold for initiation at a total strain of ~0.5%[68]. Bahrami and Woollin found similar threshold stresses for crack initiation with full size pipes[66]. There is no effect of alloy composition on the hydrogen embrittlement of duplex stainless steels, and 2205 and superduplex have been shown to have the same relative threshold stresses (actual stress/actual 0.2% proof stress) for crack initiation[69].

Light cold work has been shown to have only a minor effect on embrittlement resistance, but the effect of heavy cold work has not been well researched. However, it is expected that heavy cold work will increase the susceptibility to embrittlement[4].

Alpha prime phase has been shown to affect HISC resistance and Figure 6.30 shows the results of slow strain rate tests on lightly cold worked Z100 superduplex and material aged at 475°C[60] to produce alpha prime phase to increase the strength. Both these methods have been used to increase the strength of 25% Cr

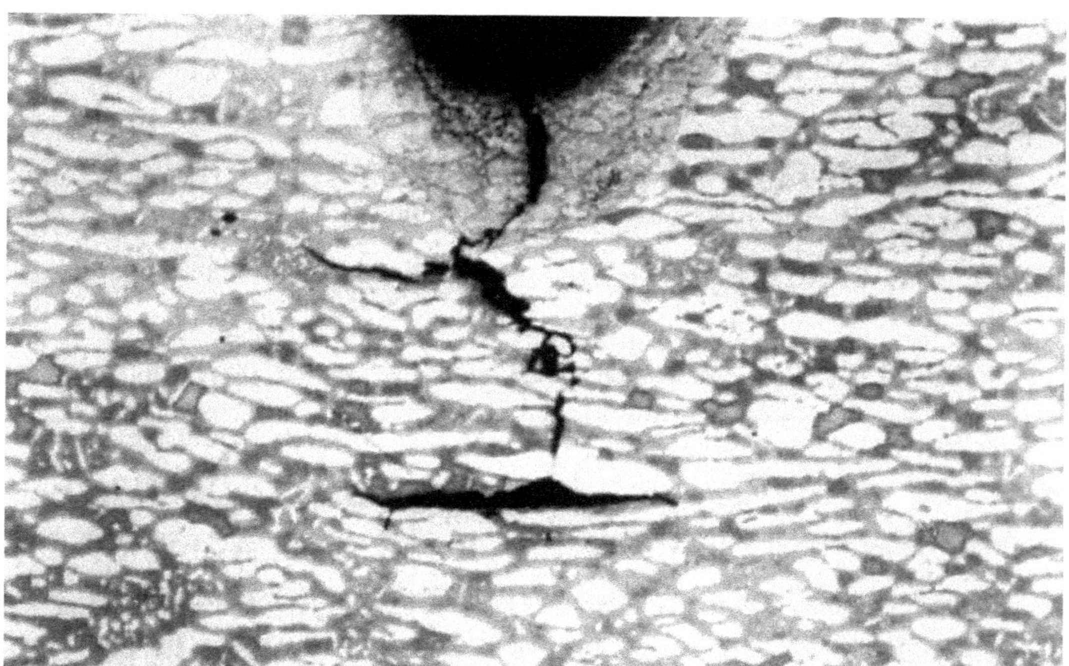

FIGURE 6.28 HISC crack from a notch being blocked by austenite[60].

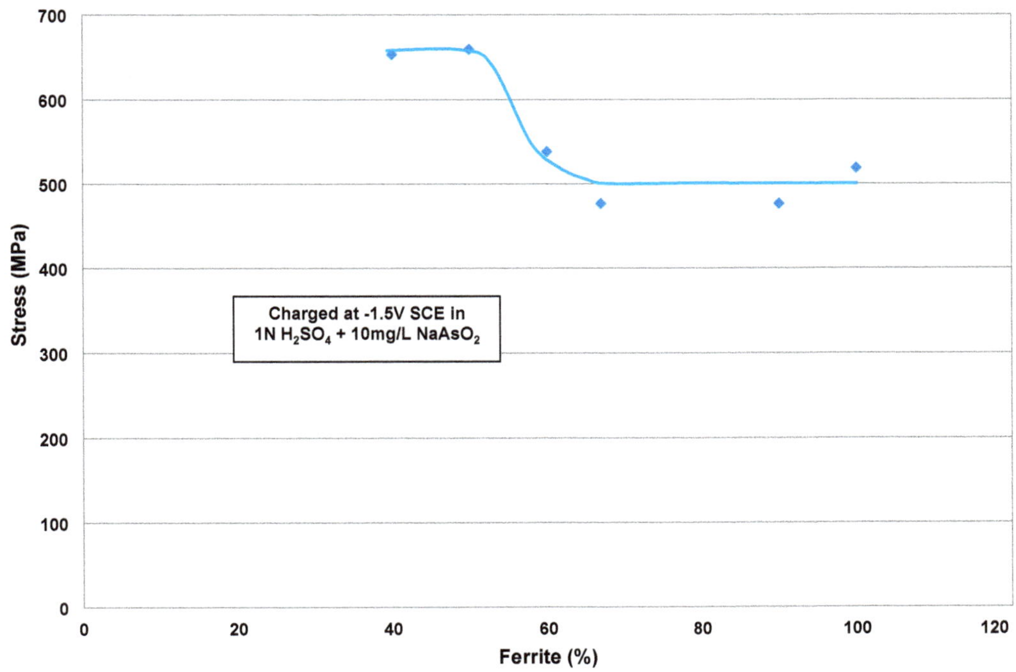

FIGURE 6.29 Effect of ferrite content on the threshold stress for embrittlement[67].

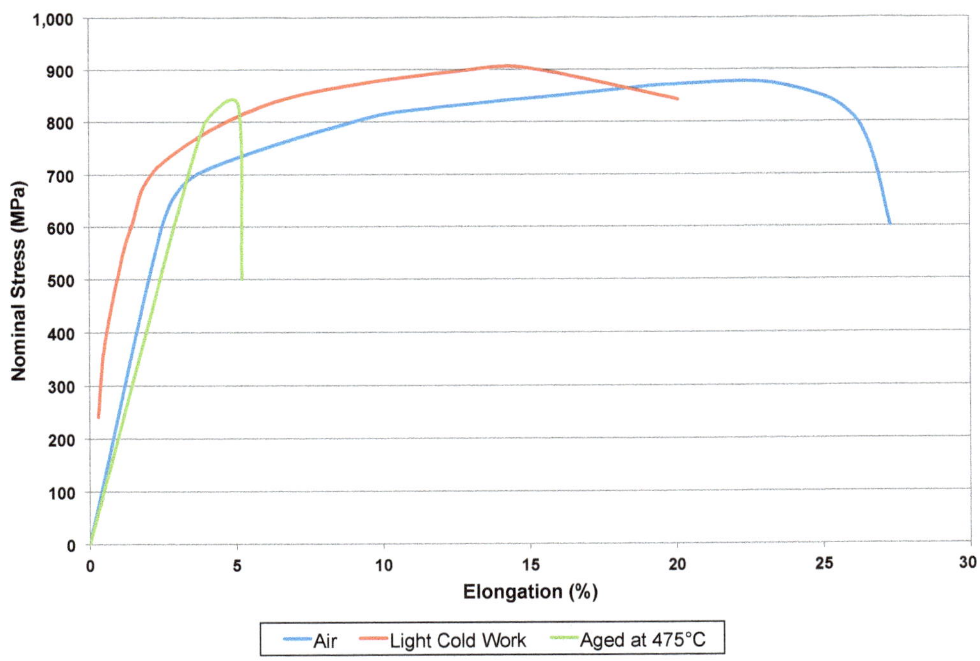

FIGURE 6.30 SSR test results for Z100 superduplex in two different conditions, while being charged at –1.04V SCE[60].

duplex alloys to that of ASTM A193 Gr B7 alloy steel (0.2% proof stress = 725 MPa). The results clearly show the drastic reduction in ductility of the aged material. The failed bolts in Figure 6.24 had been aged and these failures were followed by a company ban on aged 25% Cr duplex bolting[60]. Nitrides have also been shown to reduce the resistance to HISC, with increasing nitride content increasing the susceptibility[35].

Of critical importance are the orientation of the structure and the spacing of the austenite laths. Figure 6.31 shows a crack in the Foinaven manifold material and the orientation clearly allows the crack to propagate along the ferrite because of the orientation. Woollin found that there was a strong correlation between the stress ratio (test stress × 100/actual 0.2% proof stress) and the austenite spacing[64]. This has been confirmed by other workers (e.g., reference 70). Figure 6.32A shows the curve with the test results from several different product forms of duplex stainless steel. It can be seen that the threshold stress ratio for crack initiation is 100% or greater for material with an austenite spacing ≤ 30 μm. HIP material has a very fine microstructure with no orientation, which is why it has a high resistance to HISC. This has been confirmed by Lauvstad et al.[71].

Constant load tests on some flange forgings (top hat forgings about 0.5 m diameter and 0.5 m tall) gave the results in Figure 6.32B and they all fall above the TWI curve[72]. The reason for this is austenite spacing was devised as a research tool and is not a well-defined standard measurement. The TWI method says to ignore the small particles of austenite between the main laths when measuring austenite spacing. This "reformed" austenite precipitates when duplex stainless steel is solution annealed after hot working. However, all of the forgings had long heat treatments after forging and the reformed austenite particles were significantly large. However, at present, there is work under way to define how big austenite should be before it is counted, but it is difficult, because the structure is 3-D and a microsection is only 2-D. The alternative is to find a test that determines the actual resistance to HISC in a short time (e.g., 24 h), so that the resistance of a specific component can be measured. Research is also under way on this approach.

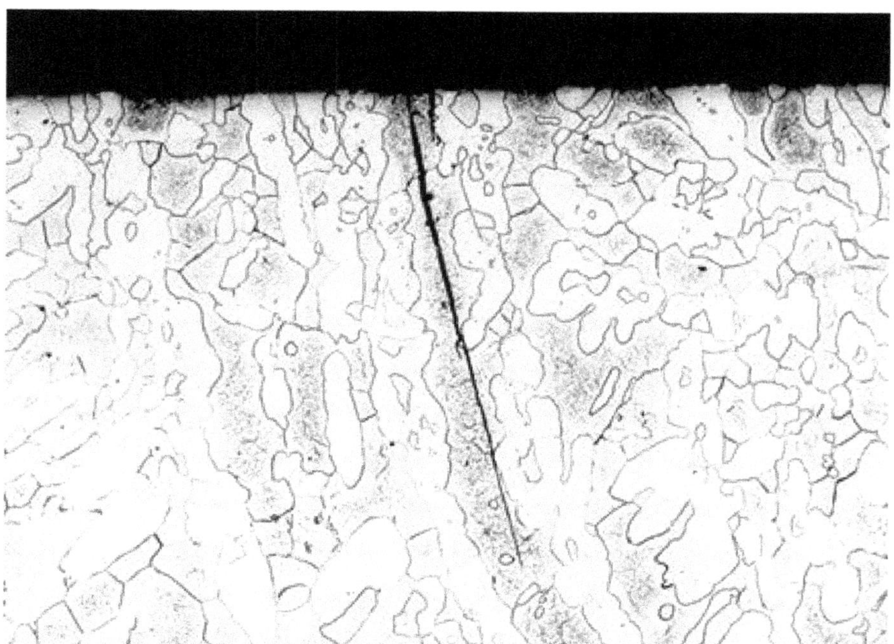

FIGURE 6.31 HISC crack propagating along the ferrite due to poor structure orientation[64].

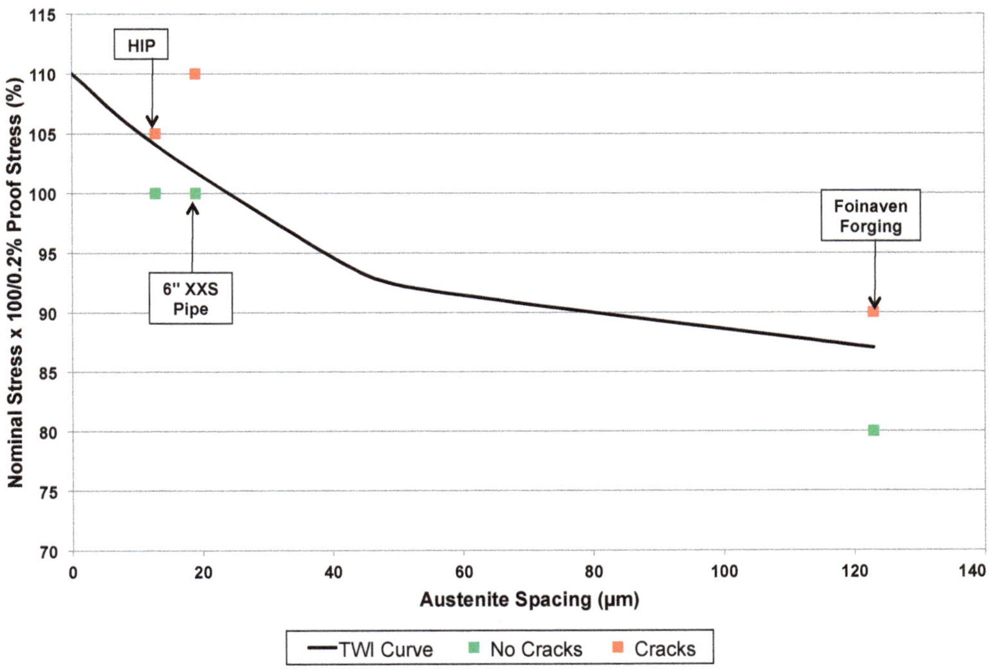

(A)

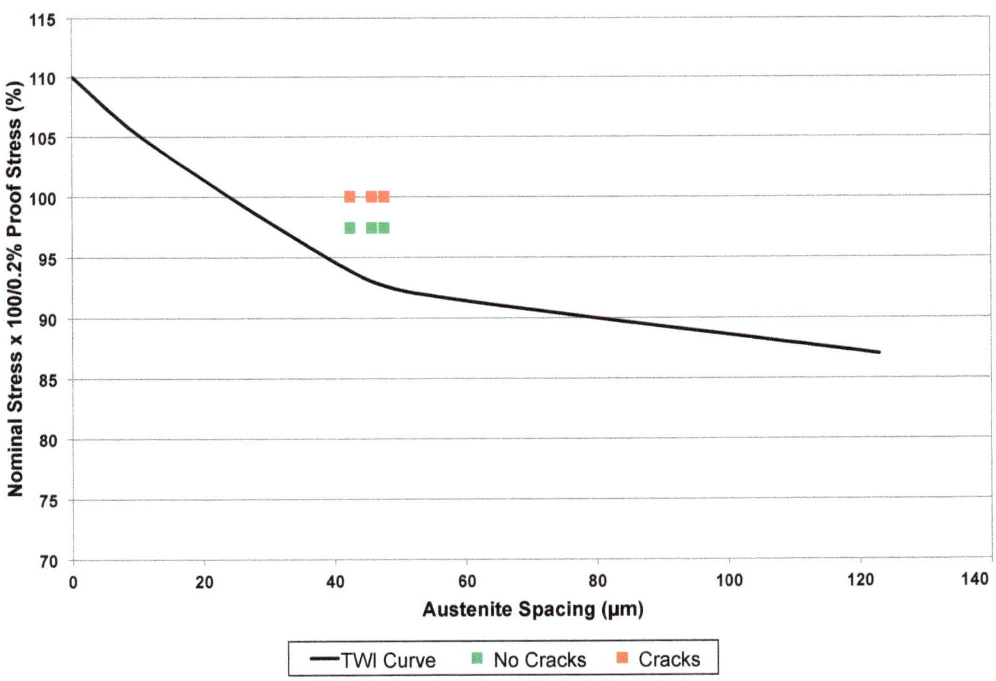

(B)

FIGURE 6.32 The effect of austenite spacing on the threshold stress for HISC at -1.1V SCE. (A) TWI data with some different product forms[80]. (B) Some results for 5 $^1/_8$" 10K lb flange forgings[72].

When designing structures subsea to resist HISC under CP, it is common to use DNV recommended practice RP-F112[73]. However, this document has been criticized as over-conservative and this, plus other methods of minimizing the risk of HISC, are discussed in reference 1.

Welds are not particularly susceptible to HISC because they tend to be stronger than parent metal and have a relatively fine structure. Some tests have shown increased susceptibility of the heat affected zone (HAZ) when its structure has coarsened significantly compared with parent metal[65]. Welds with high ferrite contents or third phases may have increased susceptibility to HISC compared with parent metal[64].

6.7 TEST METHODS

It is much more difficult to test for susceptibility to stress corrosion cracking than for pitting or crevice corrosion[2]. First, there are a lot of environmental variables, such as chloride concentration, pH, and redox potential. Second, there is stress, and the way in which the stress is applied determines the susceptibility to SCC[2]. The effect of stressing technique varies with alloy, so the relative alloy ranking can vary as well[2]. Slow strain rate testing is often regarded as the most aggressive form of testing because of the continuous plastic straining during the test, but this is not always the case. The effect of stressing method on SCC resistance for austenitic and duplex stainless steels is discussed in more detail in reference 1. The conclusion is that SCC test data needs to be viewed with some discretion when trying to apply it to a practical service application[1, 2].

There are a number of standard tests for ranking the chloride SCC resistance of alloys. One of these is the ASTM G36 test[74], in which stressed samples are exposed to boiling 45% magnesium chloride (~150°C). All stainless steels fail this test and the factor that is measured is the time to failure. This test has been criticized because high molybdenum alloys do not have an increased time to failure, while higher molybdenum alloys often perform better in many service environments. Cottis and Newman discussed this and commented that increasing molybdenum content does give increasing resistance to SCC in dilute chloride solutions, where all the alloys are initially passive[4]. However, in stronger chloride solutions, or acidic chloride solutions, the lower alloy materials can be suffering more general dissolution and this reduces their susceptibility to SCC. The higher alloy materials remain passive, but SCC may initiate where the passive film breaks down.

Another test is ASTM G123[75], which uses an environment of boiling 25% NaCl at pH 1.5. This is a much milder test and only fails low alloy materials. For this reason, boiling 40% calcium chloride is sometimes used as an alternative test, with intermediate aggressiveness. This solution is not included in any standard tests, but there is quite a lot of data for duplex stainless steels in this environment. Finally, there is also a lot of data on various alloys in 5% sodium chloride, where samples are stressed under constant load to the 0.2% proof stress at test temperature in aerated solution, and tested at various temperatures to determine the threshold temperature for SCC. This method is not incorporated into any standard.

A summary of test results from a variety of sources is shown in Table 6.5 for several of the tests described above[2, 5, 22, 76]. These results show that the ASTM G36 test in magnesium chloride is very aggressive and all the alloys failed. The 300-series austenitic alloys failed in all the tests, while the duplex alloys passed in all except the magnesium chloride test. This shows the difficulty in selecting alloys from such tests.

TABLE 6.5 Results for some common SCC tests on a variety of stainless steels[2, 5, 22, 75].

TYPE	ALLOY	RESULT (Pass/Fail)			
		Boiling 42% $MgCl_2$	Boiling 40% $CaCl_2$	Boiling 25% NaCl; pH 1.5	Boiling 25% NaCl; pH 6
Austenitic	304	F	F	F	F
	316	F	F	F	F
	904L	F	P	-	-
	6% Mo	F	P	-	-
Duplex	2101	F	P	P	P
	2003	F	P	-	P
	2304	F	P	P	P
	2205	F	P	P	P
	2507	F	P	-	P

Figure 6.33 shows the threshold temperature for SCC in aerated, neutral 5% sodium chloride for a range of stainless steels. The data was gathered from manufacturers' data sheets, but the results for the same alloy from different manufacturers are very similar. No data was available for 2101 (or the other, similar lean duplex alloys) or for 2304. However, it is expected that the threshold temperature would be at least 100°C for these alloys. The test data for 2003 showed passes at 120°C and 130°C, but tests at higher temperatures were not conducted. Superduplex alloys 2507 and Z100 did not fail at 250°C and Sandvik data sheets also show no SCC for 2507 at 300°C. Alloy Z100 was not tested at this temperature. While this data is a little more useful, it is still difficult to relate to service environments.

The ASTM C692[77] wicking test was designed to determine whether chemicals could be leached from insulation materials, and which could concentrate on hot pipes and cause SCC. The test is done at 105°C and has worked well for austenitic stainless steels. When the concern is external SCC under dripping seawater, the test technique described by Turnbull and Hinds[20] is preferred.

The standard test for sulfide SCC for many years has been NACE TM0177[78], which involves exposure in a solution of 5% sodium chloride adjusted to pH 2.7 with acetic acid and saturated with H_2S at room temperature and pressure. However, this test is unsuitable for many CRAs, because it assumes the mechanism of cracking is hydrogen embrittlement, which is true for martensitic stainless steels, but not duplex or austenitic alloys. Austenitic alloys crack by anodic dissolution due to chlorides and exacerbated by H_2S. Hence, they become increasingly susceptible to cracking as the temperature rises. Duplex stainless steels can crack by both mechanisms and experience a peak susceptibility to sulfide SCC at ~90°C (See Figure 6.11). These factors are recognized by European Federation of Corrosion (EFC) publication 17[79], which gives recommendations for suitable test temperatures for different classes of CRAs. This requires autoclave testing and careful measurement of applied stress, as described in the standard.

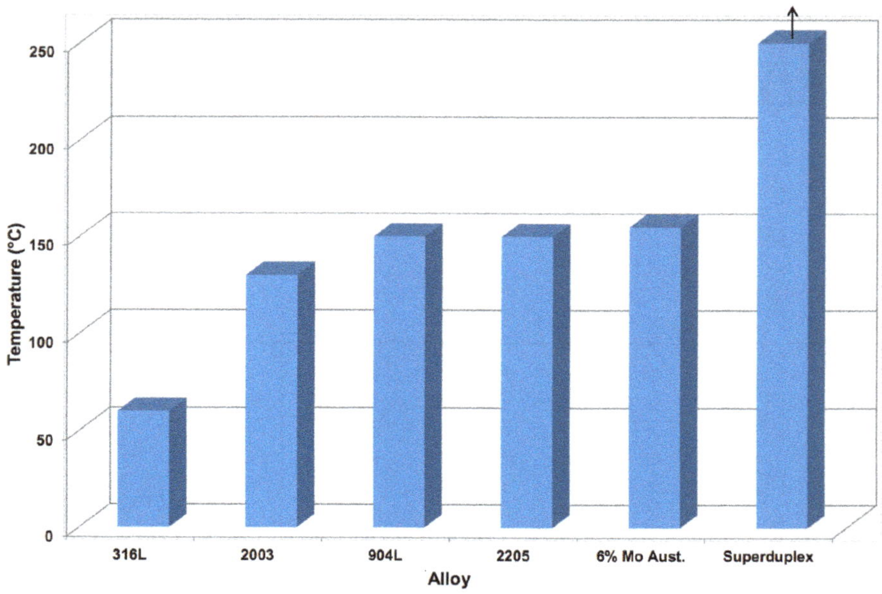

FIGURE 6.33 Threshold temperature for chloride SCC in aerated 5% sodium chloride.

There are no standard tests for caustic SCC and it is usual to use constant load, C-ring, or 4-point bent beam samples, as described in EFC 17[79], in the service environment and under service conditions.

It has been common to test for resistance to HISC using slow strain rate testing, but a number of authors have pointed out that this test is too aggressive and suggests that some materials will fail that have worked in service[62, 63, 70]. When investigating the Foinaven failure, constant strain tests were found not to give cracking as was seen in service, and the most useful test method was true constant load testing[62]. It has also been pointed out that some high dynamic straining is required to initiate HISC cracks, and this occurs early on during a true constant load test at stresses close to the proof stress[62, 80]. There are two threshold stresses, one for crack initiation and a higher one for crack propagation[64, 80]. When cracks initiate, they do so very early on in a constant load test and the crack length for non-propagating cracks was the same after 24 h and 720 h[64].

REFERENCES

1. D.R. McIntyre and R. Francis (eds), *Environmental Cracking: Guidelines for Preventing Stress-Corrosion Cracking, Hydrogen Assisted Cracking and Liquid Metal Embrittlement in the Chemical Process Industries* (St. Louis, MO: Materials Technology Institute, 2016).
2. R.F.A. Pettersson and E. Johansson, "Stress Corrosion Resistance of Duplex Grades," Duplex Stainless Steels 2010 (Zutphen, Netherlands: KCI Publishing, 2010).
3. R.J. Gerlock and J.D. Fritz, "Chloride Stress Corrosion Cracking Resistance of 6%Mo Stainless Steel Alloy," CORROSION 2002, paper no. 423 (Houston, TX: NACE International, 2002).
4. R. Cottis and R. Newman, "Stress Corrosion Cracking Resistance of Duplex Stainless Steels," Offshore Technology Report OTH 94 440 (Bootle, United Kingdom: UK Health and Safety Executive, 1994).

5. J. Sedriks, *Corrosion of Stainless Steels*, 2nd Edition (New York, NY: J. Wiley and Sons, 1996): pp. 267–325.
6. A. Laitinen and H. Hanninen, "Chloride Induced Stress Corrosion Cracking of Powder Metallurgy Duplex Stainless Steels," *CORROSION*, 52, 4 (1996): p. 295.
7. W.B. Hutchinson, K. Ushioda, and G. Runnsjo, "Tensile Behaviour in Stainless Steel Sheet," *Mater. Sci. Technol.*, 1 (1985): p. 728.
8. H. Spaehn, "SCC and Corrosion Fatigue of Martensitic, Ferritic and Ferritic-Austenitic (Duplex) Stainless Steels," in *Environment Induced Cracking of Metals*, R.P. Gangloff and M.B. Ives, eds. (Materials Park, OH: ASM International, 1991): p. 449.
9. J.M. Gras and F. Ternon, "Corrosion of Duplex Stainless Steels in Oxygenated Chloride-Containing Waters between 150 and 250°C," Duplex Stainless Steels '91 (Les Ulis Cedex A, France: Les editions de physique, 1991): p. 847.
10. J.D. Atkinson, J.R. Saithala, and H.S. Ubhi, "Stress Corrosion Cracking Potentials of Stainless Steels in Dilute Chloride Solutions at 130°C," EUROCORR 2008, European Federation of Corrosion (London, United Kingdom: Institute of Materials, Mining and Minerals, 2008).
11. J.R. Saithala, S. McCoy, J.D. Atkinson, and H.S. Ubhi, "Critical Stress Corrosion Cracking Potentials of Stainless Steels in Dilute Chloride Solutions," CORROSION 2010, paper no. 287 (Houston, TX: NACE International, 2010).
12. J.R. Saithala and J.D. Atkinson, "Stress Corrosion Behaviour of Superduplex Stainless Steel in Dilute Chloride Solutions at 130°C," Duplex Stainless Steel 2007 (Zutphen, Netherlands: KCI Publishing, 2007).
13. G. Notten, "Application of Duplex Stainless Steel in the Chemical Process Industry," Vol. I, Duplex '97 (Zutphen, Netherlands: KCI Publishing): p. 9.
14. S. Huizinga, J.G. de Jong, W.E. Like, and B. McLoughlin, "Offshore 22Cr Duplex Stainless Steel Cracking—Failure and Prevention," CORROSION 2005, paper no. 474 (Houston, TX: NACE International, 2005).
15. A. Turnbull, P. Nicholson, and S. Zhou, "Chemistry of Concentrated Salts Formed by Evaporation of Formation Water and the Impact on Stress Corrosion Cracking of Duplex Stainless Steels," *CORROSION*, 63, 6 (2007): p. 555.
16. S.P.V. Mahajanam, R.P. Case, H.E. Rincon, D.R. McIntyre, M. Joosten, and S. Gjesdal, "Effect of Sigma Phase on the Corrosion and Stress Corrosion of 2205 and 2507 Duplex Stainless Steels," CORROSION 2011, paper no. 173 (Houston, TX: NACE International, 2011).
17. J.G.A. Aerts and M.J.G. Notten, "Stress Corrosion Cracking of Duplex Stainless Steel in Chloride-Containing Media in the Chemical Process Industry: A Case Study," Duplex Stainless Steels '91 (Les Ulis Cedex A, France: Les editions de physique, 1991): p. 1141.
18. I. Oystetun, K. Johansson, and O.B. Andersen, "Stress Corrosion Cracking of Duplex SS, UNS S31803, on the Gyda HP Separator," Offshore Technology Conference, paper no. 7207 (Houston, TX: Offshore Technology Conference, 1993).
19. A. Leonard, "Review of External Stress Corrosion Cracking of 22% Cr Duplex Stainless Steel," HSE Research Report 129, ISBN 0-7176-2718-7 (Bootle, United Kingdom: UK Health and Safety Executive, 2003).
20. A. Turnbull and G. Hinds, "Stress Corrosion Cracking of Duplex Stainless Steel Under Simulated Evaporative Conditions," CORROSION 2007, paper no. 474 (Houston, TX: NACE International, 2007).
21. A. Turnbull, S. Zhou, P. Nicholson, and G. Hinds, "Chemistry of Concentrated Salts Formed by Evaporation of Seawater on Duplex Stainless Steel," *CORROSION*, 64, 4 (2008): p. 325
22. B. Ozturk, J.J. Dunn, and J.F. Grubb, "Corrosion Properties of Duplex Stainless UNS S32003 Alloy," CORROSION 2008, paper no. 186 (Houston, TX: NACE International, 2008).
23. NORSOK M-630, *Material Data Sheets for Piping*, Edition 5 (Oslo, Norway: Standards Norway, 2010).
24. NACE MR0175/ISO 15156, "Petroleum and Natural Gas Industries—Materials for Use in H_2S-Containing Environments in Oil and Gas Production," Part 3, Corrosion Resistant Alloys (Houston, TX: NACE International, 2010).

25. NACE MR0103, "Materials Resistant to Sulphide Stress Cracking in Corrosive Petroleum Refining Environment" (Houston, TX: NACE International, 2003).
26. S. O'Loughlin, M.Sc. Dissertation, "Determination of the Critical H2S Pressure and pH to Cause Sulphide Stress Cracking in Lean Duplex Stainless Steel" (Manchester, United Kingdom: University of Manchester, 2009).
27. S. Hashizume, T. Takaoka, and Y. Ishizawa, "Effect of Applied Stress on Stress Corrosion Cracking Resistance of Duplex Stainless Steel by Using Uniaxial Constant Load Test in Simulating Oil and Gas Well Conditions," Duplex Stainless Steels '91 (Les Ulis Cedex A, France: Les editions de physique, 1991): p. 921.
28. M. Schofield, R. Bradshaw, and R. Cottis, "Stress Corrosion Cracking of Duplex Stainless Steel Weldments in Sour Conditions," *MP*, 35, 4 (1996): p. 65.
29. M. Barteri, G. Rondelli, L. Scoppio, and A. Tamba, "Cold Working and Composition Effect on the Performance of Duplex Stainless Steels for OCTG," Duplex Stainless Steels '91, Beaune, France, October 1991 (Les Ulis Cedex A, France: Les editions de physique, 1991): p. 1203.
30. J-P. Audouard and M. Verneau, "Evaluation of the Corrosion Resistance of High Nitrogen Containing Stainless Steels in Chloride and H_2S/CO_2 Environments," CORROSION/97, paper no. 36 (Houston, TX: NACE International, 1997).
31. R. Francis and G. Byrne, "The Corrosion of Duplex Stainless Steels in Sour Service," CORROSION/94, paper no. 64 (Houston, TX: NACE International, 1994).
32. W. Nisbet, R.C. Newman, and G. Lorimer, "The Influence of Phase Balance on the SCC of Duplex Stainless Steels," CORROSION/94, paper no. 126 (Houston, TX: NACE International, 1994).
33. J. Oredsson and S. Bernhardsson, "Performance of High Alloy Austenitic and Duplex Stainless Steels in Sour Oil and Gas Environments," *MP*, 22, 1 (1983): p 35.
34. G. Rorvik, L.A. Marken, I.M. Kulbotten, M. Aursand, S. Olsen, and K.S. Karlsen, "Influence of Intermetallic Precipitates on the Mechanical Properties and Environmental Cracking Resistance of Duplex Stainless Steel Fittings—A Case History (Part 1)," Duplex Stainless Steels 2010 (Zutphen, Netherlands: KCI Publishing, 2010).
35. G. Byrne, R. Francis, and G. Warburton, "Variation in Mechanical Properties and Corrosion Resistance of Different Alloys Within the Generic Designation UNS S32760," Duplex America 2000 (Zutphen, Netherlands: KCI Publishing, 2000).
36. R. Francis, G. Byrne, and G. Warburton, "The Role of Environmental and Metallurgical Variables on the Resistance of Duplex Stainless Steels to Sulphide SCC," CORROSION/97, paper no. 012 (Houston, TX: NACE International, 1997).
37. T. Cassagne, J. Poultier, S. Le Menchet, and C. Duret, "An Update on the Use of Duplex Stainless Steels in Sour Environments," CORROSION 2012, paper no. 1381 (Houston, TX: NACE International, 2012).
38. L. Scoppio, M. Barteri, and C. Leali, "Sulphide Stress Cracking Resistance of Superduplex Stainless Steels in Oil and Gas Field Simulated Environments," CORROSION/98, paper no. 095 (Houston, TX: NACE International, 1998).
39. J.G. Maldonado and J.W. Skogsberg, "Cracking Susceptibility of Duplex Stainless Steel at Intermediate Temperature in the Presence of H_2S Containing Environments," CORROSION 2004, paper no. 134 (Houston, TX: NACE International, 2004).
40. B. Holmes, K. Sotoudeh, and S. Mishael, "Stress Corrosion Cracking of a Duplex Stainless Steel," CORROSION 2017, paper no. 9113 (Houston, TX: NACE International, 2017).
41. R. Francis, G. Byrne, and G. Warburton, "Application of Superduplex Stainless Steel in Geothermal Power Plants," CORROSION 2013, paper no. 2206 (Houston, TX: NACE International, 2013).
42. P. Woollin and M. Maligas, "Testing of Superduplex Stainless Steel for Sour Service," CORROSION 2003, paper no. 132 (Houston, TX: NACE International, 2003).
43. A.J. Griffiths and A. Turnbull, "Performance of Duplex Stainless Steel Coupled to Carbon Steel in Oilfield Environments," Duplex Stainless Steels '97 (Zutphen, Netherlands: KCI Publishing, 1997), p. 597.

44. S. Clarke, "Stress Corrosion Cracking of Stainless Steels in Oxygenated Caustic Environments," CORROSION 2000, paper no. 00596 (Houston, TX: NACE International, 2000).
45. S. Clarke, PAPRICAN, Vancouver, Canada, Private Communication, October (2000).
46. A. Turnbull, A. Griffiths, and T. Reid, "Corrosion and Stress Corrosion Cracking of Duplex Stainless Steels in Caustic Solutions at Elevated Temperature," *Stainless Steel World* (December 1999): p. 63.
47. G. Rondelli, B. Vincentini, M.F. Brunella, and A. Cigada, "Stress Corrosion Cracking of Austenitic and Duplex Stainless Steels in Caustic Environments," CASS 90: Corrosion–Air, Sea, Soil Conference, paper no. 37 (Auckland, New Zealand: Australasian Corrosion Association, 1990).
48. H. Leinonen, P. Pohjanne, T. Saukonen, and T. Schildt, "Effect of Selective Dissolution on Stress Corrosion Cracking Susceptibility of Austenitic and Duplex Stainless Steels in Alkaline Solutions," CORROSION 2011, paper no. 160 (Houston, TX: NACE International, 2011).
49. G. Notten, E. van den Heuvel, and H. Verhoef, "Stress Corrosion Cracking Resistance of Duplex Stainless Steels in Caustic Environments," Duplex America 2000 (Zutphen, Netherlands: KCI Publishing, 2000): p. 53.
50. E.M. Horn, S. Savakis, G. Schmitt, and I. Lewandowski, "Performance of Duplex Stainless Steels in Caustic Solutions," Duplex Stainless Steels '91 (Les Ulis Cedex A, France: Les editions de physique, 1991): p. 1111.
51. L. Wegrelius, "Duplex Solutions for Pulp and Paper Industry," 42nd Pulp and Paper International Congress and Exhibition, paper no. ABTCP 2009, Sao Paulo, Brazil, October (Sao Paulo, Brazil: Assoc Brasileira Técnica de Celulose e Papel, 2009).
52. L. Wegrelius, "Duplex Applications in Pulp and Paper Industry," TAPPI Peers Conference (Peachtree Corners, GA: TAPPI, 2015).
53. P.M. Singh, J. Mahmood, and P. Conde, "Stress Corrosion Cracking and Corrosion Susceptibility of Duplex Stainless Steels in Caustic Solutions," CORROSION 2005 paper no. 196 (Houston, TX: NACE International, 2005).
54. A. Bhattacharya, P.M. Singh, H. Leinonen, and J. Mahmood, "Effect of Welding Related Microstructure on Stress Corrosion Cracking Susceptibility in Caustic Solutions," CORROSION 2006, paper no. 497 (Houston, TX: NACE International, 2006).
55. A. Bhattacharya and P.M. Singh, "Corrosion and Stress Corrosion Cracking of Duplex Stainless Steels in Caustic Solutions," CORROSION 2007, paper no. 206 (Houston, TX: NACE International, 2007).
56. H. Leinonen, P. Pohjanne, J. Romu, and T. Saukonen, "Selective Dissolution and Stress Corrosion Cracking Susceptibility of Austenitic and Duplex Stainless Steels in Sodium Sulphide Containing Alkaline Solutions," CORROSION 2012, paper no. 1202 (Houston, TX: NACE International, 2012).
57. H. Leinonen, P. Pohjanne, T. Saukonen, and T. Schildt, "Effect of Selective Dissolution on Stress Corrosion Cracking Susceptibility of Austenitic and Duplex Stainless Steels in Alkaline Solutions," CORROSION 2011, paper no. 160 (Houston, TX: NACE International, 2011).
58. K.R. Chasse, S. Raji, and P.M. Singh, "Role of Chlorides in Stress Corrosion Cracking of Duplex Stainless Steels in Alkaline Sulphide Solutions," CORROSION 2011, paper no. 287 (Houston, TX: NACE International, 2011).
59. K.R. Chasse and P.M. Singh, "Effect of Microstructural Anisotropy on Stress Corrosion Cracking of Hot Rolled Duplex Stainless Steels," *Corros. Eng. Sci. Tech.* 47, 3 (2012): p. 170.
60. R. Francis, G. Byrne, and G. Warburton, "Effects of Cathodic Protection on Duplex Stainless Steels in Seawater," *CORROSION*, 53, 3 (1997): p. 234.
61. P. Sentence, "Hydrogen Embrittlement of Cold Worked Duplex Stainless Steel Oilfield Tubulars," Duplex Stainless Steels 1991 conference (Les Ulis Cedex A, France: Les editions de physique, 1991).
62. T.S. Taylor, T. Pendlington, and R. Bird, "Foinaven Superduplex Materials Cracking Investigation," Offshore Technology Conference, paper no. 10965 (Houston, TX: Offshore Technology Conference, 1999).

63. S. Huizinga, B. McLoughlin, M. Hannah, S.J. Paterson, and B.N.W. Sneddon, "Failure of a Subsea Superduplex Hub by HISC and Implications for Design," CORROSION 2006, paper no. 145 (Houston, TX: NACE International, 2006).
64. P. Woollin and A. Gregori, "Avoiding Hydrogen Embrittlement Stress Cracking of Ferritic Austenitic Stainless Steels under Cathodic Protection," Conference on Offshore Mechanics and Arctic Engineering, paper no. OMAE 2004-51203 (New York, NY: ASME International, 2004).
65. K.R. Chasse and P.M. Singh, "Hydrogen Embrittlement of 2205 Duplex Stainless Steel in Synthetic Kraft White Liquor," CORROSION 2010, paper no. 10084 (Houston, TX: NACE International, 2010).
66. A. Bahrami and P. Woollin, "Hydrogen Induced Stress Cracking of Duplex Stainless Steel Subsea Components," Conference on Offshore Mechanics and Arctic Engineering, OMAE 2010 (New York, NY: ASME International, 2004).
67. M. Murata, Y. Mukai, and J. Wang, "Fractographic Study of Mechanism of Hydrogen Induced Stress Cracking," 5th International Symposium of the Japanese Welding Society (Tokyo, Japan: Japanese Welding Society, 1990).
68. P. Woollin and W. Murphy, "Hydrogen Embrittlement Stress Corrosion Cracking of Superduplex Stainless Steel," *Stainless Steel World* (June 2001): p. 48.
69. P. Olsson, A.D. Bauer, and H. Eriksson, "Hydrogen Embrittlement of Duplex Grades S32750 and S31803 in Connection with Cathodic Protection in Chloride Solutions," Duplex Stainless Steels '97 (Zutphen, Netherlands: KCI Publishing, 1997): p. 607.
70. S. Ronneteg, A. Juhlin, and U. Kivisaak, "Hydrogen Embrittlement of Duplex Stainless Steels Testing of Different Product Forms at Low Temperature," CORROSION 2007, paper no. 07498 (Houston, TX: NACE International, 2007).
71. G.O. Lauvstad, R. Johnsen, I. Asbjorasen, M. Bjurstrom, and C.G. Hjorth, "Resistance Towards Hydrogen Induced Stress Cracking of Hot Isostatically Pressed Duplex Stainless Steel Under Cathodic Protection," *CORROSION*, 66, 11 (2010): p. 115004.
72. R. Francis and G. Warburton, "The Threshold Stress for Initiation of Hydrogen Embrittlement in Various Product Forms of Z100 Superduplex Stainless Steel," EUROCORR 2007, European Federation of Corrosion (London, United Kingdom: Institute of Materials, Mining and Minerals, 2007).
73. Recommended Practice DNV-RP-F112, "Design Guideline for Duplex Stainless Steel used for Subsea Equipment Exposed to Cathodic Protection" (Oslo, Norway: DNV, 2006).
74. ASTM G36, "Standard Practice for Evaluating Stress Corrosion Cracking Resistance of Metals and Alloys in a Boiling Magnesium Chloride Solution" (West Conshohocken, PA: ASTM International).
75. ASTM G123, "Standard Test Method for Evaluating Stress Corrosion Cracking of Stainless Alloys with Different Nickel Content in Boiling Acidified Sodium Chloride Solution" (West Conshohocken, PA: ASTM International).
76. E. Johansson and T. Prosek, "Stress Corrosion Cracking Properties of UNS S32101—A New Duplex Stainless Steel with Low Nickel Content," CORROSION 2007, paper no. 475 (Houston, TX: NACE International, 2007).
77. ASTM C692, "Standard Test Method for Evaluating the Influence of Thermal Insulations on External Stress Corrosion Cracking Tendency of Austenitic Stainless Steel" (West Conshohocken, PA: ASTM International).
78. NACE TM0177, "Laboratory Testing of Metals for Resistance to Sulphide Stress Cracking and Stress Corrosion Cracking in H_2S Environments" (Houston, TX: NACE International, 2005).
79. "Corrosion Resistant Alloys for Oil and Gas Production; Guidance on General Requirements and Test Methods for H_2S Service," European Federation of Corrosion publication No. 17, 2nd Edition (London, United Kingdom: Institute of Materials, Mining and Minerals, 2002).
80. P. Woollin, "Cathodic Protection of Ferritic-Austenitic Stainless Steel: Revised Final Report," TWI Report 5699/13A/03 (Cambridge, United Kingdom: TWI, 2003).

7 Corrosion Fatigue

Corrosion fatigue occurs when a material is subjected to both a cyclic stress and a corrosive environment. It is similar to stress corrosion cracking (SCC), which occurs under a nominally static load, but corrosion fatigue can occur in many more environments because the cyclic stress continually strains the material and permits corrosion more easily. SCC cracks tend to be sharp tipped and are often branched, while corrosion fatigue cracks tend to be more blunt ended and have few branches. To the eye, the fracture face of a fatigue crack is characterized by a series of "beach" marks showing the direction of propagation from the crack origin (Figure 7.1).

The stresses for corrosion fatigue may be at a low level, requiring a high number of cycles to cause failure or at a high stress level, requiring a lower number of cycles to failure. For low stress/high cycle fatigue, it is possible to define a stress level below which the material does not initiate fatigue cracking. This is called the Fatigue Limit. Ferritic steels and duplex stainless steels tend to exhibit a clear fatigue limit, and this can be used by designers to avoid fatigue failure. As a rule of thumb, the fatigue limit of ferritic and duplex steels, tested in laboratory air, is usually about 50% of the tensile strength of the steel. However, this can be reduced significantly when testing in more aggressive environments.

Austenitic alloys do not exhibit a clear fatigue limit, so designers tend to use an "endurance limit"; this is the stress level that achieves 10 million cycles of fatigue loading without crack initiation.

Fatigue loads can be high frequency, as is the case in rotating equipment such as pumps, or low frequency where structures are subjected to loading from waves in the sea. These are important factors in determining fatigue and especially corrosion fatigue resistance of a material. For a given fatigue life (number of repeats of cycle), the corrosion fatigue strength tends to be higher at higher frequencies because of the time dependency of corrosion processes.

Duplex stainless steels are much stronger than austenitic stainless steels with the same level of corrosion resistance. This means that a shaft in a seawater lift pump can be a much smaller diameter if constructed in superduplex stainless steel than would be the case if it were made from 6% Mo austenitic stainless steel. This is one reason duplex stainless steels are often chosen for rotating machinery in corrosive applications.

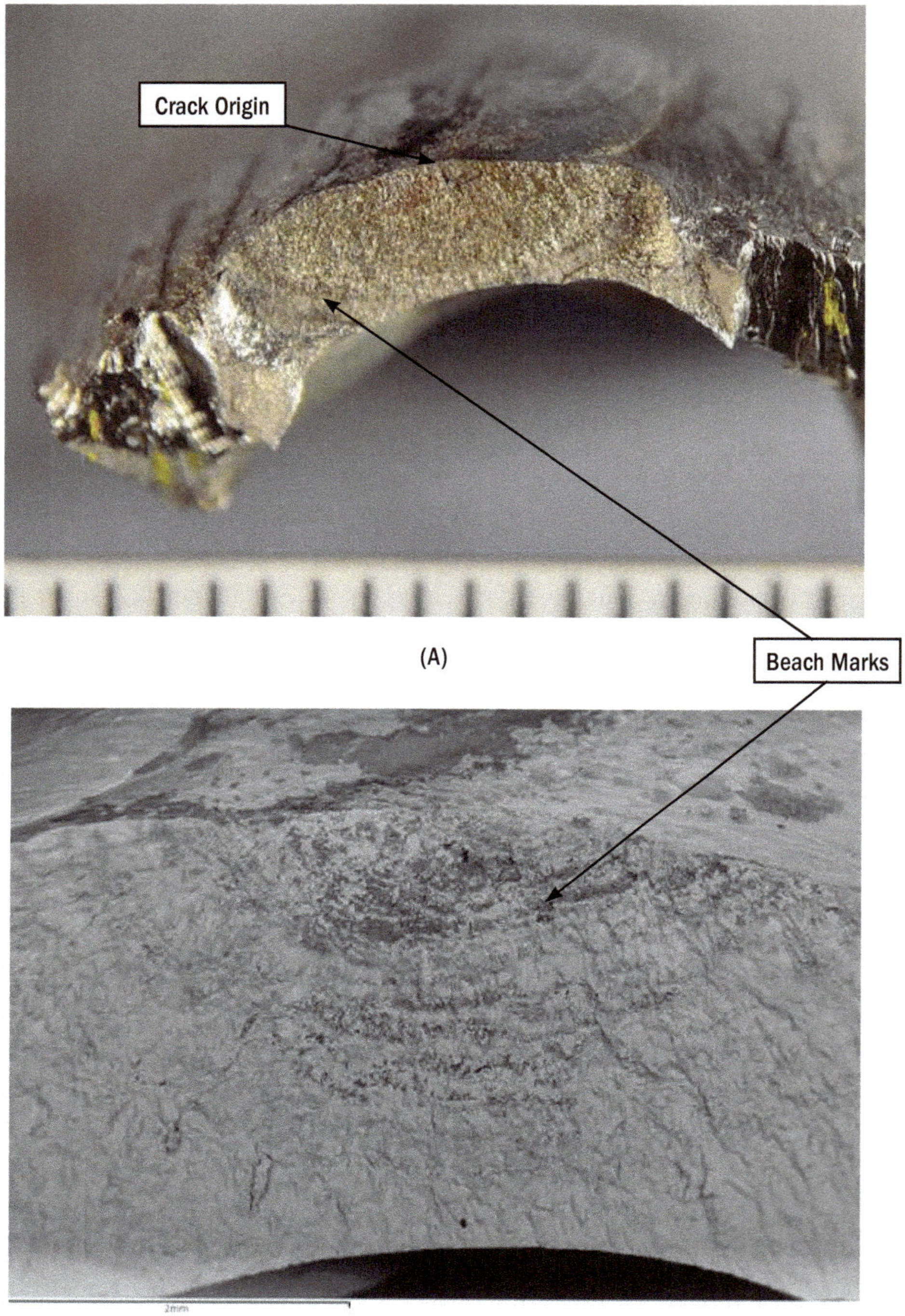

(B)

FIGURE 7.1 "Beach" marks in a classic fatigue crack on a stainless steel spool.
A) Low magnification, B) Higher magnification showing beach marks

One frequent type of service failure that has occurred on offshore platforms is failure of small branch takeoffs from large diameter, thin wall (Schedule 10S) piping, for example an NPS 2 branch from an NPS 12 main. Some of these failures have been described by Jungbauer and Smallberg[1]. Failures occur in a wide variety of piping alloys and are purely fatigue related, with no corrosive element. The failures are caused by a combination of high fluid velocities inside the pipes giving excessive vibration and the use of a standard branched connection fitting for the branch. These are thick and heavy and cause sinking if fully fillet welded, creating a local stress raiser, encouraging fatigue cracking at the weld toe, as shown in Figure 7.2a. Furthermore, if the weld preparations are only partially filled with weld metal, in an attempt to stop sinking, a stress raiser is created internally, as shown in Figure 7.2b and fatigue cracking through the weld throat can occur. The solution is to use a light weight, schedule 10S branched connection fitting, or a reduced branch tee, fully weld it up and to make sure that the branch pipe is adequately supported to reduce vibration.

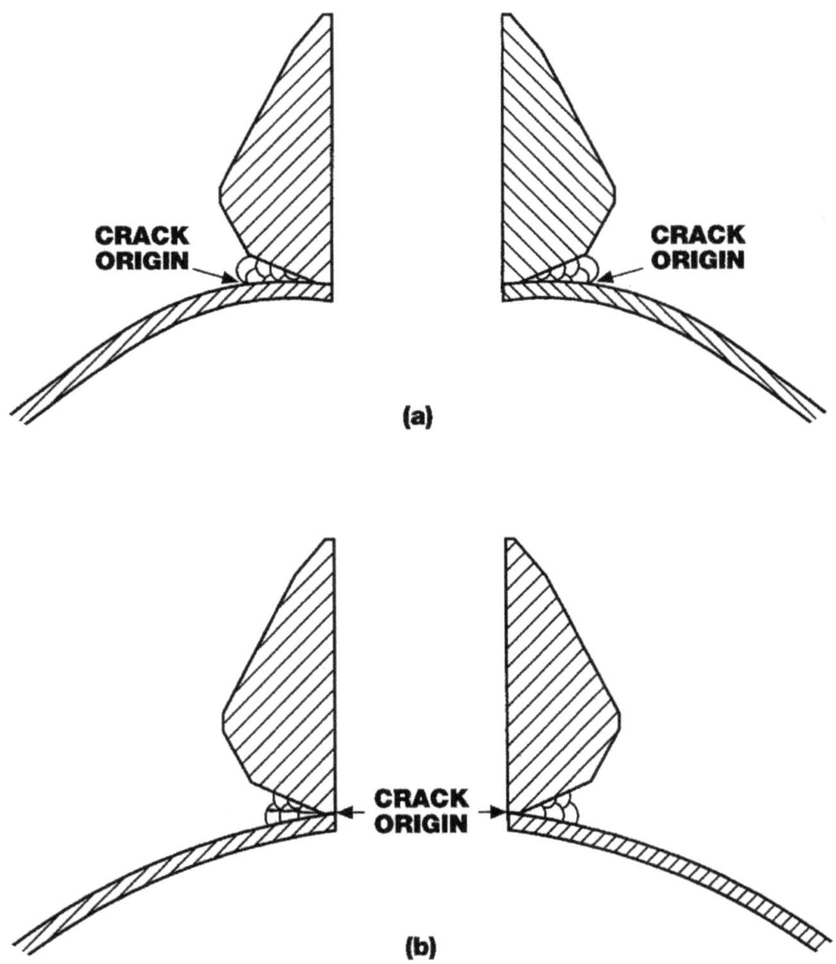

FIGURE 7.2 Schematic diagram of site of fatigue failure at a branched connection fitting on thin walled pipe. Full weld out (b) Partial weld fill

7.1 VARIABLES AFFECTING CORROSION FATIGUE

The microstructure of an alloy influences fatigue behavior in both the air and in a corrosive medium. This is because of its effect on crack path morphology and the state of the stress at the crack tip. In addition to the applied stress, the ratio of the minimum stress to the maximum stress (R) is important. Also important are the frequency of stress reversals and the range of stress intensity at the crack tip. All of these factors are critical in determining the fatigue crack growth rate.

The stress intensity (K) at the crack tip is defined by

$$K = \sigma . y . \sqrt{\pi . a}$$

where σ = stress, a = crack length, and y is a geometric factor that accounts for test specimen design.

This relationship takes into account the effects of any pre-existing defect or crack and applied stress, and represents these multiple effects in a single parameter. In the case of fatigue, where a component is subject to alternating stressing

$$\Delta K = \Delta\sigma . y . \sqrt{\pi . a}$$

where ΔK = alternating stress intensity, and $\Delta\sigma$ = alternating stress, i.e.,

$$\Delta K = K_{max} - K_{min}$$

and

$$\Delta\sigma = \sigma_{max} - \sigma_{min}$$

A higher alternating stress, $\Delta\sigma$, increases the stress intensity at the crack tip, for a given initial defect size, and similarly, larger defects produce an increase in stress intensity at the crack tip for a given stress range.

The relationship between ΔK and the fatigue crack growth rate per cycle (da/dN) is given by the Paris equation:

$$da/dN = c(\Delta K)^{m}$$

where c and m are constants.

This equation does not hold for low ΔK values, where a threshold value of ΔK is reached, i.e., ΔK_{th}. No crack growth occurs below this point, nor does it hold for high ΔK values where tensile modes of failure can be observed as maximum stresses in the fatigue cycle become high. Sometimes a step can be observed in the linear portion of the Paris regime. This can be indicative of corrosion fatigue and stress corrosion cracking, or hydrogen embrittlement occurring simultaneously, as described in reference 2.

7.2 EFFECT OF MICROSTRUCTURE

Threshold stress intensity values (ΔK_{th}) for fatigue crack growth at R=0.8 (i.e., a high mean stress) for wrought 2205 and Z100 superduplex have been determined, and were in the range from 2.3–3.3 MPa·m$^{1/2}$ for 2205[3] and 3.6–4.2 MPa·m$^{1/2}$ for Z100[4]. At a value of R=0.1, a threshold stress intensity of 7.5 MPa·m$^{1/2}$ was obtained for Z100 superduplex.

The values of ΔK_{th} associated with engineering materials are usually in the range from 2–10 MPa·m$^{1/2}$. Lower values of ΔK_{th} in this range are observed with tests carried out at high mean stress levels (high R ratios). Higher values of ΔK_{th} are usually obtained when tests are carried out at low mean stress levels (low R ratios). This is shown in Figure 7.3 where fatigue crack growth rate (da/dN) is plotted against ΔK. It can be seen that the results for Z100 superduplex fall within the normal range for engineering materials. This is also true for 2205 duplex[3].

The effects of microstructure in the near threshold region are small in relation to the effect of mean stress. Microstructural effects relate to the degree of crack closing during cyclic loading, together with grain size[5, 6]. If the structure is coarse, the opposite faces of the crack may touch and rub together. This has the effect of increasing the minimum stress intensity and, therefore, of reducing the net effective stress intensity range. A decreased stress intensity range results in a lower crack growth rate. The effect of crack closure is more pronounced in coarse grained materials[4] and is illustrated in Figure 7.3 by comparing the as-received material (solution treated with ~50% ferrite) with the 1300°C quenched material, which has a higher ferrite content and coarser grains. Marrow and King commented that the fatigue crack growth rate varies with the crack orientation in duplex stainless steel, and the growth rate decreases as the austenite spacing decreases[2].

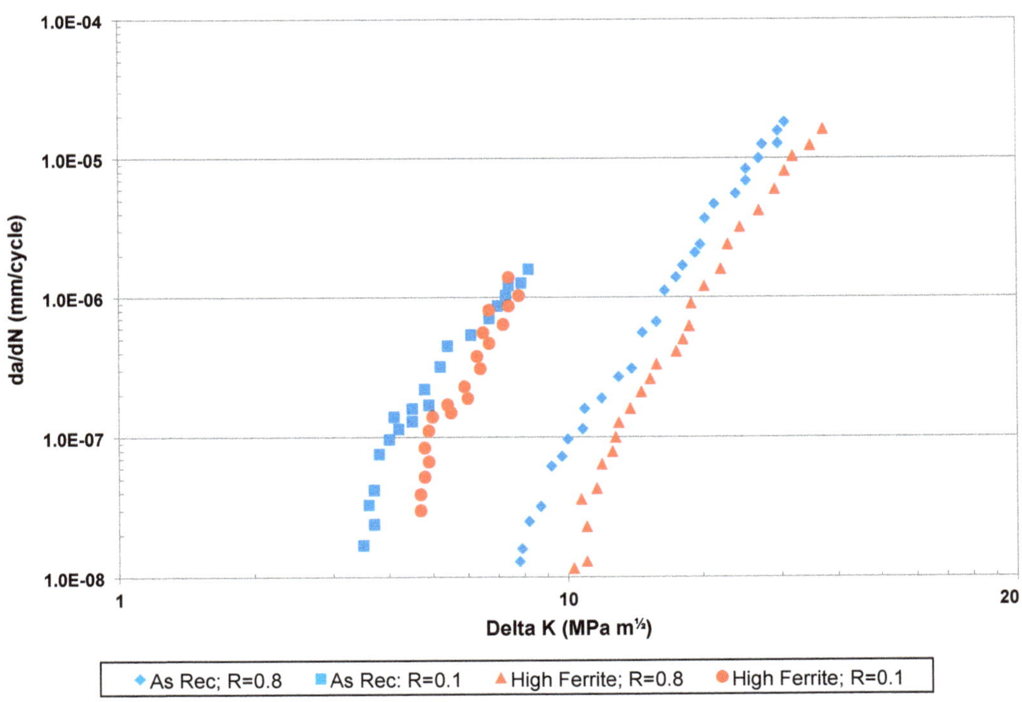

FIGURE 7.3 Fatigue crack growth data for Z100 superduplex in air at 50 Hz[4].

Perdriset et al. investigated the effect of composition on the corrosion fatigue resistance of 2205 in air and in a 3.5% sodium chloride solution[7]. They showed that the fatigue resistance increased with increasing nitrogen content, because the nitrogen strengthened the austenite. This was confirmed by Akdut, who also pointed out that the processing history and the total chemical composition could also affect the alloy morphology and, hence, the fatigue resistance[8].

7.3 EFFECT OF STRESS AND ENVIRONMENT ON CRACK GROWTH

Duplex stainless steels show good resistance to corrosion fatigue crack propagation in a range of aqueous environments including saline solution (3.5% NaCl) and 0.2M nitric acid, at ΔK values between 10 and 25 MPa·m$^{1/2}$. as shown in Figure 7.4[4] for Z100 superduplex. Within this range of ΔK, the crack growth rates for all three environments were very similar.

Woollin et al. conducted corrosion fatigue tests on 2205 duplex in 3% sodium chloride at 15°C, R=0.5 and a frequency of 1 Hz[9]. The results, in Figure 7.5, show a lot more scatter than those for superduplex in Figure 7.4. This is probably due to the different test methods and the fact that 2205 can suffer localized corrosion in this test solution. The 2205 appeared as, or more, resistant to cracking in sodium chloride compared with air at low ΔK values (<19 MPa·m$^{1/2}$), while the crack propagation rate was some 5–7 times higher at ΔK values exceeding 19 MPa·m$^{1/2}$.

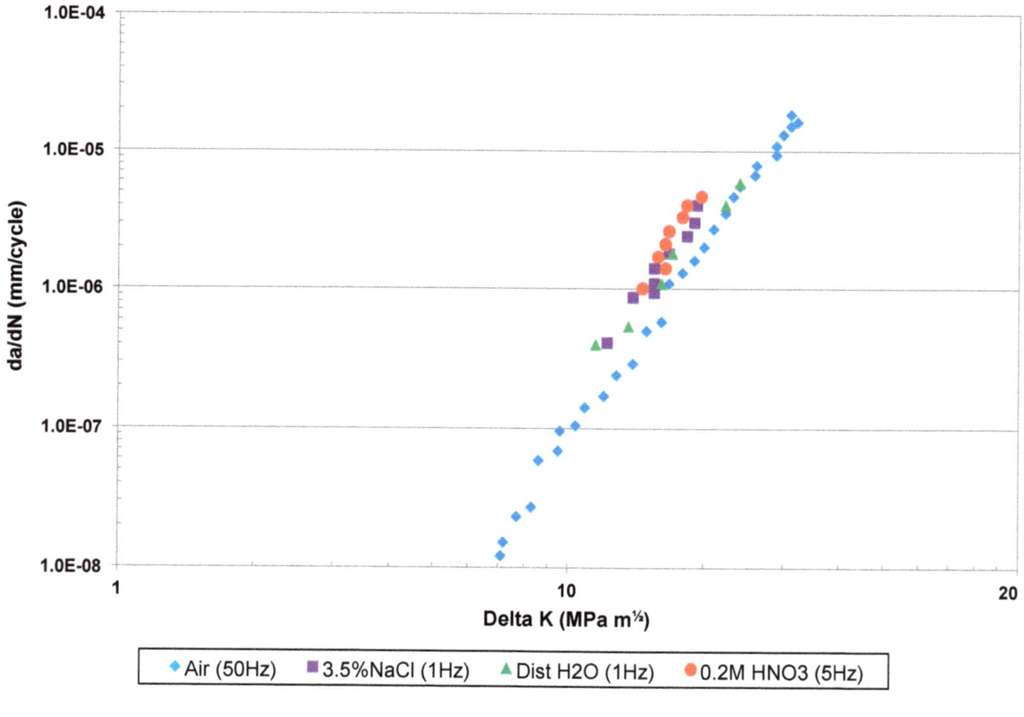

FIGURE 7.4 Fatigue crack growth data for Z100 superduplex with stress ratio R=1[4].

High corrosion resistance confers high resistance to corrosion fatigue, such that data points relating to the various test environments effectively lie on the same line given the normal scatter in such tests. A similar effect is illustrated in Figure 7.6, which shows the relative performance in air and in saline solution at a frequency of 50 Hz[4]. In each case, the threshold stress intensity of 7.5 MPa·m$^{1/2}$ is unchanged.

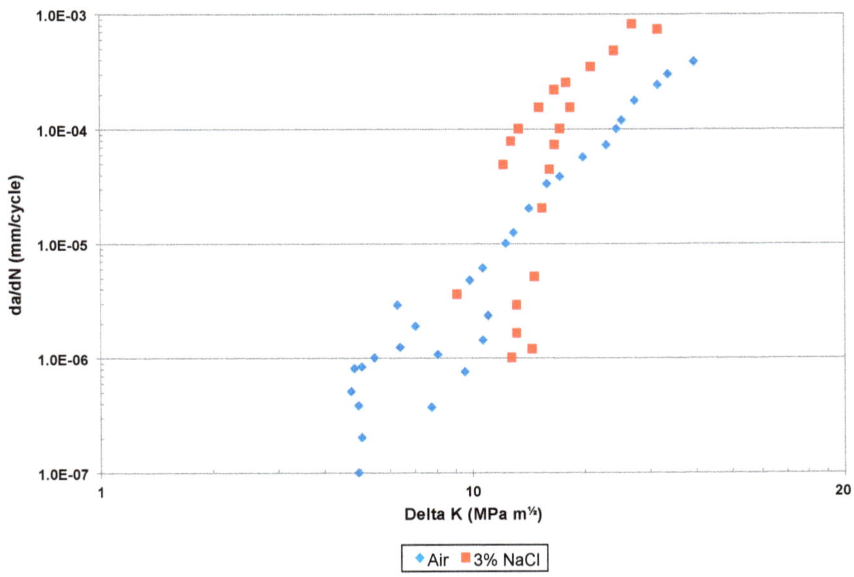

FIGURE 7.5 Fatigue crack growth data for 2205 with a stress ratio of 0.5 at a frequency of 1 Hz[9].

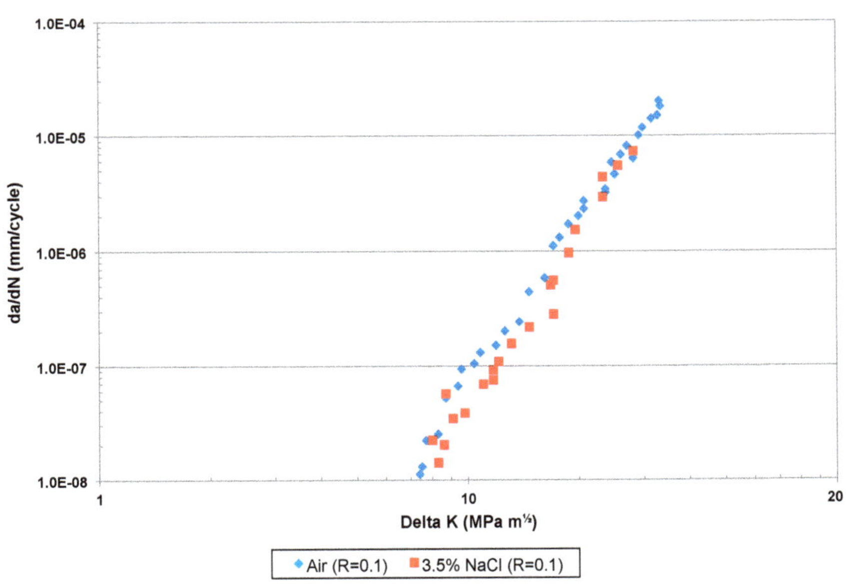

FIGURE 7.6 Fatigue crack growth data for Z100 superduplex at 50 Hz[4].

7.4 COMBINED EFFECTS OF CRACK INITIATION AND GROWTH

The effects of crack initiation and propagation in the assessment of fatigue and the performance of an alloy may be investigated in terms of the total time to initiate and propagate a crack. This differs from the previous approach described above in that the fraction of the total fatigue life required to initiate a crack is also taken into account. The data is usually presented as an *S-N* curve, plotting stress against the number of cycles to failure. In air, there is usually a stress below which fatigue failure does not occur, termed the "fatigue limit," as discussed at the beginning of this chapter. In a corrosive medium, it is more common for there to be no threshold stress, although the time to failure may be very long, for example >10 million cycles, and this is termed the "endurance limit."

7.4.1 Chloride Solutions

There is fatigue data in air for lean duplex 2003, showing a threshold stress of 345 MPa, but there is no data in any corrosive medium. There is however plenty of data for 2205 duplex and superduplex alloys.

Figure 7.7 shows the *S-N* curves for 2205 in saline (3.5% NaCl) in rotating bend tests (Wohler) at 50 Hz[10]. Also included is some data from the Sandvik data sheet at 25 Hz for comparison. The higher threshold stress at 25 Hz may be due to a finer, structure, giving somewhat higher strength.

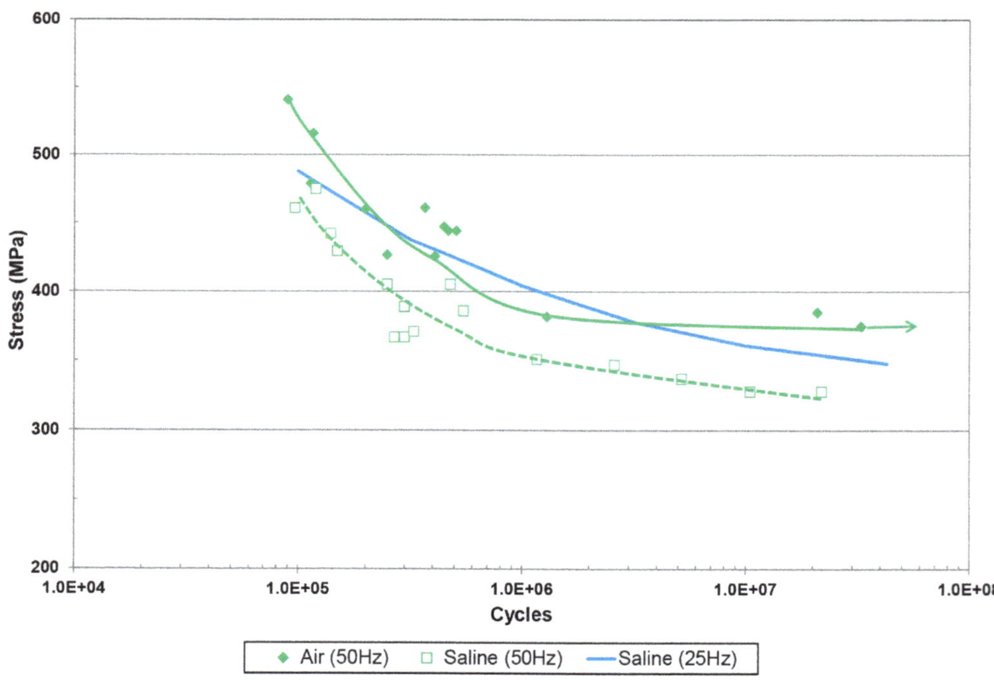

FIGURE 7.7 *S-N* curves for 2205 in rotating bend tests in air and saline[10].

Perdriset et al. investigated the corrosion fatigue resistance of 2205 in 3.5% sodium chloride solution and found that pits occurred in the austenite that helped with crack propagation across ferrite/austenite boundaries[7]. This is not too surprising as it was shown in Chapter 5 that 2205 can suffer pitting at these chloride concentrations.

Wolf et al. tested 2205 in a simulated North German Basin geothermal fluid, which was a mixture of a wide variety of chlorides (total chloride ~166,000 mg/L) plus a little sulfate at pH 5.4–6.0[11]. Normally, geothermal brines are deaerated and may also contain hydrogen sulfide. In this case, the tests were in aerated brine at 96°C. In early tests, the samples were screwed into the load frame and were, therefore, connected to electrical mains earth. The potentials were erratic and rather negative. Further tests with electrically insulated samples gave more positive potentials. The *S-N* curve in Figure 7.8 shows that the insulated samples failed quicker than earthed samples. This is a very aggressive brine at high temperature, so the relatively poor performance of 2205 is not too surprising and superduplex should perform much better.

Yang et al. tested 2205 in 26 wt.% sodium chloride at pH 2 under cyclic slow strain rate (1×10^{-6}/s) at a potential of –375 mV SCE at room temperature[12]. They investigated the effect of various peak stresses and stress ratios. They looked at three peak stresses of 110%, 120%, and 140% of yield strength. It is not clear whether this was the specified minimum yield strength (SMYS) or the actual yield strength (AYS). Because of the high maximum stresses, it is most likely to have been a percentage of SMYS. All of the cracking occurred at inclusions in the structure, except at 110% YS, where cracks initiated at pits at the inclusion/parent metal interface. At 120% YS and 140% YS, the cracks were across the inclusions, with more severe cracking at the higher stress. The composition of the inclusions suggests that they were principally oxides rather than sulfides.

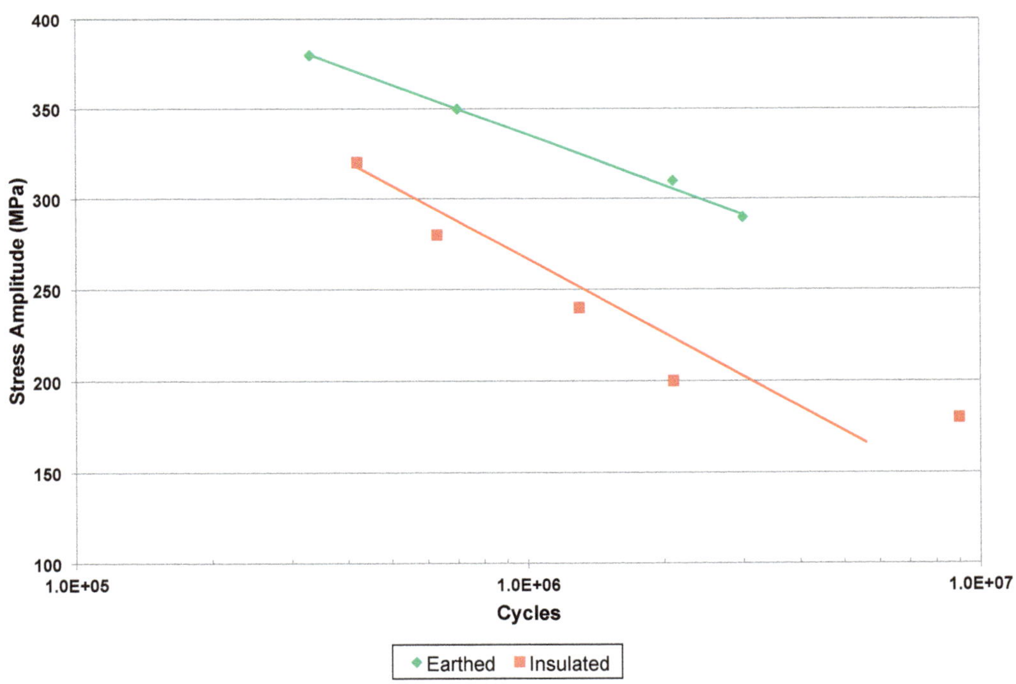

FIGURE 7.8 Effect of earthing on the *S-N* curve for 2205 in a simulated geothermal brine[11].

Yang et al. also investigated the effect of the stress ratio for 2205 at a peak stress of 140% YS, and determined the crack density at each *R* value as a measure of the severity of cracking[12]. The results in Figure 7.9 show that only short cracks occurred at *R*=0.1, but longer cracks occurred at *R*=0.5 and *R*=0.8, with slightly longer cracks at the higher *R* value. These results show the high resistance of 2205 to corrosion fatigue under the test conditions, because cracking only initiated at very high peak stresses.

Figure 7.10 shows the *S-N* curve for Z100 superduplex in air and brine, and the high corrosion resistance of the alloy in this fluid is demonstrated by the clear threshold stress in brine[10].

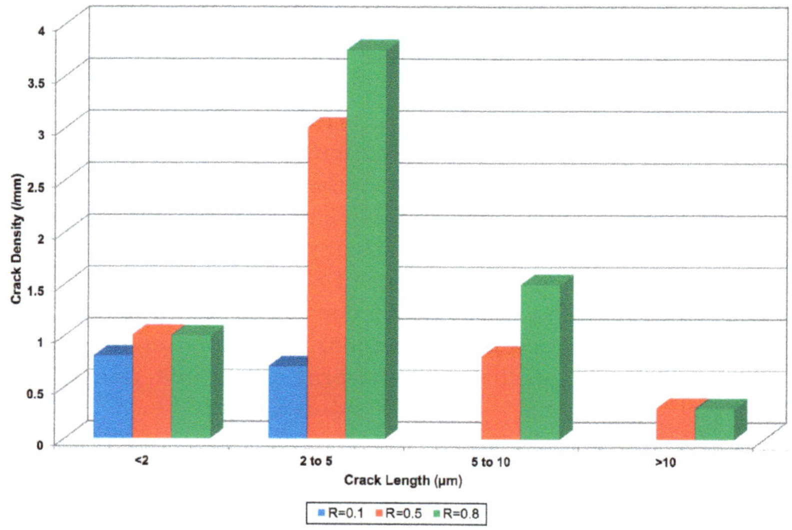

FIGURE 7.9 Effect of stress ratio on the fatigue crack density of 2205 in 26% NaCl at pH 2 and a potential of -375 mV SCE at room temperature[12].

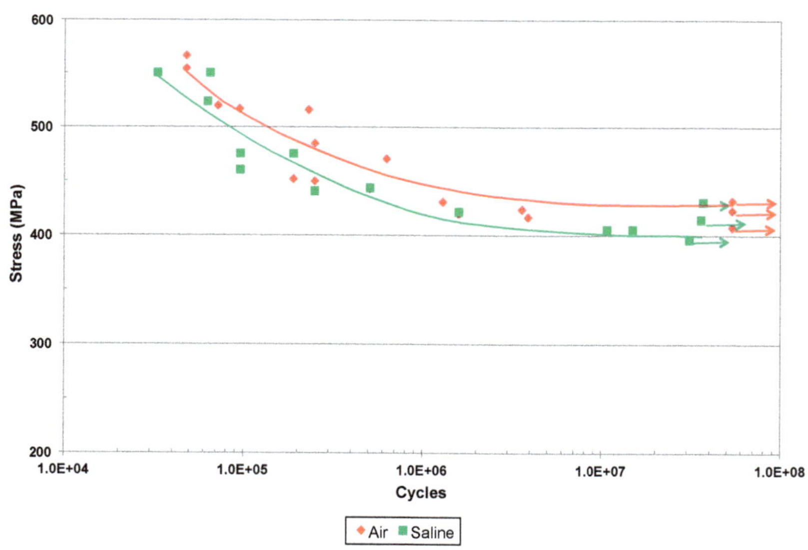

FIGURE 7.10 *S-N* curves for Z100 superduplex in air and saline at 50 Hz[10].

Girones et al. investigated the effect of chloride concentration on the corrosion fatigue properties of 2507 superduplex[13]. They found that the fatigue life decreased compared with air at 30,000 mg/L chloride, but it increased at 60,000 mg/L chloride. This was explained by dissolution of metal at slip steps by the stronger solution, making the surface smoother and crack propagation slower.

7.4.2 Pulp and Paper Industry

In the Kraft pulping process, wood chips are heated under pressure in a solution of sodium sulfide and caustic soda, which is known as "white liquor." This solution is strongly reducing. During processing, the white liquor dissolves the lignin in the wood and is converted to "black liquor." Typical compositions for simulated liquors are 150 g/L NaOH + 50 g/L Na_2S (white liquor), and 15 g/L NaOH + 50 g/L Na_2S (black liquor).

Yang et al. tested 2205 in the same manner as described above for 26% NaCl at pH 2, but the fluid was simulated white liquor at 170°C[12]. Fatigue cracking occurred in the austenite at all peak stresses. In Chapter 6, it was shown that superduplex stainless steel can crack in the austenite in hot caustic soda at reducing potentials, so the occurrence of fatigue cracks under similar conditions is not surprising. Yang et al. also examined the effect of stress ratio and only found fatigue cracking at R=0.5, while none occurred at R=0.1 and R=0.8. They proposed that this was due to the degree of strain accumulation at a constant peak stress with varying R ratios and the ability of the film to repassivate. Duplex alloy 2205 has performed well in Kraft digesters, and the results of Yang et al. show that very high stresses are needed to initiate corrosion fatigue.

The water is often removed from the pulp by pressing it continuously between two rolls, one of which is perforated and has a partial vacuum inside to facilitate white water removal. Originally, rolls were small (0.3 m diameter and 2.5 m long) and made of bronze, but as sizes increased they failed because of corrosion fatigue[14]. Martensitic or austenitic stainless steels also failed by corrosion fatigue as roll sizes continued to increase. The introduction of 2205 suction rolls has solved the fatigue problem with rolls up to 2 m diameter and 10 m long or more, and rolls are now also being fabricated in 2304[14]. The rolls are sometimes made by continuous casting, but it is more common to fabricate the rolls from plate, solution anneal to remove stresses, and then proceed to final machining.

Groth et al. summarized the results of a large program of corrosion fatigue testing of 2205 and 2304 for suction rolls[15]. The tests were in a simulated white water that was 400 mg/L chloride and 250 mg/L sulfate at pH=3.5, and tests were at 5 Hz at room temperature with a stress ratio of −1. Figure 7.11 shows the *S-N* curve data for 2205 and 2304, and the high threshold stress for both alloys is clear. Groth et al. also measured the fatigue strength at 50% failure probability for 10 million cycles at both 5 and 25 Hz[15]. The results in Table 7.1 show that 2304 appeared somewhat better than 2205 at 5 Hz, but this is from a limited data set for 2304, compared with 2205. Both alloys have high threshold stresses in this environment. The data for the 2205 and 2304 welds is discussed in Section 7.5.

Table 7.1 Fatigue strength at 50% failure probability at 10 million cycles in simulated white water at pH=3.5[15].

ALLOY	UNS No.	Fatigue Strength (MPa)	
		5Hz	25Hz
2205	S32205	208 ± 23	255 ± 20
2304	S32304	225 ± 5	Not tested

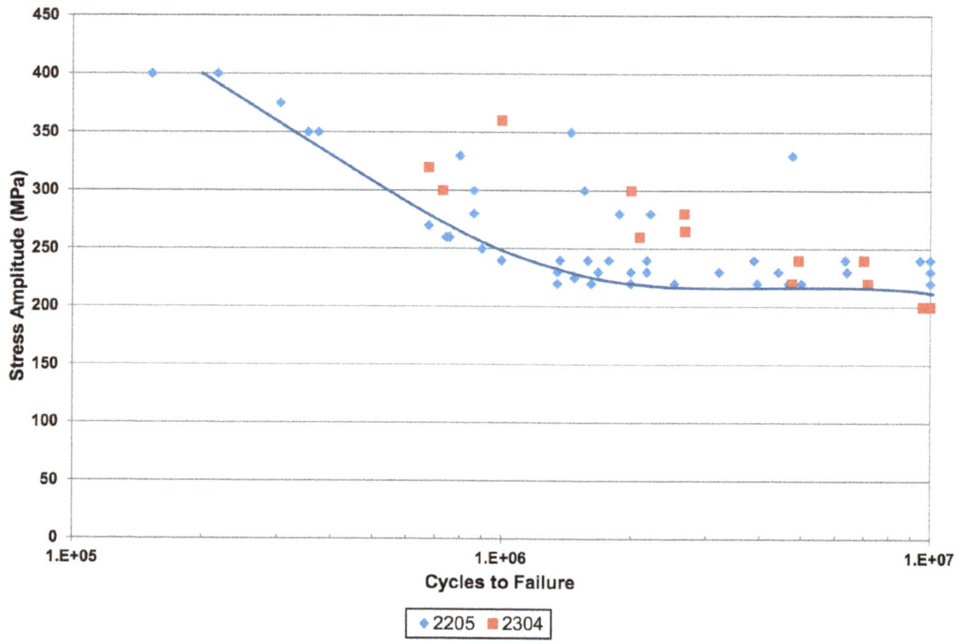

FIGURE 7.11 *S-N* curves in a simulated white water at pH=3.5 and 5 Hz[15].

7.4.3 Oil and Gas

Produced fluids in the oil and gas industry often contain carbon dioxide and hydrogen sulfide, as well as produced water that may have a high chloride concentration. This means that the in situ pH may vary from 3-5 depending on the additional presence of buffering agents, such as bicarbonate. Thodla et al. investigated the corrosion fatigue resistance of 2507 superduplex for offshore oilfield risers[16]. These pipes take the produced fluids from a subsea wellhead to the platform for processing. They can be very long so fatigue resistance is essential. Thodla et al. carried out tests in a sour brine containing 115,000 mg/L chloride, 1,000 mg/L bicarbonate, 103 bar CO_2 and 0.124 bar H_2S at 45°C[16]. This gave an in situ pH of 3.9. The corrosion fatigue tests were at 0.5 Hz and R=0.5, while the tests in air were at 10 Hz and R=0.5. The crack growth rate results are shown in Figure 7.12, where it can be seen that there was a significant increase in the fatigue crack growth rate in the sour brine compared with air. This means adopting some conservatism when designing subsea risers. It is interesting to note that superduplex stainless steel has been used for pump shafts in similar sour brines for many years without any corrosion fatigue failures. However, these shafts are typically running at around 50 Hz; 100 times the current frequency, and it was shown earlier that fatigue resistance increases at higher frequencies.

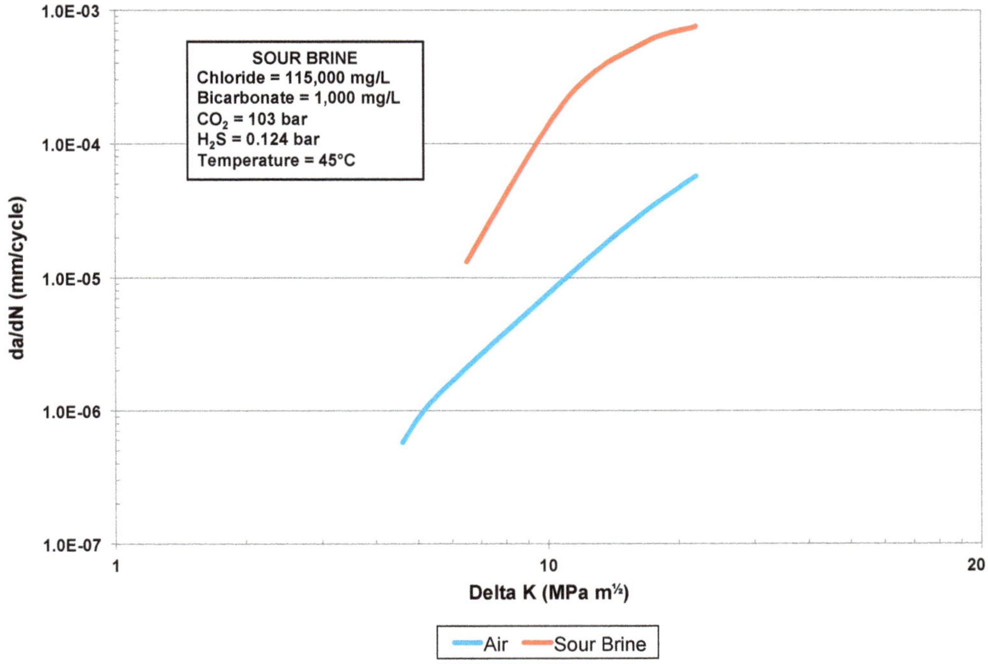

FIGURE 7.12 Crack growth rate for 2507 in a sour brine at R= 0.5 and at 0.5 Hz[16].

7.5 CORROSION FATIGUE OF WELDS

There is less published on corrosion fatigue of duplex stainless steel welds, and some of the data was reviewed by Gunn[17], but it is mostly for austenitic alloys. More recently, the corrosion fatigue resistance of duplex stainless steel welds has been examined for specific environments.

Liljas et al. carried out a wide range of fatigue tests on both 2205 and 2304 duplex with different types of weld to determine how applicable the design rules for welded carbon steel under cyclic stressing would be to fabricated duplex stainless steels used for bridges and other structures[18]. These tests were mostly in air, but they did conduct a few tests on welded 2205 and welded 316L in 3% sodium chloride at R=0.1 and a frequency of 1 Hz. They took as the endurance limit, samples that had not failed after 2×10^6 cycles, and the results are shown in Table 7.2. The welded samples were submerged arc welded, which might be expected to be the weld type with the most inclusions and hence, most susceptible to corrosion fatigue. It is clear that 2205 had a higher endurance limit compared with 316L, as would be expected from its higher strength. For both alloys the welds were stronger than the parent metal, with a higher resistance to corrosion fatigue.

Table 7.2 Results of corrosion fatigue tests of welded stainless steel in 3% sodium chloride at *R*=0.1 and a frequency of 1 Hz[18].

ALLOY	CONDITION	ENDURANCE LIMIT (R) (MPa)	R_{Corr}/R_{Air}
316L	Parent	155±127	0.766
	SAW Welded	190±155	-
2205	Parent	250±205	0.885
	SAW welded	261±214	-

The work described above of Groth et al. on duplex alloys for suction rolls in pulp and paper mills also included some data on welds[15]. The tests were at 5 Hz in the same simulated white water as described in Section 7.4.2, and the results are shown in Figure 7.13. The 2205 welds were electroslag welded, while the 2304 welds were produced by submerged arc welding. It can be seen that the data points for all the welds lie above the line for parent metal (Figure 7.9). This is most probably because duplex welds tend to be stronger than parent metal and so will be less likely to initiate corrosion fatigue.

Comer investigated the corrosion fatigue resistance of Z100 superduplex welds under a number of conditions found in seawater applications[19-21]. One weld was all gas tungsten arc (GTA) welding, while the second was a GTA root with shielded metal arc (SMA) fill, both produced by joining 10 mm plate. The results showed that the residual stresses from the welding operation were low and negligible when calculating fatigue stresses. The threshold stresses for the onset of crack growth in parent metal and welded material were similar. There were also no significant differences in the results for the two different welds. In chlorinated seawater, the potential of stainless steel is typically +600 mV SCE (see Chapter 5), and tests in synthetic seawater at this potential showed no difference in the crack propagation rates, as shown in Figure 7.14.

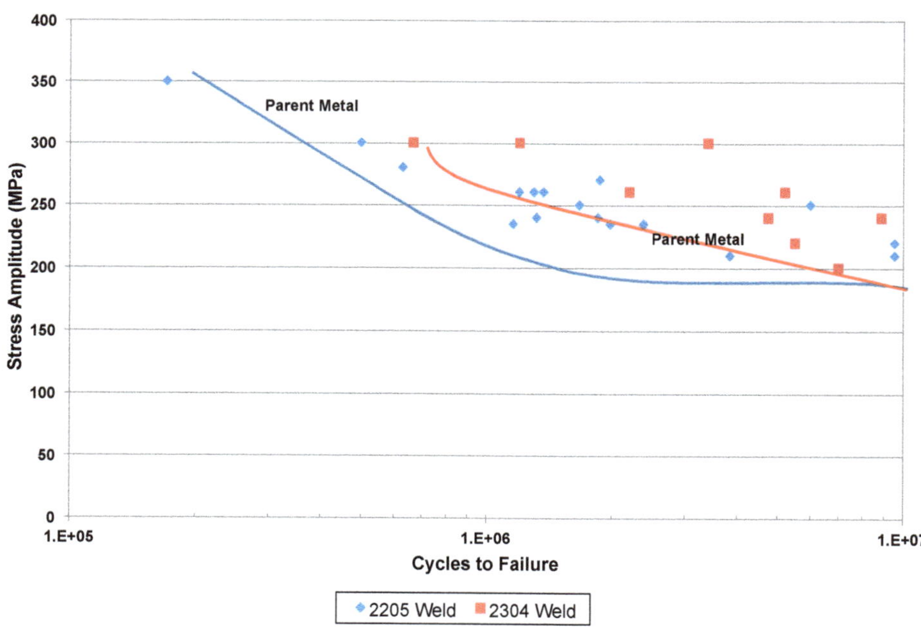

FIGURE 7.13 *S-N* curves for welded duplex in simulated white water at pH=3.5 and 5 Hz[15].

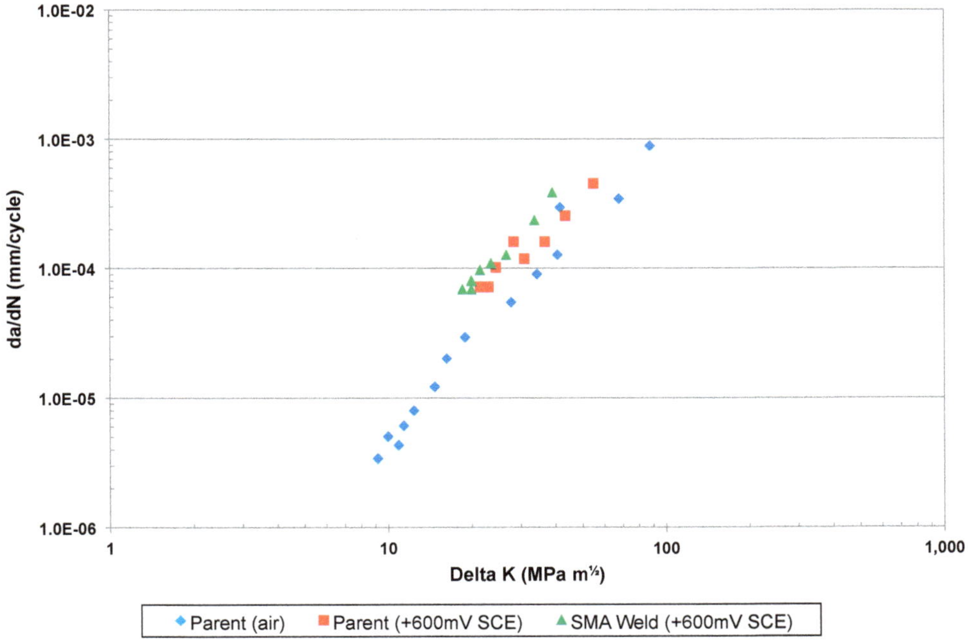

FIGURE 7.14 Crack growth rate for superduplex welds in synthetic seawater at a potential of +600 mV SCE (Frequency = 0.1 Hz and *R*=0.5)[19,21].

When designing welded components to resist fatigue, e.g., vessels to EN 13445, each part of a welded joint that could act as a site for a fatigue crack is placed in one of eight classes (B, C, D, E, F, F_2, G, and W). Comer's welds in Z100 fall into class D[19]. Figure 7.15 shows the time to failure for both base metal and weld metal as a function of the stress range. The dashed line is the class D design curve and the dotted lines show ±2 standard deviations. It is clear that the data for the Z100 parent and welds fit well within the class D envelope.

Comer also carried out tests in synthetic seawater on welded samples polarized to −1.04 V SCE, which is a typical potential for cathodic protection subsea[19-21]. His results showed that crack propagation rates increased by a factor of four for parent metal compared with air. There were no significant differences between parent metal and either of the two welds (Figure 7.16). These results indicate that the design for cathodically protected subsea items that are exposed to cyclic stresses requires a little more conservatism in design than equipment without CP.

Woollin et al. conducted corrosion fatigue tests on DP3W superduplex GTA welds made with Z100X filler in seawater with cathodic protection at approximately −1 V SCE, as might occur with material used on subsea wellheads or flowlines[9]. The results were similar to those of Comer described above, in that crack propagation rates increased under cathodic protection compared with air at high ΔK values, while there was a lot of scatter at low ΔK values (<19 MPa·m$^{1/2}$) and the differences between air and cathodically protected in seawater were small.

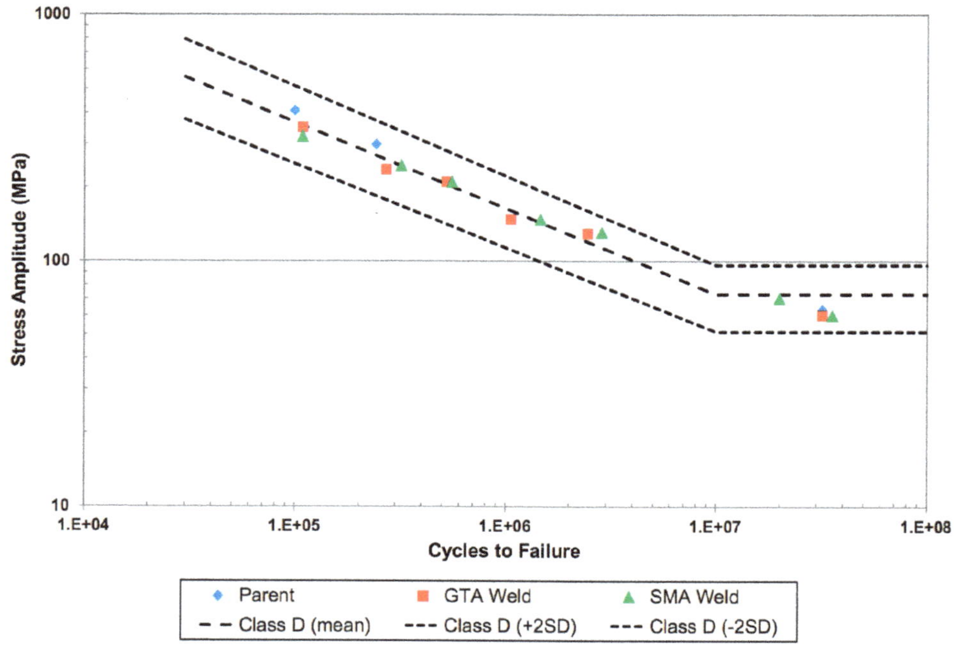

FIGURE 7.15 *S-N* curve for welded superduplex in air showing Class D line and two standard deviations (R=0.5)[19].

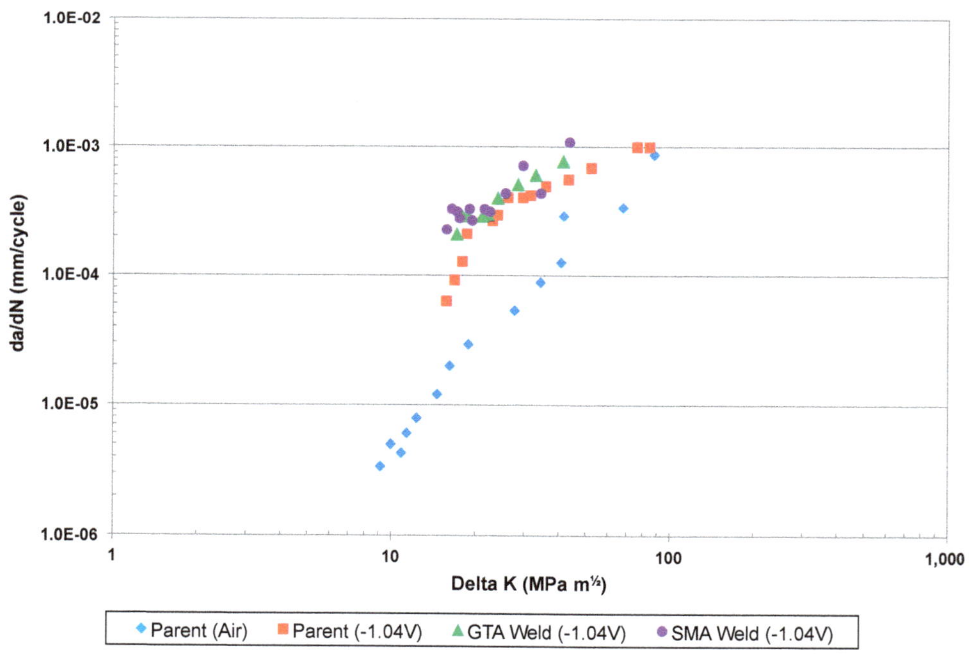

FIGURE 7.16 Crack growth rate for superduplex welds in synthetic seawater at a potential of -1.04 V SCE (Frequency = 0.1 Hz and R=0.5)[19].

Woollin et al. also tested DP3W superduplex GTA welds made with Z100X filler in a sour brine containing 14,590 mg/L chloride with 0.975 bar CO_2 and 0.025 bar H_2S, giving a pH at temperature and pressure of about 5.7[9]. The results in Figure 7.17 show a crack propagation rate up to 10 times that in air in the sour brine at both 20 and 80°C. The results suggest that crack propagation rates in the sour brine were lower at 80°C than at 20°C at low ΔK values. This is probably connected with the mechanism, which was thought to be due to hydrogen from the H_2S, produced by corrosion. Hydrogen is much more mobile at 80°C and is less likely to become concentrated at the crack tip.

Umbilicals are widely used in the oil and gas industry, either to convey power, hydraulic fluid, or chemicals to a subsea wellhead, and/or to convey process fluids to the surface. Superduplex stainless steels are widely used for umbilical tubing because of their high strength and excellent resistance to a wide range of fluids. A typical umbilical is a bundle of superduplex tubes around a core, with insulation and armor around the whole bundle. Because the tubes are hung from the platform to the sea bed and can be kilometers long, it is important that they resist fatigue. The umbilicals are so long that girth welds are required at intervals to join tubes together, and these are regarded as the weak link. Kivisaak and Chai describe the fatigue properties of 2507 superduplex and 3207 hyperduplex stainless steels with a single girth weld[22]. Figure 7.18 shows the *S-N* curves for both 2507 and 3207 in air, and the high resistance to fatigue is clear. Kivisaak and Chai claim that the corrosion fatigue resistance in seawater is the same as air, although they did not show a full *S-N* curve. However, the data for Z100 superduplex in Figure 7.8 at 50 Hz shows little difference in corrosion fatigue resistance in saline solution compared with air, so this would not be surprising at 5 Hz, as well for a superduplex alloy.

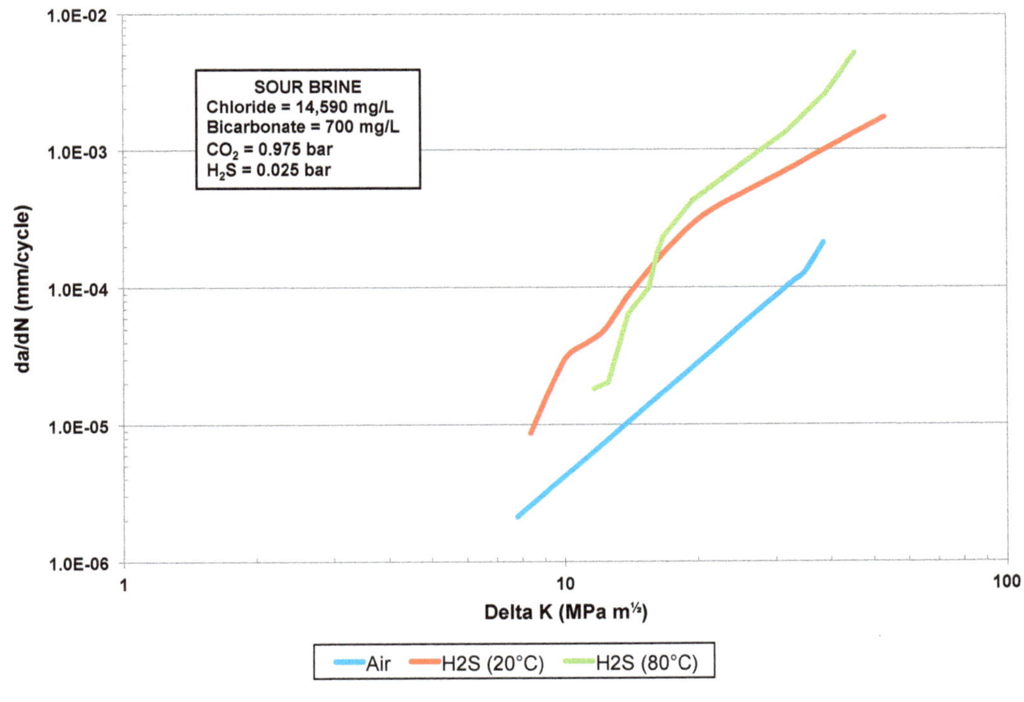

FIGURE 7.17 Crack growth rates for superduplex welds in a sour brine at R=0.5 and a frequency of 0.5 Hz[9].

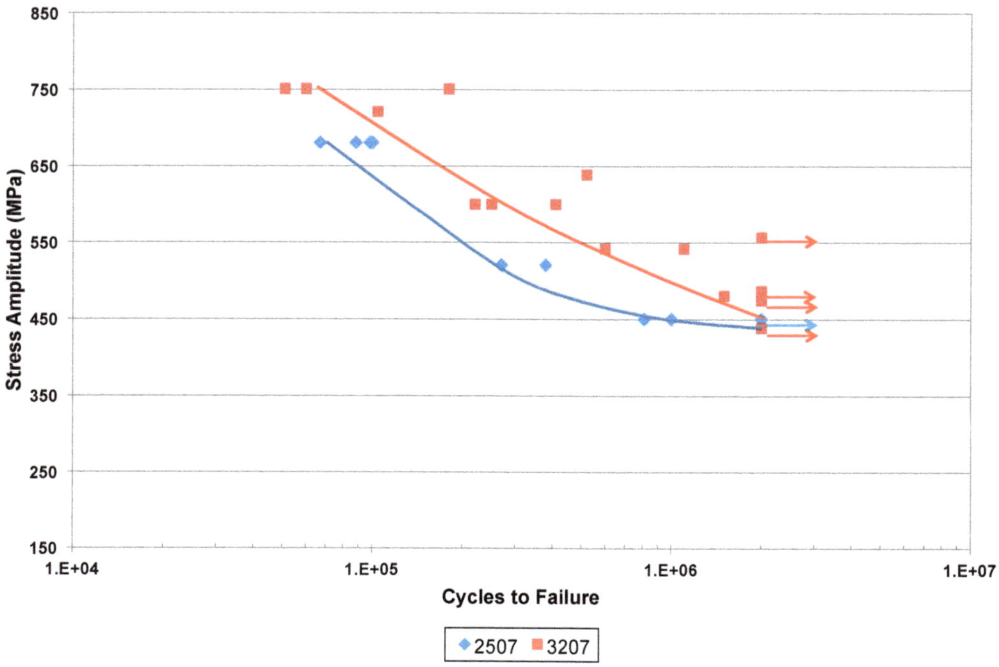

FIGURE 7.18 *S-N* curve in air at 5 Hz for 14.7×1 mm tube with a single girth weld[22].

REFERENCES

1. D.E. Jungbauer and A.J. Smallberg, "Recent North Sea Experiences Involving Fatigue Failures of Duplex Stainless Steel Piping Systems," *Stainless Steel World* (April 1996), p. 31.
2. T.J. Marrow and J.E. King, "Microstructural and Environmental Effects on Fatigue Crack Propagation in Duplex Stainless Steels," *Fatigue Fract. Eng. Mater. Struct.*, 17, 7 (1994): p. 761.
3. J. Wasen, E.U. Engstrom, and B. Karlsson, "Near-Threshold Fatigue Crack Growth in a Duplex Stainless Steel," Stainless Steels '87 (London, United Kingdom: Institute of Materials, Minerals, and Mining, 1987): p. 357.
4. M.C. Mayaki, Ph.D. Thesis, "Aspects of Fatigue Crack Propagation Behaviour of Duplex Stainless Steels" (Nottingham, United Kingdom: University of Nottingham, 1989).
5. S. Suresh, G.F. Zamiski, and R.O. Ritchie, "Oxide-Induced Crack Closure: An Explanation for Near-Threshold Fatigue Crack Growth Behaviour," *Met. Trans. A,* 12A (1981): p. 1435.
6. S. Suresh and R.O. Ritchie, "Geometric Model for Fatigue Crack Closure Induced by Fracture Surface Roughness," *Met. Trans. A,* 13A (1982): p. 1627.
7. F. Perdriset, T. Magnin, T. Cassagne, P. Hoch, and F. Dupoiron, "Environmental Effects on Low Cycle Fatigue Behaviour of Z3 CND 2205 Duplex Stainless Steel," Duplex '94, paper no. 13 (Cambridge, United Kingdom: TWI, 1994).
8. N. Akdut, "Is the Fatigue Life of Duplex Stainless Steels a Function of the Second Phase and/or the Nitrogen Content?" Duplex America 2000 (Zutphen, Netherlands: KCI Publishing, 2000): p. 79.

9. P. Woollin, S. Maddox, and D.J. Baxter, "Corrosion Fatigue of Welded Stainless Steels for Deepwater Riser Applications," 24th International Conference on Mechanics and Arctic Engineering (OMAE 2005), paper no. 67498 (New York, NY: ASME International, 2005).
10. Rolled Alloys Technical Report TN1085R3, "Corrosion Fatigue Resistance of Zeron 100 Superduplex Stainless Steel" (Temperance, MI: Rolled Alloys, 2008).
11. M. Wolf, A. Pfennig, R. Afanasiev, and T. Boellinghaus, "Corrosion Fatigue of X2CrNiMoN22-5-3 Exposed to the Geothermal Environment of the North German Basin," CORROSION 2016, paper no. 7378 (Houston, TX: NACE International, 2016).
12. D. Yang, P.M. Singh, and R.W. Neu, "Effect of Low Frequency Cyclic Stresses on Stress Corrosion Cracking of Duplex Stainless Steel in Chloride and White Liquor Solutions," CORROSION 2012, paper no. 1613 (Houston, TX: NACE International, 2012).
13. A. Girones, M. Anglada, and A. Mateo, "Chloride Content Effect on the Corrosion Fatigue Properties of Superduplex Stainless Steel," *Trans. ASME,* 129, part 4 (2007): p. 588.
14. L. Wegrelius, "Duplex Solutions for Pulp and Paper Industry," ABTCP 2009, 42nd Pulp and Paper International Congress and Exhibition (Sao Paulo, Brazil: Assoc. Brasileira Técnica de Celulose e Papel, 2009).
15. H. Groth, A. Kahonen, C. Toigerstrand, S. Ekman, M. Andersson, and D. Eyzop, "Summary of Corrosion Fatigue Test Data Duplex Suction Roll Shell Material," ACOM No. 3 (Avesta, Sweden: Outokumpu Stainless AB, 2010): p. 2.
16. R. Thodla, F. Gui, K. Evans, C. Joia, and I.P. Baptista, "Corrosion Fatigue Performance of Duplex 2507 for Riser Applications," CORROSION 2010, paper no. 312 (Houston, TX: NACE International, 2010).
17. R.N. Gunn (editor), *Duplex Stainless Steels: Microstructure, Properties and Applications* (Cambridge, United Kingdom: Abington Press, TWI, 1997): p. 152.
18. M. Liljas and C. Ericsson, "Fatigue Behaviour of Stainless Steel Welds," ACOM 1-2 (Avesta, Sweden: Outokumpu Stainless AB, 2002): p. 2.
19. A.J. Comer, Ph.D. Thesis, "Corrosion Fatigue of a Superduplex Stainless Steel Weldment" (Dublin, Ireland: Dublin City University, 2003).
20. A. Comer and L. Looney, "Crack Propagation Resistance of Zeron 100 Weld Metal Using GTA and SMA Welding Process," *Theoretical Appl. Fracture Mech.*, 45 (2006): p. 139.
21. A. Comer and L. Looney, "Corrosion and Fatigue Characteristics of Positively Polarized Zeron 100 Base and Weld Metal in Synthetic Seawater," *Int. J. Fatigue*, 28 (2006): p. 826.
22. U. Kivisaak and G. Chai, "Corrosion Fatigue Properties of Sandvik 3207 HD Umbilical Tubes," EUROCORR 2009, European Federation of Corrosion (London, United Kingdom: Institute of Materials, Mining and Minerals, 2009).

8 Erosion Corrosion and Cavitation

8.1 INTRODUCTION

Erosion corrosion occurs when the fluid flow is turbulent enough to generate a shear stress at the metal surface sufficient to disrupt the passive film. The film reforms and then is removed, causing metal loss, and this repetitive process is erosion corrosion. Because it involves film formation, it is electrochemical in nature and it can be suppressed by the application of cathodic protection (CP). Erosion is mechanical and is usually caused by particles, such as sand, in the fluid that mechanically remove the film and some metal. Because it is mechanical in nature, it cannot be suppressed by the application of CP.

In many industrial applications, there are instances where water flow is high and turbulent, such as pumps, and many of these fluids can also contain entrained solids in quantities from a few milligrams per liter to over 10 g/L. The severity of the erosion by solids depends on the angularity of the particles, as well as their size and concentration. Coarser particles are more abrasive than fine ones.

When a metal is suffering from erosion corrosion there will be a weight loss due to corrosion and another due to mechanical erosion. However, it is now well established that when the two occur together there is a synergistic effect that contributes an additional weight loss[1]. This is often expressed by the equation:

$$\text{Total Weight Loss (TWL)} = E + C + S \qquad (1)$$

Where E is the weight loss due to erosion, C is the weight loss due to corrosion, and S is the synergistic weight loss. The magnitude of these factors depends on many variables, including temperature, solids content, velocity, the alloy in question, and the corrosivity of the fluid. Stainless steels have a very protective passive film, which is difficult to break down, so they tend to have good resistance to erosion corrosion, provided they can resist corrosion in the specific environment.

He et al. investigated erosion corrosion using a rotating cylinder electrode in slurries containing alumina and various liquids, pure water, 3.5% sodium chloride, and 3.5% sodium chloride plus 0.5N sulfuric acid[2, 3]. They tested stainless steels suitable for pumps and included a ferritic alloy, a martensitic alloy, an austenitic alloy (similar to 316) and a duplex alloy, CD4MCuN. They found that CD4MCuN was the best in all three environments, particularly with higher solids contents. They ascribed this to the work hardening of the austenite phase by the solids impacts and obtained microhardness figures up to 570 HV in the austenite at the surface. This did not happen with the other alloys, despite some of them being harder than the duplex alloy to start with. Work hardening had less effect with the 316L because of the higher rate of metal loss compared with the duplex alloy.

This clearly demonstrates the advantages of duplex stainless steels in environments that can cause erosion corrosion, and they are widely used for pumps, valves, and agitators in many industries. Some of the data and applications in different media are reviewed below.

8.2 EROSION CORROSION IN DIFFERENT ENVIRONMENTS

There are two factors regarding velocity that need to be considered when selecting materials. The first of these is that nominal liquid velocities in piping rarely exceed 10 m/s, even though this can be well below the velocity limit for a stainless steel. This is because higher velocities can cause cavitation, particularly where there are a lot of sharp bends, and it can also cause excessive noise in environments where staff are operating. Higher velocities are possible with gases, as discussed below.

When assessing the maximum velocity in a centrifugal pump, it is not the speed of the outer perimeter of the impeller, because the water there is also moving and the relative speed may be low. An impeller usually consists of an upper and lower shroud with radial vanes between them. The vanes usually follow an approximate logarithmic spiral; the useful property of this shape is that as the water accelerates from the impeller eye to the perimeter, the cross sectional area increases to keep the water velocity approximately constant. Hence, the fluid velocity along the radius of the impeller is the total fluid flow divided by the cross sectional area at the perimeter of the impeller. This is often in the range from 3–10 m/s depending on the size and type of pump. However, the water flow path through an impeller means that the flow is extremely turbulent, particularly at changes in direction, so these areas can be prone to erosion corrosion under some conditions.

8.2.1 Seawater

Seawater is widely used as a cooling fluid around the world, so there is a requirement for robust pumps, valves, and piping to handle large volumes without significant corrosion. Natural seawater always contains some solids unless it is finely filtered, as in a seawater reverse osmosis (SWRO) desalination plant (see Chapter 5). The particles may be dead plankton or other creatures and vary in size from a few microns up to 50 μm or more, and sand or mud may also be present. Particles up to 50 μm diameter are usually termed silt, while coarser particles are termed sand. Seawater typically contains around 50 mg/L of fine particles (silt), although high solid burdens can occur in some areas, particularly under tidal action, e.g., River Severn estuary in the United Kingdom.

Kain et al. carried out a multi-facetted test program for the U.S. Navy to assess the seawater corrosion, galling, and wear performance of an extensive series of existing and candidate valve materials[4]. Included was a series of submerged jet-impingement tests operating with natural seawater at 30°C, and at a sustained velocity of 30 m/s. Results of these 30 day tests, shown in Table 8.1, indicate that all wrought stainless steel grades tested were far superior to M-bronze (a cast copper alloy to UNS C92200). Among the various stainless grades, only S31603 exhibited measurable attack.

TABLE 8.1 Results of impingement tests in seawater at 30 m/s and 30°C for 30 days[4].

ALLOY	MAXIMUM DEPTH OF ATTACK (mm)
316	0.06
2507	0.00
Z100	0.00
6% Mo	0.00
Ti grade 2	0.00
M-Bronze	0.81

Tests were conducted in a pin erosion test rig, as described by Dallas and McConnell[5]. The conditions were 3% sodium chloride solution at pH=8 and 55°C with 1,000 mg/L CAST sand at either 25 m/s or 30 m/s[6]. The results in Table 8.2 show that Z100 superduplex was superior to both carbon steel and 316L at both velocities, but the differences between 316L and Z100 were relatively small.

TABLE 8.2 Volume metal loss from pin erosion tests in 3% sodium chloride at 55°C with 1,000 mg/L sand[6].

ALLOY	VOLUME LOSS (ml x 0.001)	
	25 m/s	30 m/s
Carbon Steel	2.33	6.56
316L	1.50	4.85
Z100	1.30	4.18

Hu and Neville used a similar pin erosion test rig to examine the behavior of some high alloy stainless steels in 3.5% sodium chloride at 18°C and 17 m/s with various sand loadings[7]. The results after 8 h in Figure 8.1 show that there was little difference between 2507 superduplex and a 6% Mo alloy (UNS S31254). They also examined the different contributions of *E*, *C*, and *S* in equation 1. The *C* was 4 to 12% of the TWL and *C*+*S* was ~20% of TWL, which demonstrates the importance of the synergistic effect.

In further work, Hu and Neville looked at how the potential of the stainless steel changed with sand loading[8]. As the sand loading increases, the potential should go more negative because bare metal is exposed for longer. They found that the potential decreased as the sand loading increased, up to 823 mg/L, and thereafter, the potential decrease was small with further sand additions. This shows that there is a critical sand concentration to significantly depassivate the metal surface, which will result in an acceleration of erosion corrosion.

In further, work they added 316L to the tests and examined a wider range of sand loadings at both 18 and 50°C[9]. The results in Figure 8.2 show that the erosion corrosion increased with sand loading and 316L was significantly less resistant to attack, while 2507 was only marginally better than the 6% Mo austenitic alloy.

Meng et al. compared cast Z100 superduplex with cast 316 (CF8M) in 3.5% sodium chloride at two velocities (7 and 20 m/s), two sand loadings (50 and 500 mg/L), and two temperatures (20 and 50°C) in a pin erosion test rig[10]. The weight losses for 316 were greater than for Z100, as might be expected, but they also evaluated

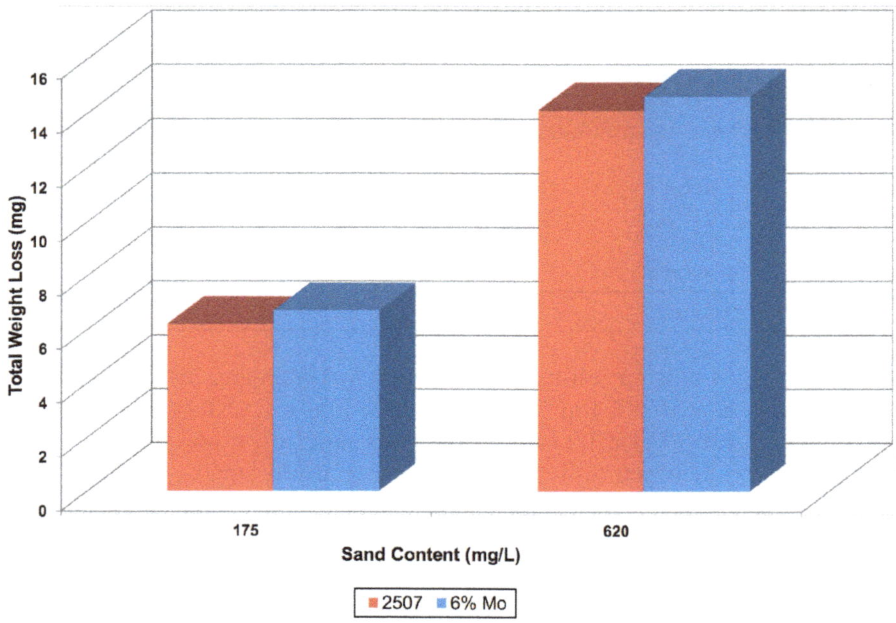

FIGURE 8.1 Total weight loss from two stainless steels in 3.5% sodium chloride at 18°C and 17 m/s at different sand loadings[7].

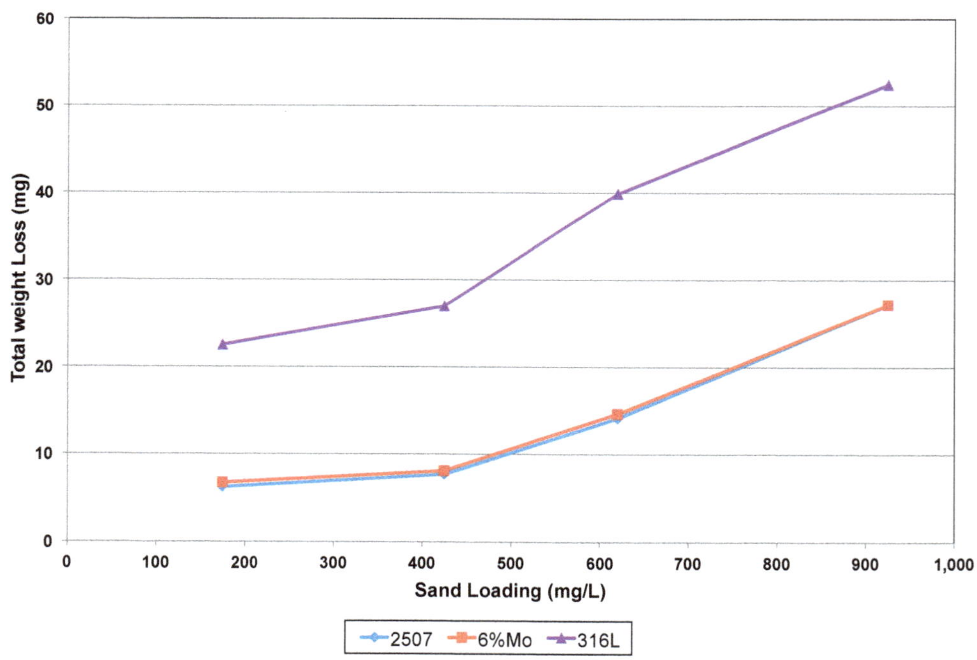

FIGURE 8.2 Total weight loss for some stainless steels in 3.5% sodium chloride at 18°C and 17 m/s at different sand loadings[9].

the effects of the three variables on the weight loss due to corrosion and the weight loss due to synergy. The results in Figures 8.3 show that the dominant factor affecting the corrosion mass loss was the velocity with smaller contributions from the sand loading and temperature, for both alloys. However, Figure 8.4 shows that while velocity and sand loading were the two chief contributors to the synergistic mass loss for 316, they were somewhat less important for cast Z100 and there was also an effect of temperature.

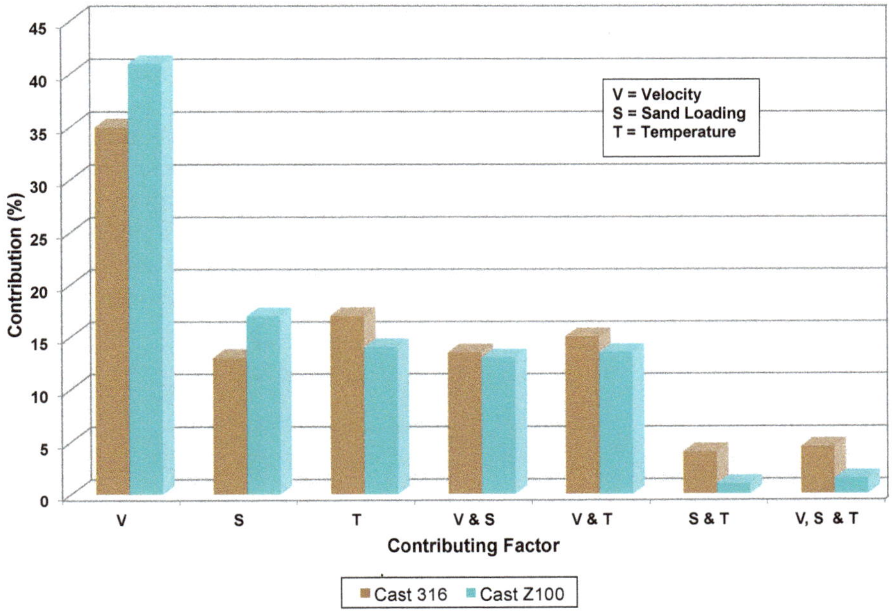

FIGURE 8.3 Effect of variables on the total weight loss in sand erosion tests in 3.5% sodium chloride[10].

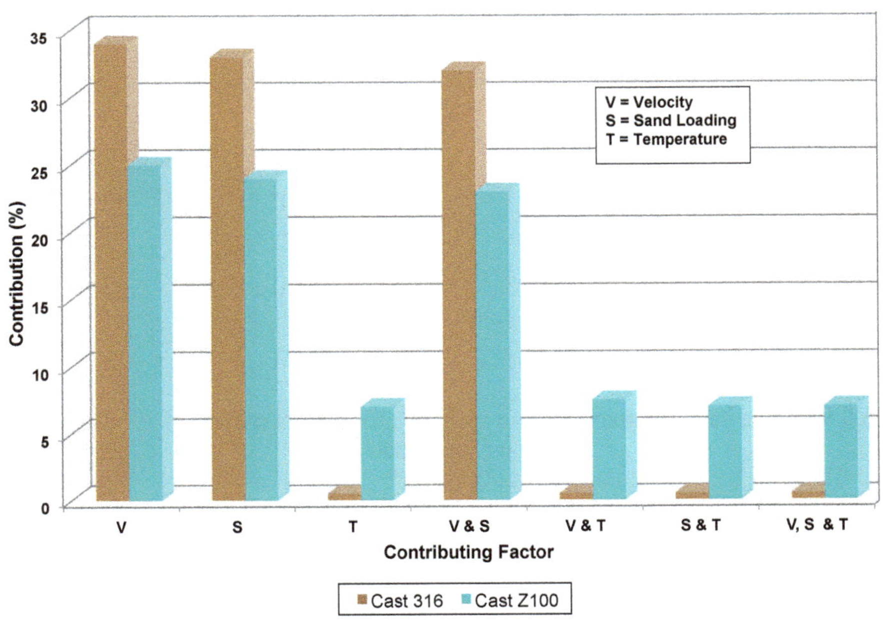

FIGURE 8.4 Effect of variables on the synergistic weight loss in sand erosion tests in 3.5% sodium chloride[10].

Hodgkiess and Mantzavinos also used a pin erosion rig to compare stainless steels and titanium alloys at 19°C and 12.6 m/s in 3.5% sodium chloride solution with varying sand loadings[11]. The results in Figure 8.5 show that there were only small differences between the stainless steels and titanium at low sand loadings, but the titanium was much better than 316L at high sand loadings. Over the range of conditions tested, there was little difference between superduplex and titanium grade 2.

Superduplex stainless steel has been used for seawater pumps since 1986 with great success, and erosion corrosion is rarely a problem unless the sand loading is very high. Even then it need not be a problem as was shown at a nitrogen gas compression plant in Mexico[6]. The gas was compressed in stages with heat exchangers to cool the gas between the compression stages. This required a lot of pumps and heat exchangers, which were originally in copper alloys, but they suffered severe erosion corrosion because of a heavy solids burden in the seawater after every tropical storm, because they stirred up the shallow bay that was the seawater source. The solids burden could peak at up to 50,000 mg/L sand and gravel. The replacement heat exchangers were in Z100 superduplex stainless steel (Figure 8.6) and they have performed well. The pumps were specially designed to cope with the sand burden and the wear rings were coated with tungsten carbide (see section 8.3) to resist wear. The pumps were so big that the cases were fabricated from plate, as shown in Figure 8.7. These have also performed very well.

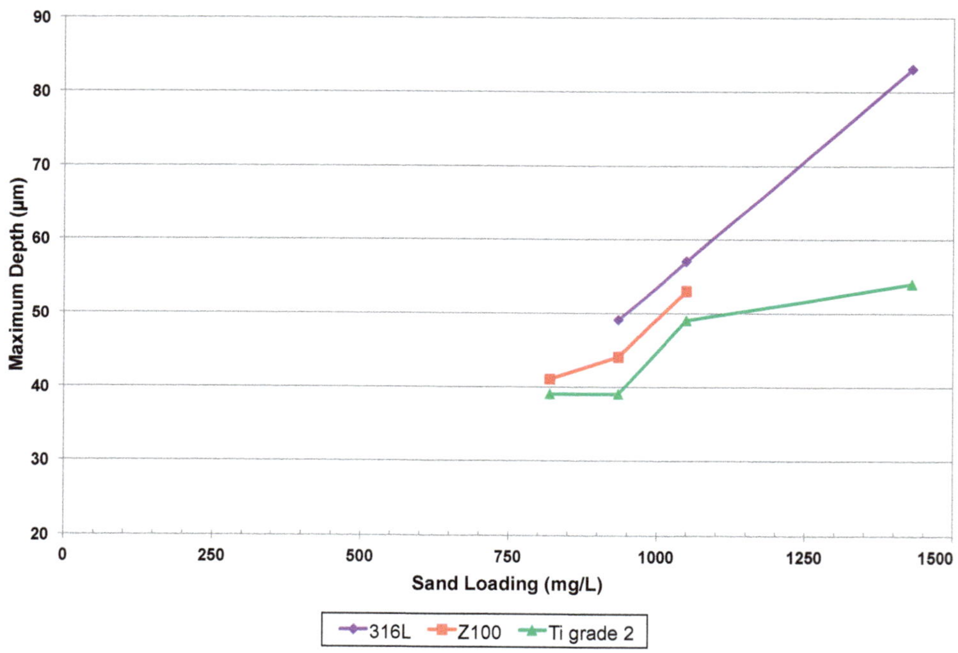

FIGURE 8.5 Maximum depth of attack on three alloys in 3.5% sodium chloride at 19°C and 12.6 m/s at different sand loadings[11].

FIGURE 8.6 A Z100 heat exchanger to cool nitrogen gas in seawater with a high solids burden.
[Photo courtesy of Rolled Alloys]

FIGURE 8.7 A Z100 pump to handle seawater with a high solids burden.
[Photo courtesy of Rolled Alloys]

8.2.2 Gas Systems

There are many situations in the oil and gas industry where methane gas must be removed and it often carries entrained water and carbon dioxide. If the gas is flowing at high velocity, then the entrained water droplets may cause erosion corrosion. This can also happen if sand is entrained in the gas stream. Erosion in gas streams has not been as well researched for duplex stainless steels as for those carrying water. A number of regimes are possible in wet gas flow, from bubble flow at low velocities, through slug flow to annular mist flow at high velocities. Annular mist flow is the most erosive in the absence of sand and in many practical situations the flow is annular mist type[12].

Castle et al. reported wellhead tests in wet gas streams on a number of alloys[13] that were conducted under sand-free conditions at 18 m/s and the results are in Table 8.3. They show that both the alloy steel and 2205 duplex were better than carbon steel, and there was essentially no significant metal loss from 2205, justifying its use at higher wet gas velocities.

TABLE 8.3 Metal loss at 18 m/s in a wet gas stream[13].

ALLOY	METAL LOSS (%)	EROSION RATE (mm/y)
AISI 1018 Carbon Steel	59	6.3
9Cr/1Mo Alloy Steel	1.1	0.08
2205 Duplex	0.01	0.001

Ikeda et al. described tests in a flow loop in a Venezuelan gas well on a range of martensitic and duplex stainless steels[14]. The gas was mostly hydrocarbons with 6.65% carbon dioxide and 5 ppm of hydrogen sulfide. They conducted two tests of about four months each. In test 1, the mean temperature was 88°C at a mean gas pressure of 130 bar, while in the second test, the mean temperature was 79°C at a mean pressure of 120 bar. The results of both tests in Figure 8.8 show that 2205 had a negligible corrosion rate in all the tests, even at a gas flow velocity of 250 m/s. The corrosion rate of 13 Cr martensitic was higher than that of 2205, but was still low enough for many practical applications.

Craig[12] looked at API RP14E[15] and variations on it. This standard is for predicting the limits of use of carbon steel under high flow annular mist conditions in oil and gas production environments. The equation takes the form:

$$V_c = C/\rho^{1/2} \qquad (2)$$

where V_c = critical velocity (for erosion corrosion), C = a constant, and ρ = fluid density.

There are values of C in the standard for some corrosion resistant alloys (CRAs), but there is no understanding of how these are derived. Craig points out that there is a better equation that uses the hardness of the oxide or sulfide film on the metal as a variable:

$$V_c = 9.78\ [P/\rho]^{3/7} \qquad (3)$$

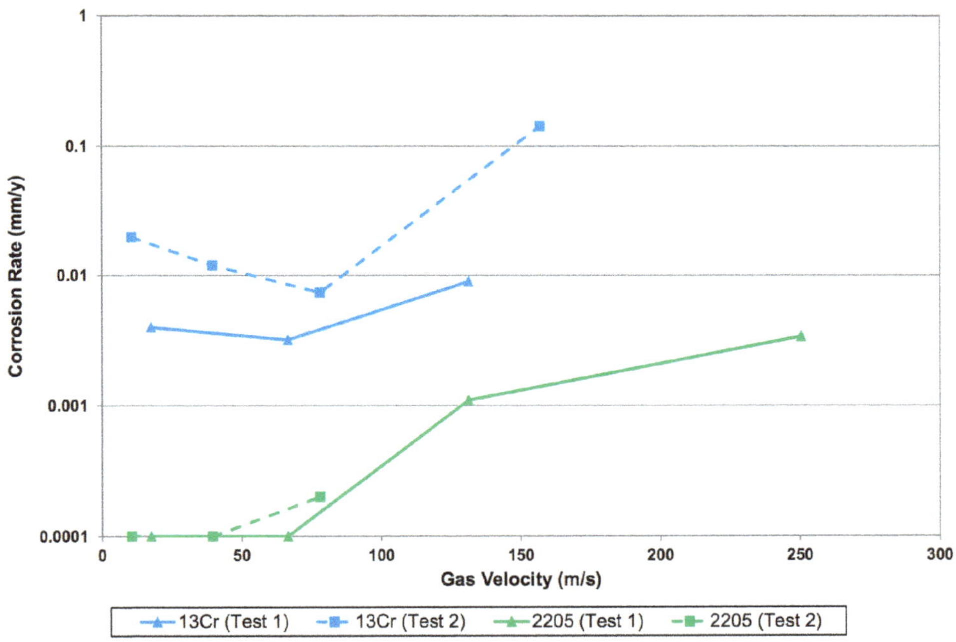

FIGURE 8.8 Corrosion rate of two stainless steels versus velocity in wet gas[14].

where P = hardness of the oxide or sulfide in kilograms per millimeter squared, and ρ = fluid density in pounds per cubic feet. Craig gives examples for different CRAs to show how the use of equation 3 gives a wide range of values for C in equation 2.

8.2.3 Oil and Produced Water

When oil and water are present together in a produced fluid, two conditions can arise. One is where the oil and water remain as separate phases with little mixing. The second case is where the oil is miscible with the water, or the oil and water are mixed in an emulsion, usually under fast flowing conditions.

In the first case, where the oil and water remain separate, corrosion can occur where there is sufficient water to wet the metal surface. This can occur with very small water fractions[16]. Exactly how much depends on the nature of the specific oil and the solubility of water in it. However, water solubility of water in most oils is low and as little as 1% water can cause corrosion[17]. Even if the water and oil form an emulsion, the oil and water can separate into their distinct phases if the flow rate decreases and a series of stratified layers can form[18]. Where a water layer contacts the metal surface, corrosion can occur. After the separators, the produced water has very little oil, but may still be carrying solids from the producing strata.

The author carried out a pin erosion test on a number of alloys to examine performance at elevated temperature (55°C) and low pH (3.5) in the presence of sand (640 mg/L) with 25,000 mg/L chloride at 40 m/s. The results in Figure 8.9 show that all the alloys were attacked, which is not surprising at this very high velocity. However, both 2205 and Z100 superduplex were more resistant than 316L.

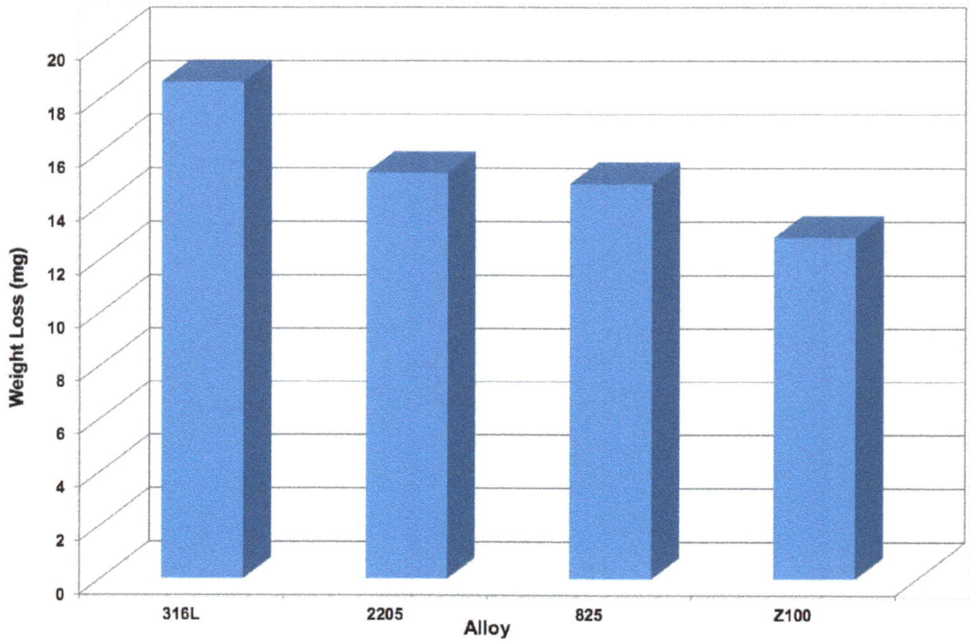

FIGURE 8.9 Weight loss in pin erosion test in 25,000 mg/L chloride at pH=3.5 and 55°C at 40 m/s.

When trying to evaluate erosion conditions in real wells, testing usually involves the presence of gas, sand, and water, without oil, as the latter tends to reduce corrosion if it has any effect. Rincon et al. carried out flow loop tests at pH 4 in a brine saturated with carbon dioxide, at both 24°C and 66°C[19]. The liquid flow velocity was relatively low (0.4 m/s) with a gas flow rate of 6.1 m/s and sand was injected at a rate of 13.6 Kg/day. They tested a 13 Cr martensitic alloy, a super 13 Cr martensitic alloy and 2205 duplex. The mass loss rates for erosion corrosion and the separate erosion and corrosion contributions are shown in Table 8.4. In these results any synergistic effect is included with the mass loss due to corrosion.

TABLE 8.4 Erosion corrosion penetration rates for some stainless steels in CO_2 and brine[19].

ALLOY	TEMPERATURE (°C)	PENETRATION RATE (mm/y)		
		Erosion Corrosion	Erosion	Corrosion
13Cr Martensitic	24	4.5	2.0	2.5
Super 13Cr		3.1	1.9	1.2
2205 Duplex		2.4	1.65	0.75
13Cr Martensitic	66	7.6	2.4	5.2
Super 13Cr		4.4	1.8	2.6
2205 Duplex		2.7	2.1	0.6

The results show that the martensitic alloys are more susceptible to erosion corrosion than 2205 duplex under these conditions at both temperatures. As the temperature increases, the mass loss due to corrosion increases significantly for the martensitic alloys, while that for 2205 decreases slightly. This demonstrates the superior performance of 2205 in these three phase flow conditions.

Haberl et al. also carried out flow loop tests in a brine with carbon dioxide and sand at 80°C for 48 h[20]. The test conditions are shown in Table 8.5 and this was calculated to give a pH of 3.4 at temperature and pressure. They tested a 13 Cr martensitic alloy (UNS S42000), 2205 duplex, and alloy 28 (UNS N08028)—a high alloy austenitic stainless steel. The results in Figure 8.10 show that the 2205 was better than the martensitic alloy at the lowest velocity, and alloy 28 was even better. As the velocity increased, first the attack on 2205 was similar to that on the 13 Cr alloy, and at the highest velocity, all three alloys were similarly attacked. This shows that corrosion resistance became less important as the severity of the erosive conditions increased.

TABLE 8.5 Conditions in erosion corrosion test loop of Haberl et al.[20, 21].

PARAMETER	TEST CONDITIONS	UNITS
Gas Liquid Ratio	~20,000	-
CO_2 gas flow	250	L/h
Average brine flow	230	mL/h
NaCl content of brine	27	g/L
Sand content	2.7	g/L
Sand grain size	≤ 150	µm
Flow velocity	10-59	m/s
Impact angle	90	degrees
Temperature	80	°C
Pressure	15	bar

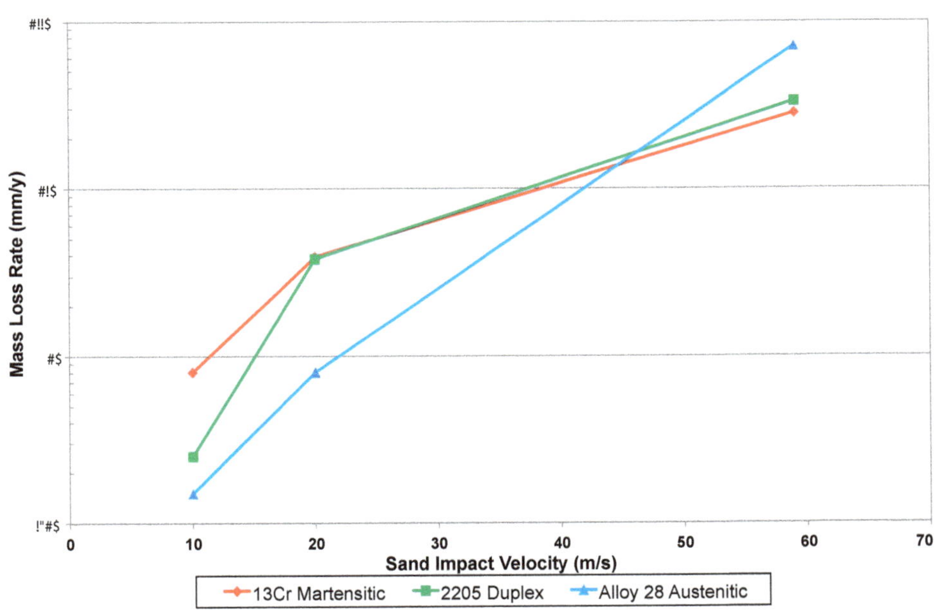

FIGURE 8.10 Erosion corrosion weight losses for some stainless steels in CO_2 saturated brine at 66°C[20].

In later work, Haberl et al. added 304L austenitic and nickel alloy 625 to the tests, as shown in Figure 8.11[21]. As might be expected, 304L was intermediate in performance to 13 Cr and 2205, while alloy 625 consistently showed the lowest attack, except at the highest velocity, where all the alloys were similarly attacked. Under such severe conditions, corrosion resistance is irrelevant and the mechanical properties of the alloy to resist mechanical damage are dominant. The properties of the alloys in Figure 8.11 are similar, so they all suffer similar attack at very high velocities.

A Norwegian North Sea platform used Z100 superduplex stainless steel for the process piping, including a manifold carrying oil, gas, and water, plus sand for the first two years of production. The operating conditions were 2.1 bar CO_2, 0.01 bar H_2S, 3 lbs/1000 bbls sand, a velocity of 30 m/s, and a temperature of 110°C.

After four years, the manifold was inspected to see whether there was any erosion corrosion damage. There was none, even on over-penetrating weld beads that were protruding into the flow.

Several authors have shown that in aqueous solutions, the rate of erosion is roughly proportional to the sand content and the cube of the velocity[5, 22].

$$\text{i.e., erosion } \alpha\ s \cdot v^3$$

where s = sand content and v = flow velocity.

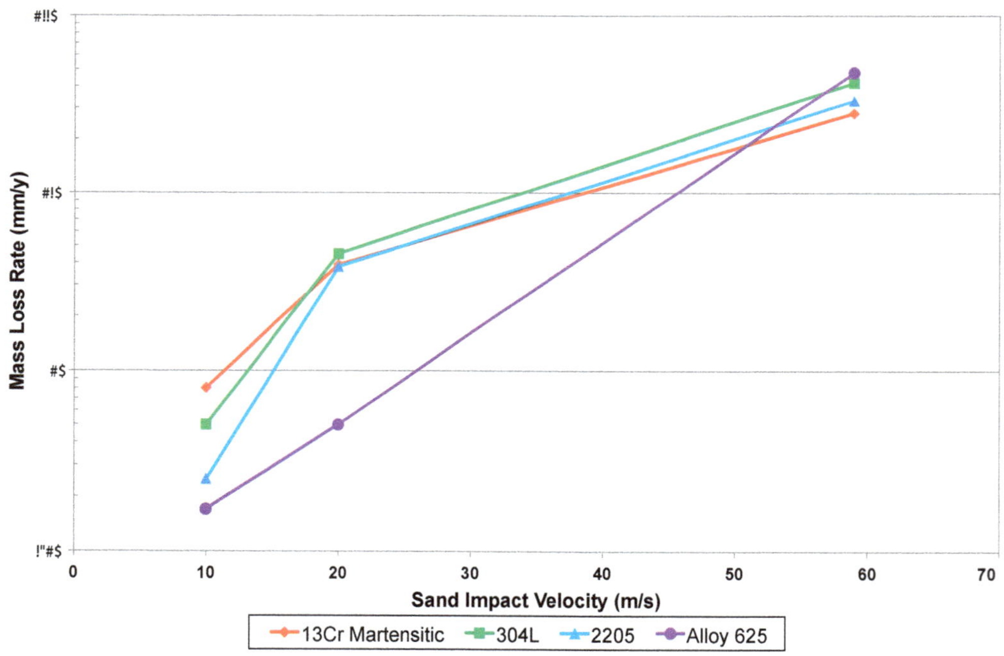

FIGURE 8.11 Erosion corrosion weight losses for some CRAs in CO_2 saturated brine at 66°C[21].

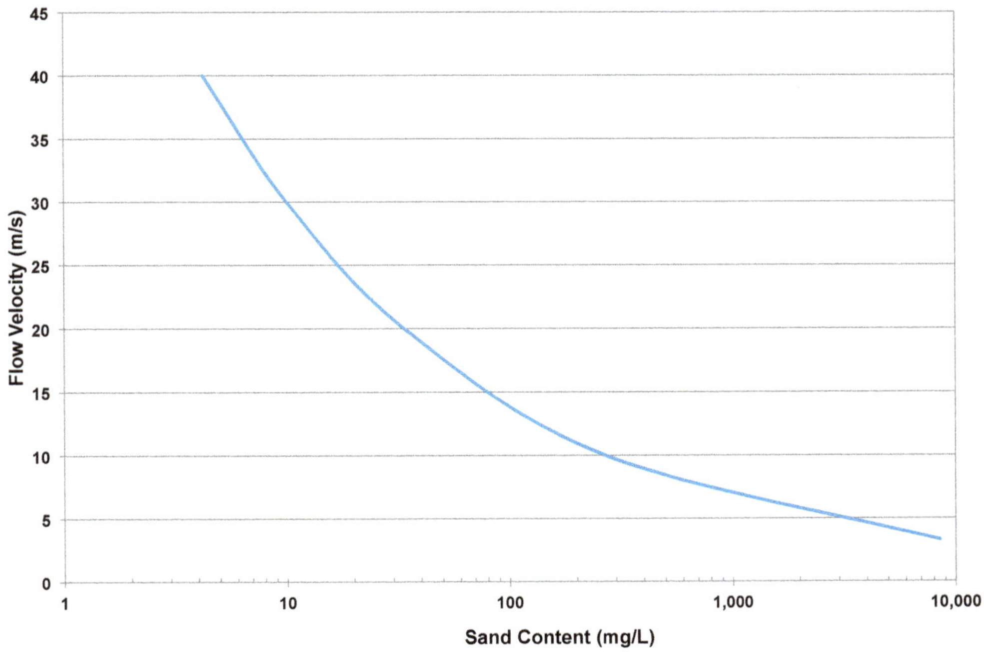

FIGURE 8.12 Conservative safe velocity limit for Z100 superduplex in brines with CO_2 and H_2S[6].

Using this equation and the data from the Norwegian platform, it is possible to construct a conservative velocity limit curve for Z100 superduplex[6] (Figure 8.12). With further data it might be possible to extend these limits further.

This curve has been used to justify the use of superduplex in other projects where sand was present, including non-oil and gas applications. An example is the nitrogen gas compression plant described in Section 8.2.1, which had intermittent sand loading up to 50,000 mg/L. The maximum flow velocity in the heat exchangers was 2 m/s and using the above equation gives a conservative maximum sand content of 34,000 mg/L. Hence, the ability of superduplex to survive intermittent sand contents of 50,000 mg/L is not surprising. It is important to note that this curve is only valid where the alloy forms a passive film and would not be valid in media where general corrosion was occurring, such as very low or high pH values.

8.2.4 Flue Gas Desulfurization

The wet limestone process for the removal of acid gases from the flue discharge of coal burning power stations was described in Chapter 5, Section 5.5, and a typical absorber tower is shown in Figure 5.22. The crushed limestone is present in large quantities, but it is only mildly abrasive, while the flue ash is in smaller quantities, but it is hard and often angular.

Dallas and McConnell carried out pin erosion tests on some candidate casting alloys for the slurry pumps in flue-gas desulfurization (FGD) plants[5]. Figure 8.13 shows the volumetric metal loss over a range of pH values. At pH 6, the differences between the alloys are small, but they increase as the pH decreases, particularly for 316L and the white iron (a cast iron with ~28% Cr and 2.5% C). The volume loss of the superduplex alloy did not change with pH, because of its high strength and corrosion resistance. The 25% Cr alloy in Figure 8.11 was a cast duplex similar to Ferralium, but without the copper and a PREN of 35.

Following the success of the pin erosion tests, Dallas and McConnell constructed a flow loop with a prototype slurry pump made of Z100 superduplex. The slurry contained 10.64 wt.% calcium sulfate, 0.26 wt.% calcium carbonate, 0.47 wt.% fly ash, and 40,000 mg/L chloride, at a temperature of 50°C.

After 5,000 hours of operation, examination of the pump components showed slight erosion corrosion of the vane tips and at the discharge, i.e., the areas of greatest turbulence. On the basis of these test results, Z100 superduplex was selected for the diffuser plates, slurry pumps, and agitator stools in the FGD plant at a U.K. power station[6]. No erosion has been seen on the diffuser plates and the pumps last 35,000 to 40,000 hours before repair of the impeller is required. Auduoard et al. also report the good performance of superduplex stainless steels in erosive FGD slurries, particularly for pumps[23].

The problems with 2205 in FGD absorber towers were described in Chapter 5, but superduplex has performed well in abrasive service, being used successfully for slurry pumps, centrifuges, and agitators in FGD plants around Europe. Superduplex piping has also been used to replace FRP pipe that failed because of erosion by FGD slurry[6].

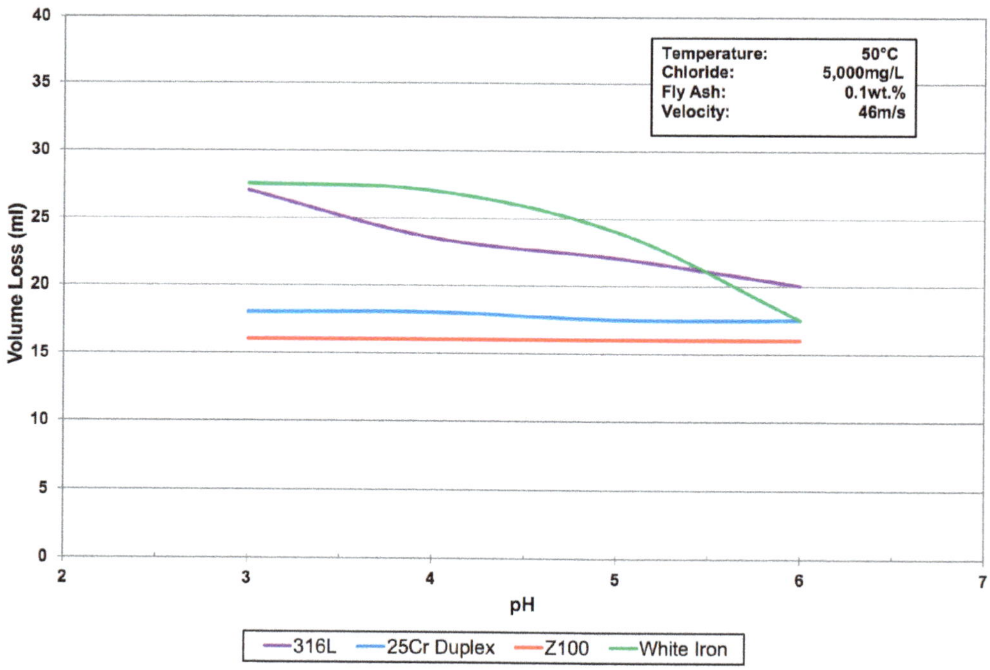

FIGURE 8.13 Effect of pH on erosion corrosion in a simulated FGD slurry[5].

8.2.5 Other Industries

When waste paper is recycled, it contains many impurities, such as plastics, metals, printing ink and netting spines from books. The austenitic stainless steel 316L offers good resistance to corrosion in this environment, which is also abrasive. Wegrelius describes a series of laboratory tests incorporating both corrosion and wear on 304L, 316L, and 2101 lean duplex[24]. The environments and test conditions are shown in Table 8.6 and the results are shown in Figure 8.14. All of the alloys showed lower weight losses in environments 1 and 2. However, the higher chloride environment with immersion before wearing showed higher weight losses. In sulfuric acid, passivity was lost and the wearing produced much greater weight losses. The lean duplex alloy showed lower weight losses than the austenitic alloys in all the tests.

TABLE 8.6 Environments and test conditions for abrasion tests of some stainless steels[24].

ENVIRONMENT	COMPOSITION	TEST	CONDITIONS
1	200 mg/L chloride	1	24 h wearing
2	1,000 mg/L chloride	2	Immersion + 5 h wearing
3	1N H_2SO_4	3	5 h wearing

The production of commercial phosphoric acid was described in Chapter 4 and it can be seen that the agitators and circulation pumps in the main reactor handle both corrosive liquid and abrasive solids. Calboreanu described the development of a modified cast duplex stainless steel to the CD4MCuN composition, but with

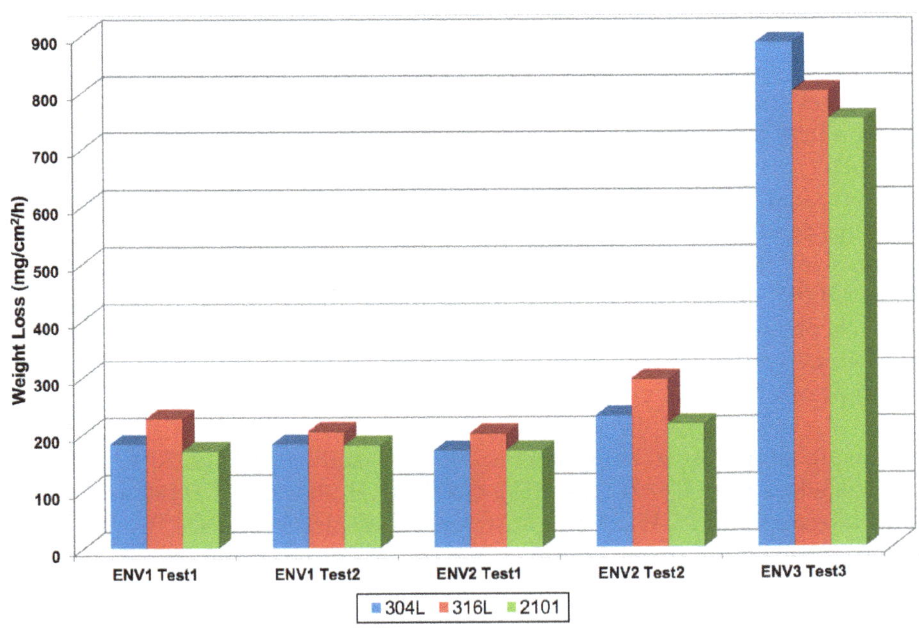

FIGURE 8.14 Results of corrosion and abrasion tests on some stainless steels[24]. (Environments and test conditions in Table 8.6)

processing to refine the grain size and increase the strength and hardness[25]. Samples of several cast duplex and austenitic stainless steels were exposed in an Asian phosphoric acid plant for 14 days. The exposure conditions and the corrosion rates are shown in Table 8.7. Although the modified CD4MCuN composition had the lowest corrosion rate, the corrosion rates of all the alloys were low compared with the values from the tests described in Chapter 4. This may be due to the low chloride content of the slurry, although the fluoride content was very high. The results in Table 8.7 show that the cast duplex alloys corroded less than the austenitic alloys. Auduoard et al. also described the successful use of cast 25% Cr duplex stainless steels for pumps in phosphoric acid plants[26].

TABLE 8.7 Results of 14 day exposure of some cast stainless steels in 22% phosphoric acid (as P_2O_5), 10–20% solids, 15,500 mg/L fluoride, and 80 mg/L chloride at 73°C[25].

ALLOY	CORROSION RATE (μm/y)
CD4MCuN	5.0
Mod. CD4MCuN	4.3
CN7M	7.0
CN3MCu (modified alloy 20)	5.5

Wischnowski and Pohl described the development of a 1.5% C, 3% Mo, 35% Cr duplex white iron, which also has nitrogen additions[27]. They claimed very good erosion corrosion performance for this duplex white iron in abrasive, corrosive slurries. However, as much of the chromium and other alloying elements would be tied up in precipitates that increase the strength and hardness, it is not clear how good such an alloy would be in an acidic slurry with chlorides at elevated temperature.

In concentrated sulfuric acid, the films on stainless steels are not as protective as those that form at near-neutral pH. Under flowing conditions, alloys such as 316L are limited to a maximum flow rate of 1.5 m/s in concentrated sulfuric acid[28] (94–100 wt.%) to avoid excessive corrosion rates. The austenitic alloys with high silicon, such as UNS S31265, can be used up to velocities of 10 m/s and higher because of the silicon in the protective film. Francis presents the results of rotating cylinder tests on Z100 superduplex in 95% sulfuric acid at 70°C, which showed a steady state corrosion rate that was less than 0.1 mm/y[29]. The rotational flow was equivalent to a flow of 2.4 m/s in an NPS 4 pipe.

Kain carried out tests on some cast stainless steels in 95 wt.% sulfuric acid at 50°C and 70°C at velocities from 0–1.8 m/s[30]. The alloys were CF8M (cast 316), CN7M (cast alloy 20), CD4MCu (an earlier version of CD4MCuN without nitrogen), and CW12M2 (a cast version of nickel alloy C-276 with high carbon). The compositions of all these alloys are shown in Tables A.3 and A.4 in the appendix. The results for 50°C and 70°C are shown in Figures 8.15A and 8.15B, respectively. At 50°C, the corrosion rate of CF8M was quite high with CN7M being lower and CD4MCu even lower. This is not surprising with a cast alloy developed for sulfuric acid service. The modern version of this alloy (CD4MCuN) would be expected to perform better, but data is not currently available. The nickel-based alloy had a very low corrosion rate under these conditions. All the alloys showed only a small increase in corrosion rate with velocity.

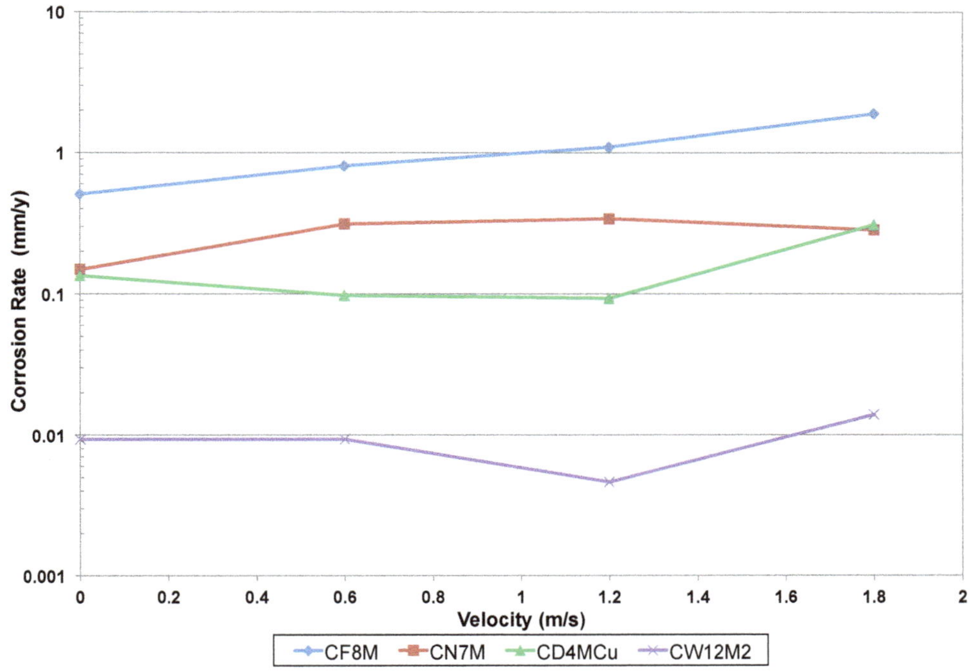

(A)

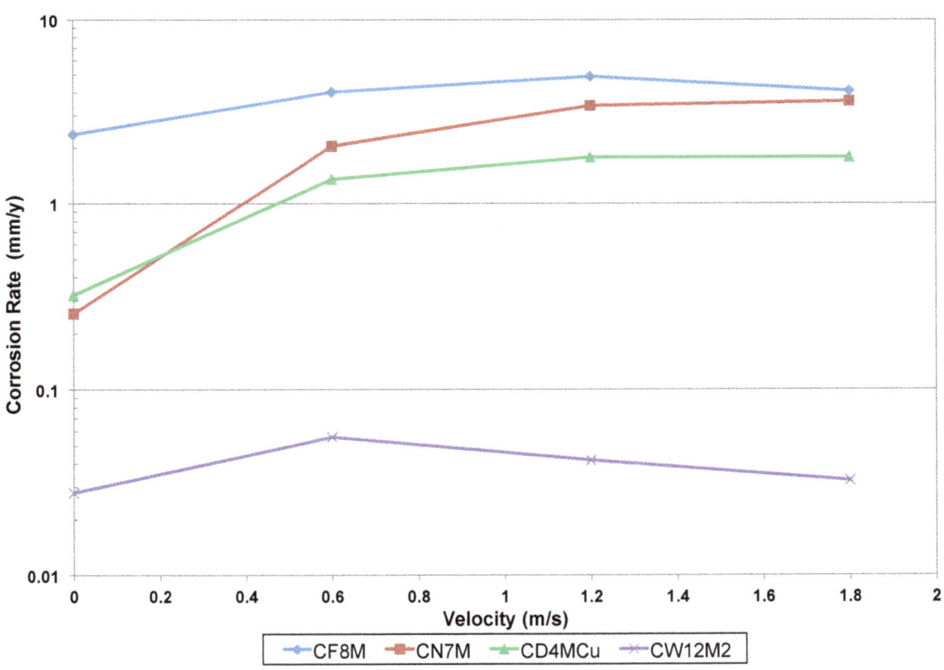

(B)

FIGURE 8.15 Corrosion rates of some cast stainless steels in 95% sulfuric acid at different velocities[30]. A) 50°C; B) 70°C.

At 70°C, the corrosion rates of all the alloys were higher and the biggest increase in corrosion rate was from 0–0.6 m/s, while the change at higher velocities was much smaller[30]. The differences between the cast stainless steels were relatively small, but the cast duplex was better than the cast austenitic alloys. Again, the cast nickel-based alloy had a very low corrosion rate. The results of Francis[29] suggest that the corrosion of Z100 superduplex would be only slightly greater than that of CW12M2 at 70°C (Figure 8.15B). Again, CD4MCuN would be expected to perform better than CD4MCu.

8.3 IMPROVING EROSION CORROSION RESISTANCE

Although duplex stainless steels generally have superior resistance to erosion compared with martensitic or austenitic stainless steels, there are times when even these alloys can be inadequate. There are three main ways in which the resistance to erosion corrosion can be increased: weld overlay, coatings, or metallurgical modification of the surface structure.

Cobalt alloys have excellent resistance to erosion corrosion and weld overlays of Stellite 6 have been used on both pump wear surfaces and valve seats[31]. There can be problems with welding Stellite 6 (UNS W73006) onto superduplex and the author has seen problems due to two different mechanisms. One is where the heat input was too high and nitrides were precipitated in the heat-affected zone (HAZ), which reduced the corrosion resistance and led to pitting in seawater. Stellite 6 contains a higher carbon content than superduplex and it is possible for carbides to form at the overlay/substrate interface. These also reduce the corrosion resistance, and the corrosion in seawater can be so extensive that the weld overlay detaches.

The solution is to use a buttering layer of a nickel alloy, such as 625. The niobium in this alloy will prevent carbon transporting to the substrate. It is still important to keep welding heat inputs and interpass temperatures within the limits for the base metal.

Nowadays, this technique has been replaced by weld overlaying directly with a cobalt alloy such as Ultimet (UNS R31233), which is low carbon, with excellent resistance to corrosion and erosion. The author has used this alloy successfully in a couple of applications where high rates of erosion were expected.

An alternative to weld overlay is high velocity oxy-fuel (HVOF) coatings, which are applied at very high velocity with essentially 100% densification. It is common to use tungsten carbide in HVOF coatings to resist erosion, so a binder alloy is needed with the tungsten carbide particles so that they adhere as a compact layer on the surface. The coating is typically 50 μm thick per pass, and a total thickness of 200 μm is commonly used, for example, on pump wear rings.

The selection of the binder alloy is critical for HVOF coatings to be successful. The binder must have both adequate corrosion and erosion resistance in the service fluid so that it remains intact and holds the ceramic particles in place. Neville and Hodgkiess examined the corrosion and erosion corrosion behavior of HVOF coatings of tungsten carbide on Z100 superduplex in seawater with two different binders[32, 33]. One was a Co-28Cr alloy and the other was a Ni-Cr-Si-B alloy. The nickel-based alloy was very susceptible to crevice corrosion and pitting at ambient temperature, while the Co-Cr alloy suffered more general corrosion[32]. At 60°C. the Co-Cr alloy suffered severe pitting within one hour[33].

Bjordal et al. carried out rotating cylinder tests on HVOF coatings of tungsten carbide on duplex stainless steels in seawater and also found that the composition of the binder was critical to the performance[34]. A bare 2205 sample performed better than one with an HVOF coating with a Co-C binder, because of corrosion of the binder alloy.

Rogne et al. looked at HVOF coatings on 2205 duplex in seawater at 23 m/s with 2,500 mg/L sand (diameter ~250 μm)[35]. They tested binders of Co-Cr, Ni-Cr, Co-Ni-Cr-Mo, Co-Cr-Mo, and C-276 (UNS N10276). They found that cobalt-based binders were better than Ni-Cr ones, but high chromium is needed for corrosion resistance in seawater. The best performance was where the tungsten carbide particles were small and a similar size to the binder metal particles. They also found that spraying the binder as a powder alloy was better than a powder made up of the different alloying elements.

The overall conclusion is that it is important to select a binder alloy with both good corrosion and erosion resistance in the service fluid. HVOF coatings have been used to increase the life of superduplex pump wear rings, and the pumps (described in Section 8.2.1) handling seawater with high solids in a nitrogen gas compression plant used this technology with good results.

Finally, the surface of stainless steels may be modified by diffusing nitrogen or carbon into the metal to modify the structure without forming precipitates. This is the principle of the Bodycote Kolsterizing process. It is claimed that when carbon is diffused into the surface region at moderate temperatures, carbides do not form, but the carbon atoms sit in the interstitial positions in the lattice and strengthen it, increasing both corrosion and erosion corrosion resistance. The Kolsterized layer is 10–40 μm thick and has a surface hardness >1,000 HV.

There does not appear to be any detailed data comparing the erosion corrosion resistance of coated with uncoated duplex stainless steels, but Dent reported some corrosion and toughness data[36]. In the ASTM G48 ferric chloride test, he showed that the CPT increased by 5°C for Kolsterized 316L and 2205, while that of Z100 superduplex decreased by 5–10°C. In Charpy impact toughness tests at -50°C, the toughness of 2205 was significantly increased, while that of Z100 was significantly decreased. The reason is that the carbon is diffused in for a few hours at 400–450°C and at this temperature, alpha prime phase forms in superduplex stainless steel fairly quickly. The time to form alpha prime is longer for 2205 because of its lower alloy content. It is believed that it was the formation of alpha prime that decreased the pitting resistance and toughness of superduplex. Hence, it could only be considered as a method of surface modification for 2205 and lower alloyed duplex stainless steels.

8.4 CAVITATION

Cavitation occurs when there is a sudden local pressure drop such that vapor cavities can form. These migrate along the pressure gradient until the external pressure causes them to collapse. Where they collapse on a metal surface, the force can exceed the metal strength such that metal is lost. This is essentially a mechanical process and the resistance to cavitation is governed by the strength and hardness of the metal. For two materials of similar strength the one that is more ductile will suffer less cavitation damage. This is because the high stresses generated by collapsing cavitation bubbles can remove metal more easily from a brittle alloy. Where the cavitation is mild and the fluid is corrosive then the corrosion of the base metal may also be important, with corrosion resistant alloys also resisting cavitation better than less corrosion resistant ones.

Cavitation may occur in pumps operated away from their design curve, after partially throttled valves, after orifice plates or on marine propellers. It has also been seen on hydrofoil blades.

Cavitation damage tends to appear rather rough compared with the smooth water swept pits of erosion corrosion. It also occurs in different locations. On a marine propeller, cavitation bubbles can initiate near the hub and they then migrate along the blade, often imploding about a third of the way to the tip. Erosion corrosion is most likely at the blade tips, where water velocities and turbulence are highest.

The severity of cavitation is sometimes measured using the dimensionless cavitation number *Ca* so that

$$Ca = \frac{P - P_v}{\frac{1}{2}\rho v^2}$$

where P is the local pressure, P_v is the vapor pressure, ρ is the density of the liquid, and v is the fluid velocity.

In fluids at high velocities, the cavitation number can be small (<1), but the severity of attack can be acute (see below).

There are three main ways of testing for cavitation resistance and it is important to realize the drawbacks to each one. The first is ultrasonic cavitation testing to ASTM G32, where a small metal sample is fastened to the end of a piezoelectric oscillator, running at approximately 20 kHz, and it is then immersed in the fluid of interest[37]. This is not true hydrodynamic cavitation, but it has the advantages of being simple to carry out and is quick. All metals are attacked, so it is possible to rank materials fairly quickly. However, these results cannot be directly compared to any particular service application.

The second method is ASTM G134, which uses a submerged cavitating liquid jet, or series of jets and samples that are held a few millimeters from the jet, and the severity of attack can be determined as a function of time[38]. The severity of the cavitation can be adjusted by varying the jet velocity and nozzle diameter, so that materials can be compared under conditions approximating service.

The third method is to use a restricted flow channel or venturi carrying the fluid of interest (often seawater) and inserting a cylindrical or triangular former across the flow to create cavitation. Samples may be fixed to the channel walls or to the reverse side of the former, depending on where the cavitation bubbles impact. This is expensive to set up but can closely model service conditions where cavitation is occurring.

There is rather limited information on the cavitation resistance of duplex stainless steels and it is mostly in seawater or chloride solutions of a similar salinity. Kwok et al. tested some austenitic and duplex stainless steels in 3.5% sodium chloride using the ASTM G32 test method[39]. The results at 23°C are shown in Figure 8.16 as a function of time; it can be seen that the stronger 2205 duplex is better than 316L, and Z100 superduplex is even more resistant. The unusual result is for 304 austenitic stainless steel. The attack on 304 is much less than on 316L, and Kwok et al. ascribed this to the strain induced transformation of austenite to martensite, or work hardening of the 304 under cavitation impacts. Heathcock and Protheroe showed that the work hardening rate of 304 is higher than that of 316L, which explains the superior performance of 304 in these tests[40].

Kwok et al. went on to compare the cavitation resistance of 304 and Z100 as a function of both pH and temperature under the same conditions[39]. Figure 8.17 shows the effect of temperature at pH 7. The Z100 superduplex is superior under all conditions, but the differences were small at 10°C and 70°C. The increase in cavitation damage from 10–50°C was ascribed to a decreasing stability of the passive film with temperature.

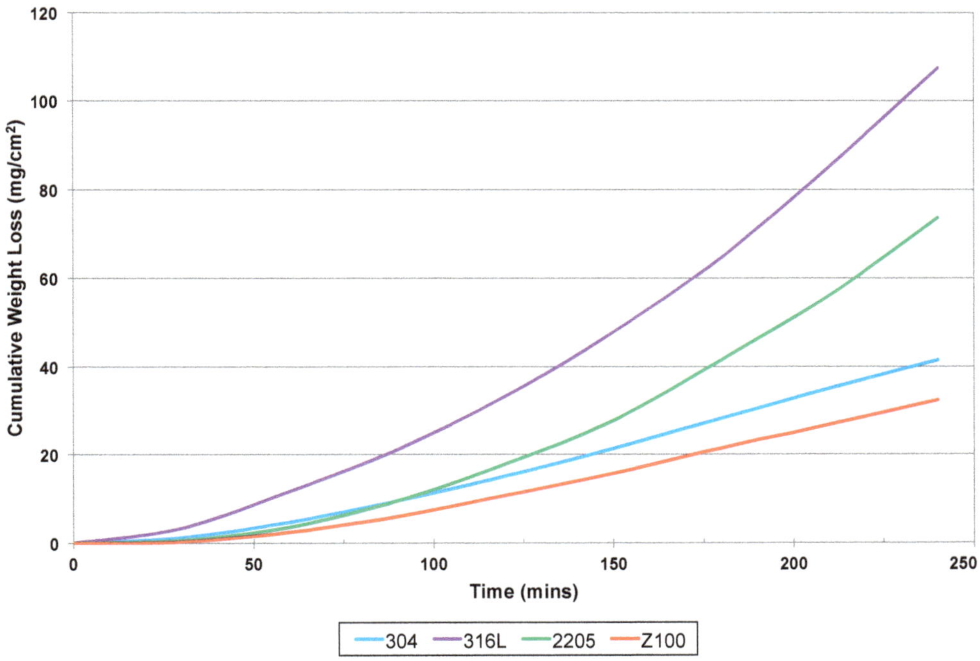

FIGURE 8.16 Cumulative cavitation weight loss for some stainless steels in 3.5% sodium chloride solution at 23°C[39].

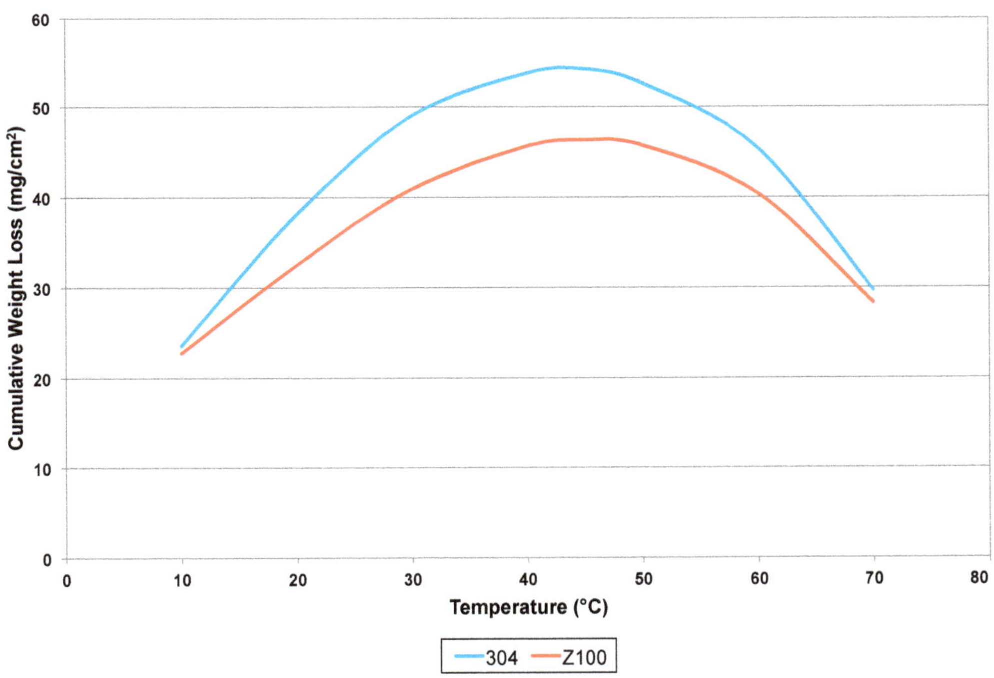

FIGURE 8.17 Cumulative cavitation weight loss in 3.5% sodium chloride as a function of temperature at pH 7[39].

The lower metal loss for the superduplex was thought to be due to its higher corrosion resistance in chloride solutions compared with 304. Above 50°C, the cavitation damage decreases and this was due to the increase of vapor pressure with temperature and the cushioning effect under cavitation bubble impacts.

In Figure 8.18, the effect on cavitation of pH at two temperatures is shown. At 23°C, there is little change for 304, while Z100 shows a marked decrease in cavitation damage from pH 1 to pH 9. This was ascribed to the increase of passive film stability on superduplex as the pH increases. Above pH 9, the stability of the film decreased and the cavitation damage also increased. It is well known that in alkaline solutions, the composition of the passive film changes significantly and the nickel content of the alloy becomes more important. At 50°C, the results for both alloys were similar to those at 23°C, but with more metal loss because of the decreased stability of the passive film.

The previously cited U.S. Navy valve material study also included a series of multiple duration cavitation tests[4]. As originally reported, these tests were performed in accordance with ASTM Standard Method G-32 using substitute ocean water (ASTM D1141). Test results, as assessed by accumulated mass loss, shown in Table 8.8 indicate that all the stainless steels were superior to M-bronze, a copper alloy commonly used for valves in seawater systems. As was the case in related jet-impingement tests, the higher Mo-containing austenitic and superduplex alloys again exhibited much greater resistance to cavitation than conventional 316L. However, the relative ranking of performance of the high alloy materials shown in Table 8.8 varies from alloy to alloy of the same nominal composition. These differences were reproduced in replicate tests and the author considers that they could be due to differences in the hardness and strength of the particular heats that were tested.

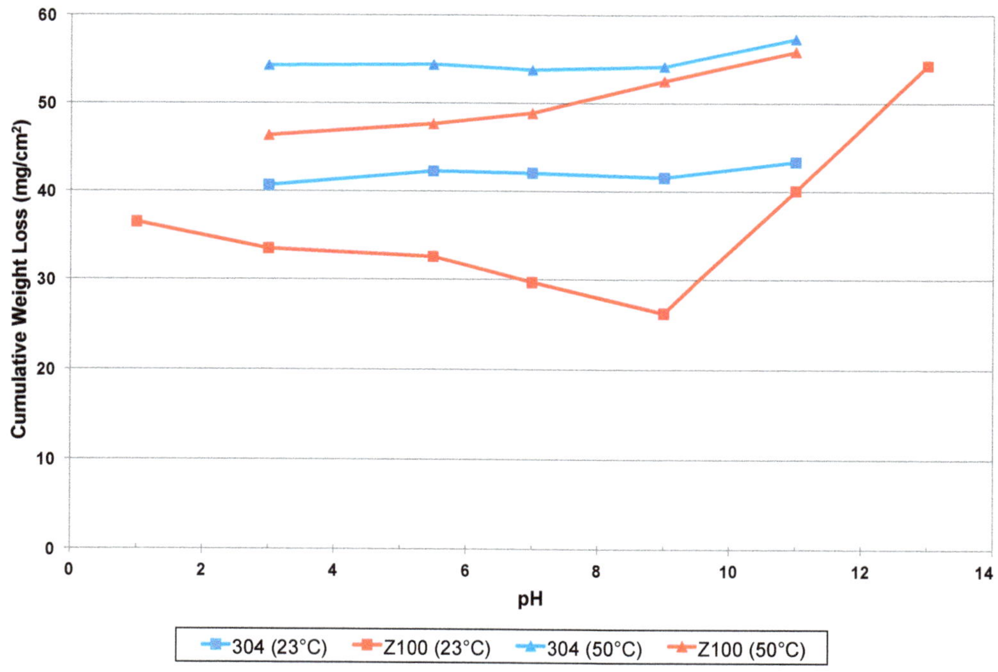

FIGURE 8.18 Cumulative cavitation weight loss in 3.5% sodium chloride as a function of pH at two temperatures[39].

TABLE 8.8 Metal Loss due to cavitation in synthetic seawater[4].

ALLOY	UNS No.	Cumulative Metal Loss (mg)		
		2 hours	6 hours	10 hours
316L	S31603	5	20.5	43.4
6%Mo	S31254	<1	<1	<1
6%Mo	N08367	<1	3.9	9.1
2507	S32750	0	1.4	5.4
Z100	S32760	<1	2.3	6.4
M-Bronze	C92200	25.8	102.7	180

Lush discusses all the factors affecting cavitation and presents the results of a series of extended tests by the U.K. Royal Navy in a cavitation tunnel with natural seawater at ambient temperature[41]. A large number of alloys were tested and the results are shown in Figure 8.19, where the results for different alloy types are grouped together. Nickel aluminum bronze (NAB) has been used for many years by the Royal Navy and it can be seen that most of the other alloys offer superior cavitation resistance under the test conditions: 40 m/s and a cavitation number of 0.08. A large number of nickel alloys were tested and the softer, lower strength ones were inferior to NAB. Similarly, the titanium alloys offer a range of strengths and hardnesses with varying cavitation resistance. The austenitic stainless steels were better, but the 25% Cr duplex alloy gave even less attack. There was a range of metal losses for the martensitic stainless steels, covering a range of strengths and hardnesses and the best of these were superior to the 25% Cr duplex alloy. However, martensitic stainless steels generally have poor resistance to localized corrosion in seawater and would not normally be used for seawater pumps or valves.

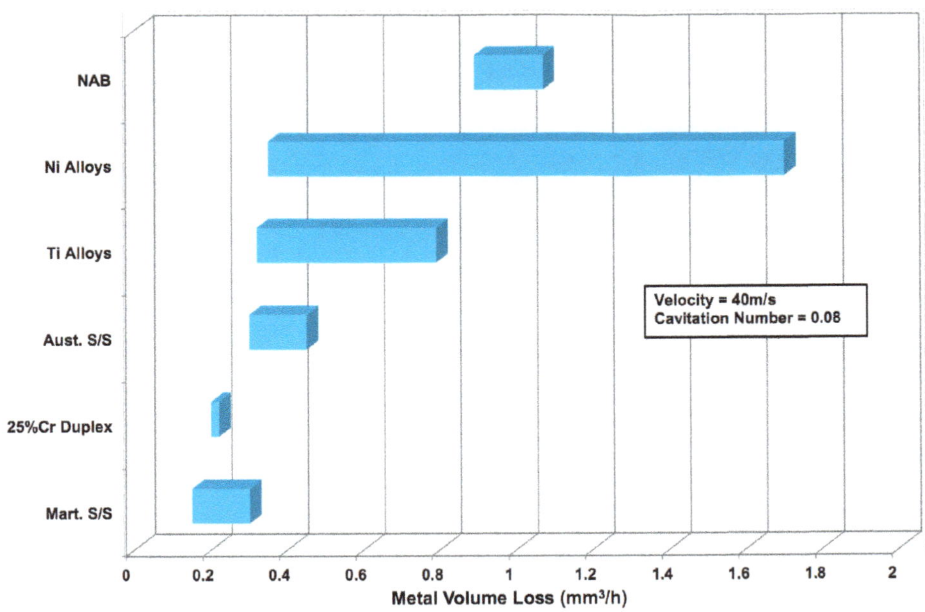

Figure 8.19 Cavitation metal loss for some different alloy groups in natural seawater at ambient temperature[41].

Superduplex stainless steel pumps have been used for seawater since the mid-1980s and cavitation damage is rare and is generally mild even when it occurs. When the large seawater pumps at a multi-stage flash (MSF) desalination plant are started up, they take a while to build up to normal operating pressure. It is common to connect them to waste through a pressure discharge vessel, so that the pumps are operating in the correct part of the design curve. The valves to feed water to the plant can then gradually be opened as the ones to the pressure discharge vessels are closed. A pressure discharge vessel is essentially a large diameter horizontal pipe with a series of orifice plates along the length to gradually reduce the pressure prior to discharge. At one MSF plant in the Middle East, two pressure discharge vessels, operating in parallel, were fabricated from FRP. They rapidly suffered severe damage because of cavitation, and replacement vessels were fabricated in Z100 superduplex stainless steel (Figure 8.20). The superduplex vessels have operated without problems for over 20 years.

FIGURE 8.20 Pressure break vessel in Z100 superduplex for a Middle East desalination plant.
[Photo courtesy of Rolled Alloys]

REFERENCES

1. A. Neville, T. Hodgkiess, and J.T. Dallas, "A Study of the Erosion-Corrosion Behaviour of Engineering Steels for Marine Pumping Applications," *Wear*, 186–187 (1995): p. 497.
2. D. He, X. Jiang, and H. Guan, "Erosion and Erosion-Corrosion Behaviors of Several Stainless Steels in a Dual Phase Liquid," *CORROSION* 58, 3 (2002): p. 276.
3. D. He, X. Jiang, S.Z. Li, and H. Guan, "Erosion-Corrosion of Stainless Steels in Aqueous Slurries–A Quantitative Estimation of Synergistic Effects," *CORROSION* 61, 1 (2005): p. 30.
4. R.M. Kain, L.S. Marshall, D.G. Melton, D.M. Aylor, R.A. Hays, and R.J. Ferrara, "Corrosion, Galling and Wear Testing of Duplex and Austenitic Stainless Steels for Seawater Valve Applications," *Stainless Steel World* (December 2000): p. 28.
5. J.T. Dallas and T.A. McConnell, "Design and Testing of a Prototype Slurry Pump for FGD Plants," Solids Pumping Conference 1991 (London, United Kingdom: Institution of Mechanical Engineers): p. 9.
6. Rolled Alloys Technical Report TN781R4 "The Erosion Corrosion Resistance of Zeron 100 Superduplex Stainless Steel" (Temperance, MI: Rolled Alloys, 2008).
7. X. Hu and A. Neville, "Assessing the Role of Corrosion in Erosion Corrosion of High Grade Alloys in Marine Environments," CORROSION 2002, paper no. 189 (Houston, TX: NACE International, 2002).
8. X. Hu and A. Neville, "In-Situ Measurements of Corrosion Characteristics of High Alloy Stainless Steels in Saline Liquid-Solid Impingement Conditions," CORROSION 2003, paper no. 264 (Houston, TX: NACE International, 2003).
9. X. Hu and A. Neville, "Stainless Steels for Controlling Erosion Corrosion," Stainless Steels America 2004 (Zutphen, Netherlands: KCI Publishing, 2004): p. 296.
10. H. Meng, X. Hu, and A. Neville, "Study of Corrosion in Erosion Corrosion of Two Stainless Steels in Simulated Marine Conditions via Experimental Design," CORROSION 2007, paper no. 254 (Houston, TX: NACE International, 2007).
11. T. Hodgkiess and D. Mantzavinos, "Some Comparisons of the Erosion Corrosion Behaviour of Titanium and Stainless Steel," Stainless Steel World Conference, paper no. P5099 (Zutphen, Netherlands: KCI Publishing, 2005).
12. B.D Craig, "Predicting Critical Erosion-Corrosion Limits of Alloys for Oil and Gas Production," *MP*, 37, 9 (1998): p. 59.
13. M. Castle and D. Teng, "Extending Gas Well Velocity Limits: Problems and Solutions," Asia Pacific Conference (Richardson, TX: Society of Petroleum Engineers, 1991).
14. A. Ikeda, M. Ueda, J. Vera, A. Viloria, and J.L. Moralez, "Effect of Flow Velocity on CO_2 Corrosion Performance of 13Cr, Super 13Cr and Duplex Stainless Steels," 12th International Corrosion Congress (Houston, TX: NACE International, 1993).
15. API Recommended Practice 14E, "Recommended Practice for Design and Installation of Offshore Production Platform Piping Systems" (Washington, DC: American Petroleum Institute, 1991).
16. N.G. Galindez Ruiz and B.D. Craig, "Effect of Crude Oil on Corrosion of Alloys in Severe Environments," *MP*, 34, 4 (1995): p. 50.
17. B.D. Craig, "Predicting the Conductivity of Water-in-Oil Solutions as a Means to Estimate Corrosiveness," *CORROSION*, 54, 8 (1998): p. 657.
18. C. de Waard, L M. Smith, and B.D. Craig, "The Influence of Crude Oils on Well Tubing Corrosion Rates," CORROSION 2003, paper no. 629 (Houston, TX: NACE International, 2003).
19. H.E. Rincon, J.R. Shadley, K.P. Roberts, and E.F. Rybicki, "Erosion-Corrosion of Corrosion Resistant Alloys Used in the Oil and Gas Industry," CORROSION 2008, paper no. 08571 (Houston, TX: NACE International, 2008).
20. J. Haberl, G. Mori, P. Wagner, W. Havlik, and P. Hosemann, "How do Microstructure and PREN Affect the Erosion Corrosion Behaviour of Stainless Steels?" CORROSION 2009, paper no. 09480 (Houston, TX: NACE International, 2009).

21. J. Haberl, G. Mori, P. Wagner, W. Havlik, and P. Hosemann, "Erosion Corrosion Behaviour of Materials Used in Oil and Gas Production," EUROCORR 2009, European Federation of Corrosion (London, UK: Institute of Materials, Mining and Minerals, 2009).
22. E. Rabinowicz, "The Wear Equation for Erosion of Metals by Abrasive Particles" (Cambridge, MA: Massachusetts Institute of Technology, 1979).
23. J.P. Audouard, M. Verneau, and J. Grocki, "Corrosion Performance and Field Experience with Superduplex and Super Austenitic Stainless Steels in FGD Systems," CORROSION 2000, paper no. 828 (Houston, TX: NACE International, 2000).
24. L. Wegrelius, "Duplex Solutions for Pulp and Paper Industry," 42nd Pulp and Paper International Congress and Exhibition, paper no. ABTCP 2009 (Sao Paulo, Brazil: Assoc. Brasileira Técnica de Celulose e Papel, 2009).
25. G.A. Calboreanu, "Duplex Stainless Steel with Superior Erosion Corrosion Performance Utilized in Pumping Phosphoric Acid," Phosphoric Acid-Fertilizer Conference (New York, NY: American Institute of Chemical Engineers, 2003).
26. J.P Audouard and P. Soulignac, "Duplex Stainless Steels for Phosphoric Acid Plants; Corrosion Properties and In-Service Experiences," Duplex Stainless Steels '91 (Les Ulis Cedex A, France: Les editions de physique, 1991): p. 1121.
27. F. Wischnowski and M. Pohl, "High Carbon Ferritic-Austenitic Duplex Stainless Steels for Use in Hydroabrasive Media," Duplex Stainless Steels '97 (Zutphen, Netherlands: KCI Publishing, 1997): p. 705.
28. D.K. Louie, *Handbook of Sulfuric Acid Manufacturing*, 2nd Edition (Toronto, Canada: DKL Engineering, 2008).
29. R. Francis, "The Performance of Stainless Steels in Concentrated Sulfuric Acid," *Stainless Steel World*, 22 (January 2010): p. 53.
30. R.M. Kain, "Velocity Effects on Several Wrought and Cast Stainless Steels in Concentrated Sulfuric Acid at Elevated Temperatures," Corrosion Research Symposium, CORROSION/83 (Houston, TX: NACE International, 1983).
31. P.A. Smith, "Development of Main Oil Line and Seawater Injection Pumps for Oil Industry Applications," 5th European Congress on Fluid Machinery for the Oil, Petrochemical and Related Industries (London, United Kingdom: Institution of Mechanical Engineers, 1993): p. 1.
32. A. Neville and T. Hodgkiess, "Corrosion Behaviour and Microstructure of Two Thermal Spray Coatings," *Surf. Eng.*, 12 (1996): p. 303.
33. A. Neville and T. Hodgkiess, "The Corrosion Behaviour of a Cermet Coating Applied by HVOF Spraying," 10th ASM National Thermal Spray Conference (Materials Park, OH: American Society of Metals, 1997).
34. M. Bjordal, E. Bardal, T. Rogne, and T.G. Eggen, "Erosion and Corrosion Properties of WC Coatings and Duplex Stainless Steel in Sand-Containing Synthetic Seawater," *Wear*, 186–187 (1995): p. 508.
35. T. Rogne, J. Berget, and T. Solem, "Corrosion, Erosion-Corrosion and Wear Resistance of HVOF Sprayed WC Type Coatings with a Corrosion Resistant Binder," CORROSION/99, paper no. 48 (Houston, TX: NACE International, 1999).
36. P. Dent, Exova Corrosion Centre, Private Communication (2016).
37. ASTM G32, "Standard Test Method for Cavitation Erosion Using Vibratory Apparatus" (West Conshohocken, PA: ASTM International).
38. ASTM G134, "Standard Test Method for Erosion of Solid Materials by a Cavitating Liquid Jet" (West Conshohocken, PA: ASTM International).
39. C.T. Kwok, H.C. Man, and L.K. Leung, "Effect of Temperature and pH on the Cavitation Erosion of Superduplex Stainless Steel," CORROSION/96, paper no. 514 (Houston, TX: NACE International, 1996).
40. C.J. Heathcock and B.E. Protheroe, "Cavitation Erosion of Stainless Steels," *Wear*, 81 (1982): p. 311.
41. P.A. Lush, "Materials for Minimum Cavitation," *Chartered Mech. Eng.* (1987): p. 31.

9 Underground Corrosion

Corrosion in soils is complex and depends on many variables. A wide range of corrosivity is also possible in soils, i.e., failures of buried components have occurred within a year, while archaeological samples have been preserved for hundreds of years. Although a wide range of equipment may be buried, the most common items are pipes (for water, sewage, gas, etc.), cables (electricity and telecom), tanks, pilings, and foundations. These are often required to last for periods from 50–100 years.

The most common materials used underground are steel and galvanized steel, although cast iron, copper alloys, lead, and stainless steel are also used. Avoidance of corrosion involves an understanding of the factors affecting the corrosivity of soils.

9.1 FACTORS AFFECTING CORROSION UNDERGROUND

Soil has essentially three components: particles, which may be inorganic (clay, sand, etc.) or organic (peat, mud, etc.); moisture, which includes the solids dissolved in the water; and gas. Both nitrogen and oxygen can diffuse into soils from the air, but other gases may be produced by local organic activity, e.g., carbon dioxide and hydrogen sulfide. A number of soil properties influence corrosion:

1) Soil Moisture: This includes the dissolved solids and varies seasonally
2) Oxygen: The quantity depends on the ability of the gas to penetrate the soil
3) Redox Potential: This is generally determined by the oxygen content of the soil
4) pH: This is influenced by the dissolved solids and any chemicals being produced by local organic activity
5) Conductivity: This affects corrosion in a similar way to that in liquids
6) Microbial Activity: In microbially active soils a range of chemical by-products may be produced that influence corrosion, e.g., H_2S and H_2SO_4

Soil conditions can vary considerably, even over a small distance, and this can cause large changes in corrosivity. When corrosion in soils is severe, it is similar to aqueous corrosion in seawater or fresh water, while when corrosion is mild, it is similar to atmospheric corrosion. Because of the variability of soil conditions, the anode and cathode in uncoupled corrosion can be widely separated in underground corrosion[1].

Although conductivity is used when assessing aqueous solutions, it is more common to use resistivity when considering underground corrosion. This is the inverse of conductivity i.e., 400 µS/cm = 2,500 Ωcm. Typically, soils with a resistivity below 1,000 Ωcm are regarded as aggressive, while those with a resistivity greater than 5,000 Ωcm are mildly corrosive.

To assess the corrosivity of a soil, a number of methods have been proposed. One of the best known is that used by the German Gas Industry[2] for assessing the risk of corrosion of buried pipelines. This involves an index calculated from 10 parameters of the water and soil, and the more negative the index, the more corrosive the soil.

Stray currents from electrical conductors are well known to affect corrosion underground–sometimes dramatically. This is a specialized subject and outside the scope of this book. For more detailed information, the reader is referred to specialist texts on the subject, while Mattsson[1] gives an introduction to the subject.

Note that when equipment is buried and the soil is replaced, the conditions around the buried item can be very different to that in the adjacent, undisturbed soil[3]. The moisture and oxygen content around the buried item can be much higher, leading to greater corrosion, than would be expected from general testing of the soil.

9.2 UNDERGROUND EXPOSURE DATA

There have been a large number of studies of corrosion underground, comparing the same materials in a variety of soils. There are fewer that have addressed stainless steels. One is that of Gerhold et al.[4], which investigated a range of stainless steels in six soils in the United States for up to 8 years. The details of the different soils are shown in Table 9.1, and the corrosion of carbon steel in these soils is shown in Figure 9.1. The data show that the soils represent a range of aggressivity, but general corrosion is not the main problem with stainless steels, and severe corrosion would show as pitting.

Table 9.1 Description of North American soils investigated by Gerhold et al.[4].

DESIGNATION	NAME	DESCRIPTION	RESISTIVITY (Ωcm)	pH	WATER COMP$^{n.}$ (mg/L)	
					TDS	Chloride
A	Sagemoor sandy loam	Well drained, alkaline soil	400	8.8	7,080	330
B	Hagerstown loam	Well drained, slightly acid soil	12,600-34,760	5.3	+	+
C	Clay	Clay pit subject to flooding	400-1,150	4.3	14,640	3,529
D	Lakewood sand	White loose sand near coast	13,800-57,500	5.7	+	+
E	Coastal sand	Beach sand, always damp	1,320-49,500	7.1	11,020	5,765
G	Tidal marsh	Poorly drained marsh	400-15,500	6.0	11,580	3,259

TDS = Total dissolved solids

+ Analysis not carried out because of low dissolved solids at these sites.

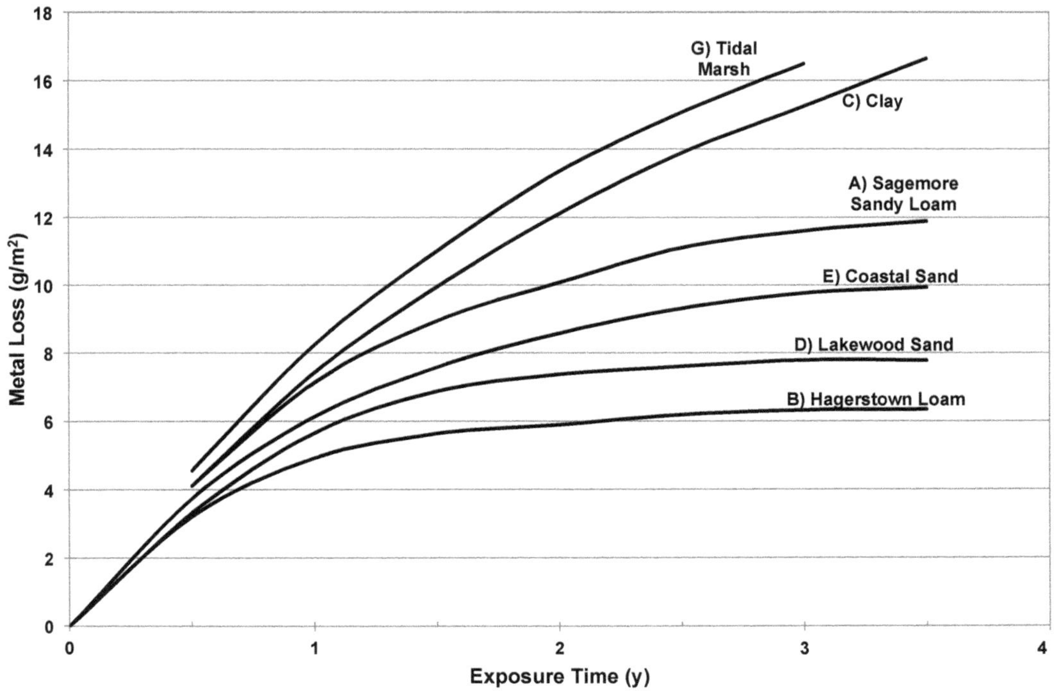

Figure 9.1 Corrosion of carbon steel at the six North American sites[4] used by Gerhold et al.

The study included ferritic, martensitic, and austenitic stainless steels, plus one duplex alloy. Unfortunately, the duplex alloy was 26Cr/6.5Ni with no nitrogen or molybdenum. This means the alloy would be ferritic with poor partitioning of elements between the two phases (see Chapter 2). Hence, its corrosion resistance would be expected to be rather poor.

The corrosion results are summarized in Table 9.2, and as expected, the duplex alloy did not perform very well at all. All the corrosion was pitting attack. The 6% Mo austenitic alloy showed no corrosion at any site. The 304 alloy performed satisfactorily at sites A, B, and D, which were mildly aggressive, but suffered severe attack at the other three sites. Alloy 316 was a little better at site C, but otherwise was similar to 304 in that it perforated at the more aggressive sites. The exposure results show that the ranking of the sites with regard to corrosivity was somewhat different for the stainless steels compared with carbon steel. The sites causing pitting of 300 series alloys (C, E, and G) all had high chloride contents.

TABLE 9.2 Results from 8-year underground exposure tests of some stainless steels[4].

SITE	ALLOY	RESULT
A	304	No Corrosion
	316	No Corrosion
	6% Mo Aust.	No Corrosion
	Duplex*	No Corrosion
B	304	No Corrosion
	316	No Corrosion
	6% Mo Aust.	No Corrosion
	Duplex*	No Corrosion
C	304	Perforated
	316	Slight Attack
	6% Mo Aust.	No Corrosion
	Duplex*	Perforated (<1 y)
D	304	No Corrosion
	316	No Corrosion
	6% Mo Aust.	No Corrosion
	Duplex*	Perforated (>3 y)
E	304	Perforated
	316	Perforated
	6% Mo Aust.	No Corrosion
	Duplex*	Perforated (<1 y)
G	304	Perforated
	316	Perforated
	6% Mo Aust.	No Corrosion
	Duplex*	Perforated (<1 y)

* Low PREN. Not to any current standard (see text).

Gerhold et al. also exposed stressed samples at all six sites, and none of the austenitic alloys or the duplex alloy suffered chloride stress corrosion cracking[4].

Boulton and Wilson exposed a range of stainless steels including three duplex alloys, 2304, 2205 and 2507, at 8 sites on North Island, New Zealand[5]. These covered a range of soil conditions, and the basic parameters at each site are shown in Table 9.3. In addition, soil resistivity measurements were taken to ASTM G57, the results of which are shown in Table 9.4. The corrosion assessments in Table 9.4 used the criteria of Chaker[6].

TABLE 9.3 Soil conditions at eight sites in New Zealand[5].

SITE No.	SOIL TYPE	pH	pH* (H_2O_2)	CONDUCTIVITY (µS)	ORG. MATTER (%C)	SOL. SULFATE (meq/100 g)	SOIL CORROSIVITY
1	Clay	6.9	6.4	80	0.36	0.84	Corrosive
2	Sandy Soil	8.0	6.1	70	0.54	0.06	Noncorrosive
3	Peat Soil	4.8	4.2	150	16.4	0.35	Corrosive
4	Reclaim Soil	7.5	7.3	100	1.6	0.17	Corrosive
5	Volcanic Soil	6.6	6.9	40	1.4	0.17	Noncorrosive
6	Pumice	6.2	6.3	20	0.23	0.01	Noncorrosive
7	Geothermal Soil	4.5	2.8	50	0.04	0.05	Corrosive
8	Peat	4.0	0.72	390	52	593	Corrosive

* Value ≤3 indicates oxidizable sulfides in soil.

TABLE 9.4 Resistivities of New Zealand soils from Table 9.3[5].

SITE No.	SOIL TYPE	RESISTIVITY (Ωcm)	SOIL CORROSIVITY*
1	Waltakere clay	900	Very Corrosive
2	Sandy soil	500,000	Noncorrosive
3	Reclaimed soil	3,400	Corrosive
4	Peaty soil	14,400	Mildly corrosive
5	Volcanic soil	11,800	Mildly corrosive
6	Pumice soil	286,500	Noncorrosive
7	Geothermal soil	20,100	Mildly corrosive

* From assessment criteria of Chaker[6].

Samples were examined after two years, and none of the higher alloy duplex stainless steels or 316L showed any signs of corrosion at any of the sites. There was some slight crevice corrosion and corrosion associated with mechanical damage on 304L and 2304 at the more aggressive sites[5]. Boulton and Wilson do comment that most of the soils were not aggressive and this explains the good performance of the lower alloys.

Peultier and Jacques reported the results of three underground exposure trials of stainless steels in Japan, Saudi Arabia, and Europe[7]. Only the latter trial included some duplex stainless steels (types 2304, 2205, and 2507), as well as austenitic grades. There were four sites in Sweden and three in France covering a range of soil types, with exposure being for three years. They do not give a detailed breakdown of the soil conditions, but Table 9.5 summarizes the seven sites. They concluded that the severity of corrosion at these sites depended chiefly on soil resistivity, pH, and chloride concentration. Based on the severity of corrosion observed after three years, they recommended different alloys, depending on the soil conditions, as shown in Table 9.6. It can be seen that higher alloyed duplex stainless steels are recommended as the chloride content of the soil increases. They also commented that even in apparently noncorrosive soils, it is quite common to select 316L or 2304 rather than 304L, because of possible soil variability over longer distances. Extrapolating this data to the newest duplex stainless steels suggests that the lean duplex alloys, with a 24<PREN<26, will perform similarly to the 316L austenitic alloy.

TABLE 9.5 Soil conditions for European underground exposure trials[7].

SITE No.	LOCATION	SOIL TYPE	COMMENTS
1	Goteborg (sea side)	Clay	Marine clay
2	Kramfors (inland)	Alluvial clay	Sulfates and chlorides
3	Varberg (sea side)	Sand	Patches of quicksand with chlorides
4	Laxa (inland)	Fibrous alluvial peat	Acidic, organic soil
5	Fos (sea side)	Alluvial sand	Warm with chlorides
6	Mallemort-Durance (inland)	Alluvial clay	Moderate chlorides in humid soil
7	Mallemort-Fumades (inland)	Sand and pebbles	Patches of coarse particles

Xunji et al. reported on the use of 2205 gas-gathering lines in China, which were carrying wet gas with a high CO_2 content, and high chlorides in the water[8]. While 2205 would be resistant to the conditions inside the pipeline, they were concerned about external corrosion, because the pipeline passed through a very aggressive soil with a resistivity from 5.15–15.08 Ωcm and a soluble salt content from 0.35–6.08 wt.%

They exposed samples of 2205 for up to three years without CP and some of the samples were scratched to remove the passive film. Additional samples also had CP at different potentials. After three years, none of the samples showed a corrosion rate of more than 0.001 mm/y with no pitting and this demonstrated there was no need for CP with 2205 in this soil[8]. These results appear to contradict Table 9.6, which suggests that superduplex stainless steel would be the best option in such an aggressive soil. However, Table 9.6 does not take the redox potential into account. Soils where oxygen can readily access the buried metal will have a higher redox potential than soils where access is limited. At lower redox potentials, it is quite likely that 2205 will be acceptable, and this is probably the reason that 2205 did so well in the Chinese trials[8].

TABLE 9.6 Recommendations for alloy selection for underground[7].

RESISTIVITY (Ωcm)	CHLORIDE (mg/L)			
	200	1,000	2,000	15,000
>5,000	304/304L			
2,000-5,000	316/ 316L/ 2304		2205	Superduplex
1,000-2,000	2205		Superduplex	
<1,000	Superduplex			

London Underground Limited (LUL) was experiencing severe corrosion of cast iron tunnel linings in one part of the system. The ground in this area had been a salt marsh so chlorides were high. In addition, there were pyrites (FeS) in the soil and sulfur-oxidizing bacteria were able to produce sulfuric acid because of oxygen entering the soil from the tunnels (see Chapter 10). The LUL consultants determined that only an acid and chloride resistant duplex stainless steel would be satisfactory. LUL lined two sections a quarter of a mile long with castings in Z100 superduplex made to the top of the copper specification, to increase resistance to sulfuric acid (as discussed in Chapter 4), and they used wrought Z100 fasteners to bolt the segments together (Figure 10.6 in Chapter 10). The tunnel in this area is designed with a life of 400 years.

Stainless steels are not often used for major underground components, although they may be used for pipelines or storage tanks for aggressive fluids. More often, they are used as pipe clamps, for monitoring equipment, submerged switches, and other ancillary equipment. As such, the main tank or pipeline may be carbon steel with CP and/or a coating. If a duplex stainless steel is used as part of such a unit with CP, there is the risk that hydrogen embrittlement may occur. There is no significant risk of failure unless the component is stressed close to or above its actual 0.2% proof stress (see Chapter 6). In most practical applications this is highly unlikely and a lean duplex could be used rather than 316L, particularly if the higher strength is useful, or 2205 if the soil is very aggressive.

REFERENCES

1. E. Mattsson, *Basic Corrosion Technology for Scientists and Engineers*, 2nd edition (London, United Kingdom: Institute of Materials, Minerals, and Mining, 1996), p. 75.
2. DVWG, "German Gas and Water Works Engineer's Association Standard," DVWG GW9 (Frankfurt, Germany: German Gas and Water Works Engineer's Association, 1971).
3. L.L. Shreir, R.A. Jarman, and G.T. Burstein, *Corrosion*, 3rd ed. (London, United Kingdom: Butterworth Heinemann, 1994), p. 2:80.
4. W.F. Gerhold, E. Escalante, and B.T. Sanderson, "The Corrosion of Selected Stainless Steels in Soil Environments," NBS IR 81-2228 (Gaithersburg, MD: National Institute of Standards and Technology, 1981).
5. L.H. Boulton and P.T. Wilson, "Underground Corrosion of Engineering Alloys in the Water Industry," presented at The New Zealand Waterworks Association Conference (1995).
6. V. Chaker, "Simplified Method for the Electrical Soil Resistivity Measurement," ASTM STP 741 (West Conshohocken, PA: ASTM International): p. 61.
7. J. Peultier and S. Jacques, "Les aciers inoxydables dans les sols," presented at the ID INOX Water Symposium, Marseille, France (2007).
8. L. Xunji, Z. Lizhi, C. Zeliang, M. Xiangjuan, and S, Wenwen, "Corrosion Behaviour of UNS S31803 Duplex Stainless Steel in Western China's Yardang Soil," *MP*, 55, 8 (2016): p. 36.

10

Microbially Influenced Corrosion

10.1 INTRODUCTION

For about 100 years, it has been postulated that the presence of bacteria in natural waters could affect metallic corrosion. However, it was only in the latter half of the 20th century that this was clearly demonstrated. One important factor is that the bacteria do not consume the metal, but their metabolic byproducts can change the local environment sufficiently to cause corrosion, when it would not otherwise occur. Hence, the name "microbially influenced corrosion," or MIC as it is commonly known. In 1983, MIC was estimated to be costing the world economy 30–40 billion U.S. dollars annually[1].

All natural, surface-derived waters (both seawater and fresh water) contain bacteria, although the types and concentrations vary with the location and temperature. Borehole waters in the United Kingdom are usually well aerated and low in bacteria, although they often contain sulfate reducing bacteria, which will become active if the dissolved oxygen content becomes very low. This will then result in significant quantities of hydrogen sulfide in the water, produced by the bacteria.

Costerton et al.[2] reviewed the formation of biofilms on metal surfaces. Once the first bacteria colonize a surface, they produce extracellular material that typically consists of polysaccharide polymers. As the biofilm grows and thickens, the aerobic bacteria consume the oxygen near the metal–biofilm interface and anaerobic bacteria become active. In the outer layer of the biofilm, oxygen can still penetrate and aerobic bacteria are active. This is shown schematically in Figure 10.1. The secondary colonization of the biofilm can promote or inhibit the activity of the primary bacteria depending on the type[2]. Corrosion only occurs once an effective microcolony is established and an electrochemical potential difference is established between different areas on the metal surface. The growth and development of biofilms are affected by the temperature, pH, water velocity, and surface roughness[3]. Most bacteria thrive best within a limited range of temperature and pH, although these can vary significantly from species to species. Bacteria do not adhere well to metal surfaces if the water flow is high or the surface is very smooth.

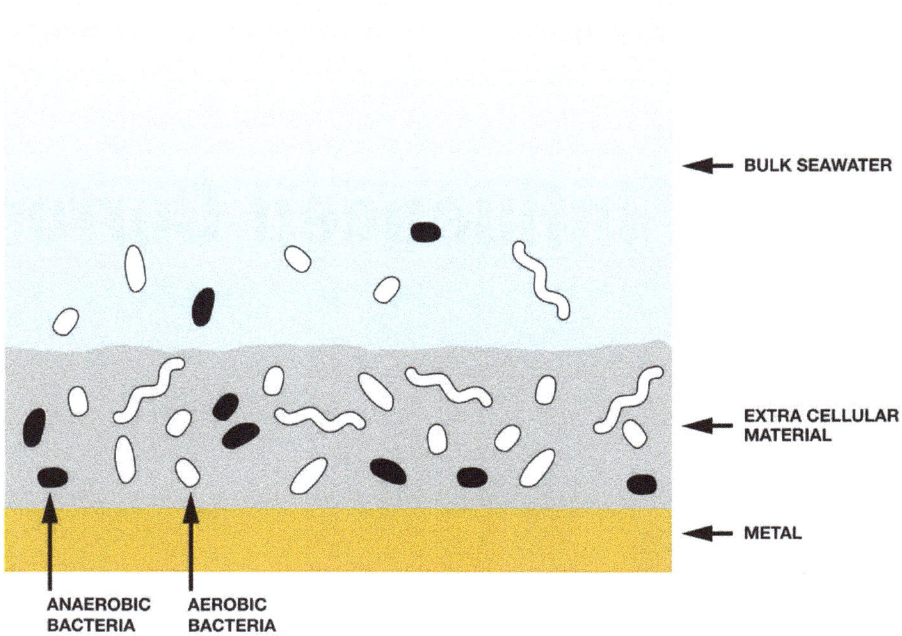

FIGURE 10.1 Schematic diagram of a biofilm on an immersed metal surface.

10.2 TYPES OF BACTERIA

Microbes in water can influence corrosion in one of five ways[4]:

1. Production of aggressive metabolic products, such as sulfuric acid, or chelating agents.
2. Cathodic depolarization associated with anaerobic growth.
3. Changes in oxygen potential, salt concentration, pH, etc., which establish local electrochemical cells.
4. Removal of corrosion inhibitors or protective coatings.
5. The biomass itself stimulates attack, for example by creating an occluded cell.

Where the water is not chlorinated, or otherwise treated to inhibit biological activity, biological colonization occurs rapidly. Colonization starts within hours and becomes well established in periods from a few days to a few weeks, depending on local conditions.

There are cathodically depolarizing biofilms that form on stainless steels in natural waters, and these change the local redox potential and, hence, the corrosion behavior (e.g., Chapter 5). There are other bacteria that thrive under oxidizing conditions, i.e., in aerated water. One type is the iron oxidizing bacterium. This works by creating a differential aeration cell and it is usually a problem with cast iron and carbon steel[4]. Stainless steels are more resistant to this type of attack, but problems have been reported with stainless steels in the presence of manganese oxidizing bacteria. These bacteria oxidize manganese in solution in the water to manganese dioxide, which deposits on the metal surface. Manganese dioxide is very efficient at reducing dissolved oxygen—the most common cathodic reaction in aerated waters. Stainless steels have a thin protective film that is essentially chromium oxide, which is continually breaking down and re-forming in service. The presence of an efficient cathode, like manganese dioxide, can greatly increase the chances of localized corrosion initiating and propagating after a film breakdown event[5]. Note that the addition of a strong oxidizer, such as chlorine, to water to control MIC, may result in the exact same problem, because chlorine readily oxidizes manganese in solution to manganese dioxide. This is more likely to happen in waters with high dissolved manganese concentrations.

Another type of bacterium that thrives under aerated conditions is the sulfur-oxidizing bacteria (SOB) species, which creates sulfuric acid as a byproduct. This will cause severe attack of carbon steel and may cause corrosion of lower alloy stainless steels. Hence, alloys that are resistant to sulfuric acid at all concentrations should be resistant to this type of attack.

When the water is stagnant, the aerobic bacteria will consume the available oxygen and then the anaerobic bacteria will become active. The most well known are the sulfate reducing bacteria (SRB), which produce H_2S as a byproduct. Stott[6] commented that the effect of H_2S is to reduce the potential for pitting/crevice corrosion of stainless steels, and that chloride is still required to initiate attack. The susceptibility to attack because of SRB increases as the chloride concentration and temperature increase[6].

It can be seen that MIC is not really a new sort of corrosion, but bacteria can change the local environment such that a type of corrosion can initiate that would not do so in their absence.

10.3 EFFECTS OF BACTERIA ON CORROSION

It is well documented that when a stainless steel is immersed in an aerated natural water, the potential increases over a period of a few days to weeks, finishing in the range +250 to +400 mV SCE. This has been found for both seawater and fresh waters. The ennoblement is due to the formation of a biofilm that cathodically depolarizes the reduction of dissolved oxygen. The exact mechanism has not been determined, but it is currently believed that hydrogen peroxide is produced as part of the metabolic processes[7]. This is an oxidizer and additions of peroxide to chloride solutions result in the same kind of potential increases for stainless steels as those seen in natural waters.

Figure 10.2 shows the cathodic polarization curves on stainless steel under different conditions[8]. In natural seawater (after the biofilm forms), only a small change in potential results in high currents, while without the biofilm, an equal potential change results in much smaller currents. The greater the current, the greater is the corrosion and if this occurs at a site of film breakdown, it increases the chances of this becoming propagating corrosion rather than repassivating. If a biocide, such as chlorine, is added, the biofilm does not form, but the higher redox potential results in a higher open circuit potential. The cathodic reaction is now the reduction of hypochlorite to chloride, but this is nowhere near as efficient as the reduction of dissolved oxygen with a biofilm, so currents are lower than when the biofilm is present.

These electropositive potentials can mean that an alloy is taken past its pitting or crevice potential. The potential of such materials then decreases and this is often used as an indicator that crevice corrosion has initiated (Figure 10.3).

With SOB, it is the sulfuric acid, produced as a metabolic byproduct that causes the corrosion. Resistance to this type of attack can be assessed by examining an alloy's resistance to sulfuric acid over a wide range of concentrations, at the temperature of interest.

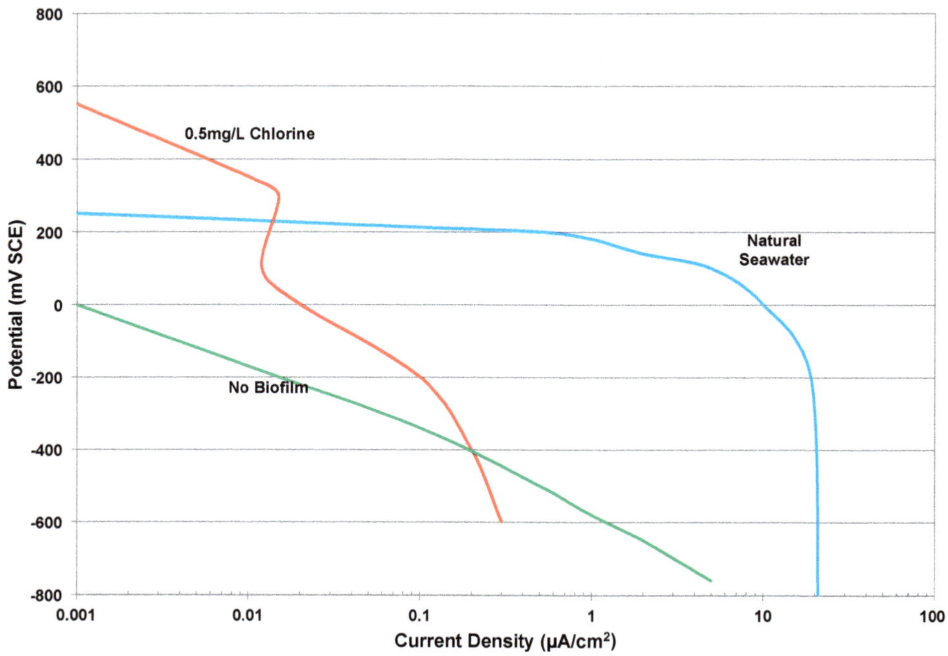

FIGURE 10.2 Cathodic polarization curves for stainless steel in seawater[8].

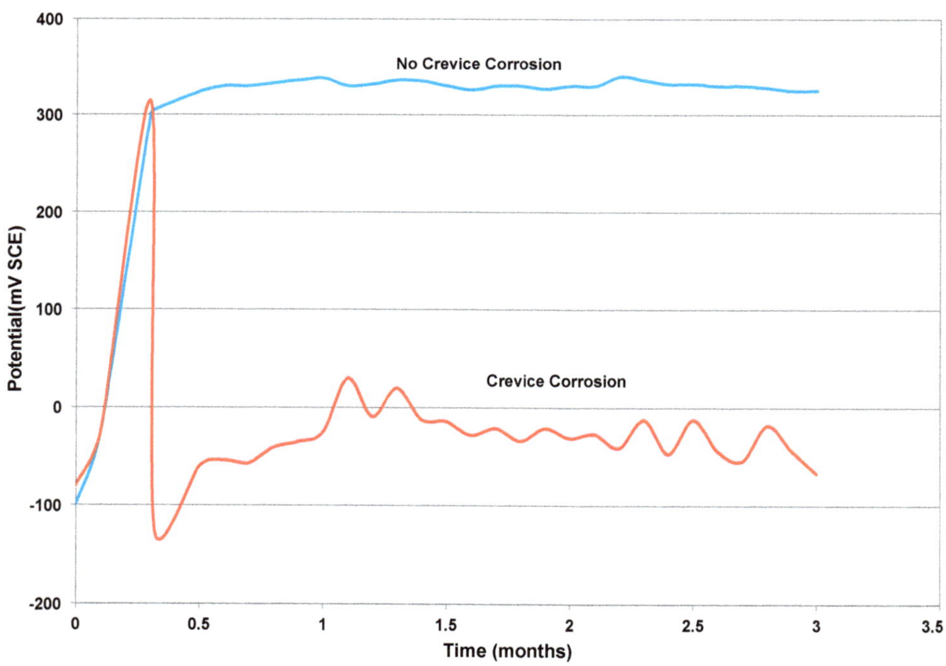

FIGURE 10.3 Potential vs. time curves for stainless steel in seawater with a biofilm, showing the effect of the initiation of crevice corrosion[19, 20].

With SRB, the H_2S produces local reducing conditions and negative potentials. Iversen[9] demonstrated this by exposing stainless steels in some Swedish wastewater plants. In the early stages, a biofilm formed that produced electropositive potentials, similar to those seen in seawater. However, over time the potential slowly decreased to very negative values. This was ascribed to the formation of a thicker biofilm such that SRB became active beneath it. With this electronegative potential, no localized corrosion initiated. However, some wastewater plants add oxidizing chemicals, such as potassium permanganate, as part of the waste treatment process. This can change the local redox potential such that corrosion can occur because sulfide can lower the threshold potential for localized corrosion, as described by Stott[6].

10.4 ENVIRONMENTAL VARIABLES

Figure 10.4 shows the critical chloride concentration for the initiation of crevice corrosion in sterile chloride solutions for some duplex stainless steels with 316L for comparison. This was discussed in Chapter 5. When MIC occurs, it is often at temperatures and/or chloride concentrations where the alloy might be thought to be satisfactory.

There are a number of variables, such as flow velocity, that might be thought to affect MIC. One factor that is well documented is surface finish, in that bacteria find it easier to attach to rough, rather than smooth, surfaces. It is not that attachment does not occur on smooth surfaces, but it is more difficult and usually takes longer.

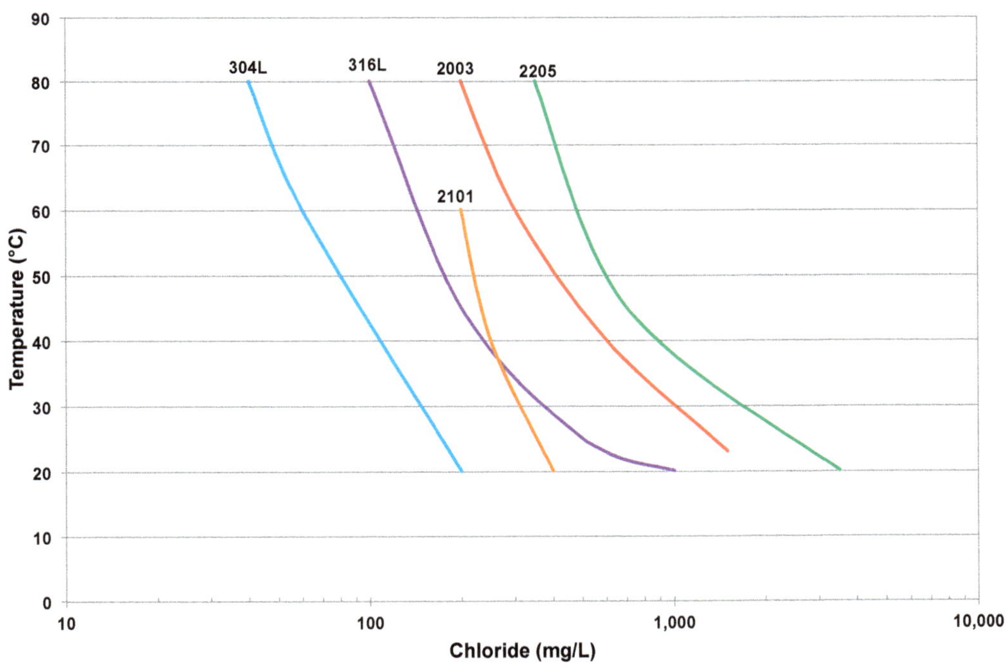

FIGURE 10.4 Threshold temperatures for crevice corrosion for some stainless steels as a function of chloride concentration in sterile solutions.

Walsh et al.[10] conducted exposure tests on a number of stainless steels in a fresh water and found that the rougher surfaces were more easily colonized than polished surfaces. They also found that welds and the HAZ were also favored colonization sites. A similar conclusion was reached by Amaya et al.[11], who examined MIC of 304L stainless steel welds in a water containing 30 mg/L chloride. It was concluded that the rougher surface of the weld and the heat tint on the HAZ created preferred colonization sites compared with the parent metal. Corrosion can also initiate more easily at the HAZ because the heat tint leaves a thin, chromium-denuded layer on the metal surface beneath it. The lower chromium content decreases the PREN (as discussed in Chapter 2) and increases the risk of localized corrosion initiating there. Welds that were pickled or ground showed no corrosion.

Felder and Stein[12] tested a number of stainless steels in a fresh water for 4 years. Continuous flow rates varied from 0.2 m/s to 1.6 m/s and loops with intermittent flow were also included. They found the same bacterial colonization level at all flow rates. However, they did find a difference in colonization level with alloy content. There was most colonization on 304L stainless steel (100%), with less on 316L stainless (20%) and hardly any on a 6% Mo austenitic stainless steel (<0.1%). A similar finding was made by Neville and Hodgekiess[13], who exposed a range of stainless steels in seawater with and without SRB cultures. The cultures also contained significant quantities of nutrients to ensure rapid growth. The finding was that all alloys were more susceptible to localized corrosion in the presence of SRB than without them. However, the more highly alloyed stainless steels, Z100 superduplex and 6% Mo austenitic, were more resistant than 316L and 2205. These results suggest that the more corrosion resistant an alloy is, the more resistant it is to MIC.

Machuca et al.[14] tested 2205 duplex stainless steel in stagnant natural seawater for various times, with tests in filtered, sterilized water for comparison. They were evaluating the effect of biofilms when piping is hydrotested with seawater and is then left full of stagnant seawater before start-up. Even after 12 weeks exposure, there was no crevice corrosion on the test specimens or the control in sterile seawater. Only by anodically polarizing the samples could crevice corrosion be induced. This is not surprising because the potential throughout the tests was approximately –150 mV SCE or less. Referring back to Chapter 5, 2205 is resistant to crevice corrosion up to high temperatures at this potential.

One factor that should be noted is that just because bacteria are present in a water does not mean that they are active. They must have a suitable supply of nutrients and a hospitable environment. This means that it is difficult to ascribe a particular failure categorically to MIC. It also means that reproducing MIC is difficult because small changes in the conditions may not favor proliferation of a particular species.

10.5 SERVICE EXPERIENCE

There are many published experiences of failures of stainless steel believed to be due to MIC, but most of these relate to 300-series austenitic stainless steels. There are far fewer relating to higher alloyed duplex or austenitic alloys. Below is a summary of experiences from the literature.

The preference of bacteria to attach at welds and the HAZ was described above. Figure 10.5 shows MIC of a 316L vessel, which operated in fresh water in Europe. The attack on the seam welds is clearly visible.

What is important is to separate data where bacterial cultures and nutrients have been deliberately added, from that due to simple exposure. The former can exacerbate conditions sufficiently to induce corrosion of CRAs that is not observed in service.

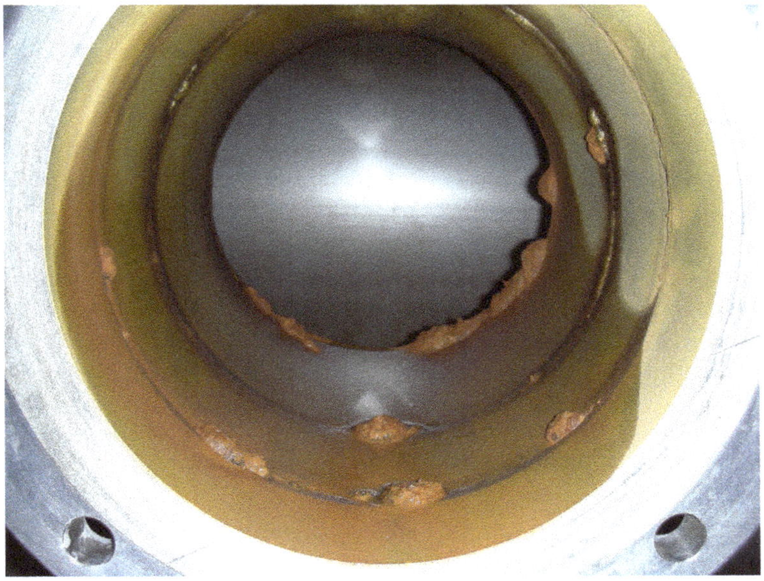

FIGURE 10.5 MIC of the welds on a 316L vessel exposed in fresh water.

Ohashi et al. carried out crevice corrosion tests on a range of austenitic and duplex stainless steels in seawater[15]. Samples were exposed for three weeks in airtight jars containing a mixed bacterial culture containing SRB from the Arabian Gulf, so they would be accustomed to warm environments. The samples were then exposed to natural aerated seawater at 30°C for six months. Parallel tests were conducted on samples exposed only to sterile seawater. The results in Table 10.1 show that the prior exposure in water that supports biofilm development can greatly increase the susceptibility to crevice corrosion and pitting for lower alloyed material. Even superduplex stainless steel suffered attack in these tests, while it is normally accepted that it is suitable up to around 40°C in natural seawater. The results also show the importance of the biofilm in assisting in crevice corrosion initiation, because 2205 suffered no attack in the sterile seawater. These tests might be said to represent a worst-case scenario where a system is left full of stagnant seawater for several weeks with lots of nutrients for bacterial growth prior to operation in aerated warm seawater.

TABLE 10.1 Results of crevice corrosion tests on some stainless steels after prior exposure to seawater with a bacterial culture[15].

ALLOY	BIOFILM ENVIRONMENT	MAX. DEPTH (mm)	WT. LOSS (mg)	STERILE ENVIRONMENT	MAX. DEPTH (mm)	WT. LOSS (mg)
304L	Very severe and extensive attack	1.15	469	Limited crevice corrosion	0.065(C)	2
316L	Very severe and extensive attack	1.29	218	Minor crevice corrosion, some pitting	0.059(P) 0.012(C)	2
2101	Very severe and extensive attack	>2	497	Very minor pits	0.021(P)	2
2205	Severe crevice corrosion	0.54	15	No Corrosion		1
2507	Some crevice corrosion	0.26	17	No Corrosion		1

(C) = Crevice Corrosion
(P) = Pitting Corrosion

Kovach and Redmond reviewed the literature on laboratory exposures versus service experience for 6% Mo stainless steels[16]. They concluded that the 6% Mo alloys only showed superficial corrosion in tests with bacterial cultures and nutrients, while 304L and 316L suffered pitting. They could find no reported service failures of 6% Mo alloys due to MIC. They reported one case where 6% Mo piping had replaced 304L, which had failed by pitting due to MIC; the 6% Mo alloy was in good condition when inspected after two years. They also report a failure of 904L heat exchanger tubes attributed to MIC after just 14 days in service. This was attributed to SRB in conjunction with chlorine additions to the water.

Tuthill reviewed MIC failures of CRAs in an unpublished report for the Nickel Development Institute[17]. He concluded that 304L, 316L, and alloy 400 (UNS N04400) were all susceptible to MIC, although the role of MIC was not conclusively demonstrated in some cases. For 6% Mo stainless steels, high nickel alloys, such as C-276 (UNS N10276), and titanium, no failures were reported and these alloys were all judged resistant to MIC. There was one report of MIC of a 25% Cr duplex alloy (UNS S32550) with a PREN of ~37, which was reported to have corroded because of MIC, but Tuthill judged that this mode of failure was not proven. Tuthill also reported a failure of 904L heat exchanger tubes, which he regarded as clearly due to MIC[17].

Stainless steels are frequently used to handle wastewater, where bacteria are usually present and are often active. Tuthill and Lamb described the use of 304L and 316L stainless steels in wastewater plants and described a few failures due to MIC[18]. They also describe how to get the best out of these alloys by good design, pickling all welds, and removing any buildup of deposits.

Iversen described the results of exposure tests of 304L, 316L, and 2205 at six Swedish wastewater plants[19, 20]. The results showed that 304L suffered MIC at four plants, all where the chlorides exceeded 200 mg/L, which is above the threshold for crevice corrosion of 304L at room temperature (Figure 10.4). The 316L corroded at one plant, where the chloride exceeded 500 mg/L and there were many deposits on the metal. This is below the chloride threshold for crevice corrosion of 316L at room temperature in sterile solutions. Alloy 2205 suffered no corrosion in any of the plants. Iversen concluded that chloride alone could not predict the likelihood of MIC[19]. There appears to be no data for lean duplex stainless steel under such wastewater conditions.

A review of MIC of stainless steel in water used for cooling or hydrostatic testing concluded that 304L and 316L could both suffer from MIC when active bacteria were present[21]. In one case, they described the replacement of 304L stainless steel, which had suffered MIC in lake water, with a 6% Mo stainless steel, and this gave excellent performance.

Felder and Stein presented the results of a four-year exposure trial of several stainless steels in fresh water containing up to 600 mg/L chloride[12]. As described above, they used several flow rates including intermittent flow, and found no corrosion on 6% Mo austenitic under any conditions, although there was little bacterial colonization, compared with 304L and 316L.

Renner described three case histories of MIC from chemical plants based on the River Rhine in Germany[22]. In all cases, the chloride content was a maximum of 100 mg/L. The first was firewater piping in 316L, which leaked after 180 days in service, due to pitting at the welds and HAZ. This was repaired with a higher alloy filler (317L type) and it leaked again after another 90 days. No corrosion occurred on solution-annealed seam welds and the failure was attributed to SRB. A second failure was of 321 stainless steel (304 stabilized with titanium) heat exchanger tubes with carbon steel tube plates. These suffered severe pitting after four years and substantial manganese-rich deposits were found; the failure was attributed to manganese oxidizing bacteria[22]. The third failure was 316 in an open surface cooler running at a high (unspecified) temperature.

After 11 years, there were leaks at nozzle welds in the dead zone due to pitting and also in areas of the base metal. Pitting was up to 10 mm deep in the vessel walls and both SRB and manganese oxidizing bacteria were found[22].

Although Iversen saw no attack of 2205 in wastewater treatment plants with chlorides less than 1,000 mg/L[19, 20], MIC of 2205 has been seen in higher chloride waters. Hesselmans reported the failure of 2205 wastewater piping, handling high chloride water at 35°C, where a 4 mm wall thickness was penetrated in less than three months[23].

Two new 2205 duplex coolers were installed in a petrochemical plant in Holland, with the process fluid (150°C) on the inside of the tubes and brackish water on the shell side[24]. The water contained up to 13,000 mg/L chloride and varied from 10–20°C during the year. In addition, the water was dosed with hypochlorite at about 1 mg/L for 30 minutes every 8 hours. After six weeks one cooler started leaking, and soon after that the other cooler also developed leaks. On inspection, barnacles and pitting were observed in areas where the water temperature was 30–40°C. There was no attack in cooler or hotter areas. There was only superficial corrosion under the barnacles, but through-wall pitting occurred beneath light brown-yellow volcano-like deposits. Figure 10.6 shows a typical tube after cleaning off the corrosion products. When the corrosion products were tested with hydrochloric acid, there was a strong smell of hydrogen sulfide suggesting that SRB were active. Investigation revealed that the hypochlorite dosing system had been down for maintenance when the 2205 coolers were installed and for some time afterwards. The system was cleaned, a new state-of-the-art anti-fouling system (pulse chlorination) was fitted, and new 2205 bundles were installed in the coolers. Since that time, the coolers have performed well with no further failures.

FIGURE 10.6 Appearance of 2205 cooler tube after removing the corrosion products[24].

Z100 superduplex and 316L austenitic stainless steels were exposed in a biologically active marine mud with a high SRB count[25]. Samples were exposed both fully and partially immersed for five years. At the end of the test, there was pitting on the 316L pipe up to a depth of 0.37 mm, while there was no attack of the Z100 parent metal, welds, or HAZ.

London Underground Limited (LUL) was experiencing severe corrosion of cast iron tunnel linings in one part of the system. The ground in this area had been a salt marsh so chlorides were high. In addition, there were pyrites (FeS) in the soil and SOBs were able to produce sulfuric acid because of oxygen entering the soil from the tunnels. The LUL consultants determined that only an acid and chloride resistant duplex stainless steel would be satisfactory. LUL lined two sections a quarter of a mile long with castings in Z100 made to the top of the copper specification, to increase resistance to sulfuric acid (as discussed in Chapter 4), and they used wrought Z100 fasteners to bolt the segments together (Figure 10.7). The tunnel in this area is designed with a life of 400 years.

FIGURE 10.7 Z100 superduplex castings and bolting for tunnel linings under test. [Photo courtesy of Rolled Alloys]

An oil carrier in the Far East suffered severe MIC of carbon steel transfer piping between the port and starboard ballast tanks. This was because they contained a mixture of seawater and sour crude. The pipe crossed inside a crude tank, which also contained high-chloride produced water and this was where the worst MIC occurred (Figure 10.8A). Various water treatments were tried to correct the problem, but the problem was finally solved by replacing the carbon steel with Z100 piping, which performed without problem (Figure 10.8B).

(A)

(B)

FIGURE 10.8 Transfer piping between the ballast tanks on an oil carrier: A) Carbon steel corroded by MIC, B) Z100 superduplex replacement piping.

A 6% Mo alloy (UNS N08367) was used to replace 304L piping handling low-chloride service water in a U.S. power station, which had failed by MIC at welds and elsewhere[26]. It has now been in service for over 20 years without problems.

Stainless steels with a PREN>40.0, such as superduplex and 6% Mo austenitic, appear resistant to a range of MIC types and have given good service in environments that have caused severe corrosion of lower alloyed materials.

REFERENCES

1. J.A. Hardy, "Biological Corrosion," *CORROSION*, 18 (1983): p. 190.
2. J W. Costerton, G.G. Geesey, and P.A. Jones, "Bacterial Biofilms in Relation to Internal Corrosion Monitoring and Biocide Strategies," *MP*, 27, 4 (1988): p. 49.
3. T.R. Bott, "Biological Fouling of Heat Transfer Surfaces," *Effluent and Water Treatment J*, 19 (1979): p. 453.
4. L. Shreir, R.A. Jarman, and G.T. Burstein, *Corrosion*, 3rd Edition (Oxford, United Kingdom: Butterworth-Heinemann, 1994): p. 2:87.
5. R. Newman, Toronto University, Canada, Private Communication.
6. J.F.D. Stott, "Assessment and Control of Microbially-Induced Corrosion," *Metals and Materials* (April 1988): p. 224.
7. H. Amaya and H. Miyuki, "Microbially Influenced Corrosion of Stainless Steels in Natural Seawater," *Welding International* 9, 12 (1995): p. 941.
8. T. Rogne and U. Steinsmo, "Practical Consequences of the Biofilm in Natural Seawater and Chlorination on the Corrosion Behaviour of Stainless Steels, in Seawater Corrosion of Stainless Steels—Mechanisms and Experiences," EFC Publication No. 19, European Federation of Corrosion (London, UK: Institute of Materials, Mining and Minerals, 1996): p. 55.
9. A. Iversen, "MIC on Stainless Steel in Waste Water Treatment—Anaerobic and Aerobic Treatments Influence on Enoblement and the Passive Surface," paper no. 562, CORROSION 2003 (Houston, TX: NACE International, 2003).
10. D. Walsh, Q. Qiong, J. Seagoe, and L. Williams, "Factors Affecting Microbiologically Influenced Corrosion of Stainless Steel," International Trends in Welding Science and Technology, 3rd International Conference on Trends in Welding Research (Materials Park, OH: ASM International, 1992): p. 673.
11. H. Amaya, H. Miyuki, Y. Takeishi, and M. Ozawa, "Effects of Shape of Weld Bead on Bacterial Adhesion and MIC Occurrence at Stainless Steel Welded Joints," CORROSION 2002, paper no. 556 (Houston, TX: NACE International, 2002).
12. C.M. Felder and A.A. Stein, "Microbiologically Influenced Corrosion of Stainless Steel Weld and Base Metal—4 Year Field Test Results," CORROSION/94, paper no. 275 (Houston, TX: NACE International, 1994).
13. A. Neville and T. Hodgekiess, "A Comparative Study of the Corrosion Behaviour of Duplex and Austenitic Stainless Steels in Marine Environments Containing Sulphate Reducing Bacteria," Duplex '94 (Glasgow, United Kingdom: TWI, 1994).
14. L.L. Machuca, S.I. Bailey, R. Gubner, E.L.J. Watkin, M.P. Ginige, and A.H. Kaksonen, "Crevice Corrosion of Duplex Stainless Steels in the Presence of Natural Marine Biofims," CORROSION 2012, paper no. 01486 (Houston, TX: NACE International, 2012).

15. K. Ohashi, R. Kobayashi, J.F.D. Stott, and M.J. Schofield, "Marine Crevice Corrosion of Stainless Steels Under Biofilmed and Sterile Conditions," CORROSION 2016, paper no. 7109 (Houston, TX: NACE International, 2016).
16. C.W. Kovach and J.D. Redmond, "High Performance Stainless Steel and Microbially Influenced Corrosion," Stainless Steel '96 (Dusseldorf, Germany: Verein Deutscher Eisenhüttenleute, 1996).
17. A. Tuthill, "Base Metal Resistance of Alloys to Microbiological Influenced Corrosion," Unpublished report (Toronto, Canada: Nickel Institute, 1992).
18. A. Tuthill and S. Lamb, "Stainless Steel in Municipal Waste Water Treatment Plants," NiDI Technical Publication No. 10076 (Toronto, Canada: Nickel Institute, 1998).
19. A. Iversen, "MIC on Stainless Steels in Wastewater Treatment Plants," CORROSION/99, paper no. 171 (Houston, TX: NACE International, 1999).
20. A. Iversen, "MIC on Stainless Steels in Wastewater Treatment Plants–Field Tests and a Risk Assessment," CORROSION 2002, paper no. 451 (Houston, TX: NACE International, 2002).
21. G. Kobrin, S. Lamb, A.H. Tuthill, R.E. Avery, and K.A. Selby, "Microbiologically Influenced Corrosion of Stainless Steels by Water Used for Cooling and Hydrostatic Testing," NiDI Technical Publication No. 10085 (Toronto, Canada: Nickel Institute, 1998).
22. M.H.W. Renner, "Corrosion Engineering Aspects Regarding MIC Related Failures of Stainless Steels," CORROSION/98, paper no. 285 (Houston, TX: NACE International, 1998).
23. J. Heselmans, "Performance of Stainless Steel in Marine Applications," *Stainless Steel World* (December 2006): p. 2.
24. J. Houben, ExxonMobil, Netherlands, Private Communication (2017).
25. R. Francis, G. Byrne, and H.S. Campbell, "Performance of Some Stainless Steels in a Marine Mud," CORROSION/99, paper no. 313 (Houston, TX: NACE International, 1999).
26. Strauss, S.D., "SWS Pipe Replacement Addresses MIC Problem," *Power*, 135 (1991): p. 59.

11 Atmospheric Corrosion

Engineering equipment frequently has to operate in the open without protection from the weather so resistance to corrosion is important. While paints and coatings are often used to control corrosion, uncoated metals are also frequently used. Some atmospheres are particularly corrosive and under the most adverse conditions premature failure of coatings is not uncommon. Use of stainless steels can offer an attractive alternative, particularly in environments where maintenance of coatings is costly or difficult. Atmospheric corrosion is somewhat different from full immersion corrosion in liquids, such as water. In the next section, the important factors influencing atmospheric corrosion are discussed.

11.1 FACTORS AFFECTING ATMOSPHERIC CORROSION

The most important factor is water. Corrosion in direct contact with air at ambient temperature is generally not significant from an engineering point of view. For corrosion to occur, water must condense on the exposed metal surfaces or precipitate as rain. This will be followed by evaporation as the components dry out and, thus, a most important parameter for atmospheric corrosion is the time of wetness (sometimes abbreviated as TOW). Condensation is controlled by the relative humidity and metal surface temperature. It is generally accepted that a relative humidity of 80% or more, plus a temperature greater than 0°C are required for significant condensation[1]. Rainfall is dependent on the local weather and can vary greatly, even in a small country like the British Isles.

Corrosion in the atmosphere is generally not as great as that on fully immersed metal surfaces in water. As the water evaporates, any dissolved salts in it will concentrate. Hence, when these species are aggressive to a metal, the corrosion rate will increase as the water evaporates. The common species found in condensation and rainwater are briefly discussed below.

1. Oxygen: This is always present in the atmosphere and oxygen diffuses rapidly through the thin water film. Hence, depolarization of the cathodic reduction of dissolved oxygen is strongly favored.

2. Carbon dioxide: This gas is common in the atmosphere and dissolves readily causing the water to be mildly acidic because of the formation of carbonic acid. However, it is not considered to be a major contributor to atmospheric corrosion.[2]
3. Ammonia: This is common in rural areas, for example, because of decaying organic matter and/or use of fertilizers. Ammonia is readily soluble in water and increases the pH, which is generally beneficial for most alloys, although stress corrosion cracking of some copper alloys is possible. Ammonia may also cause slight dissolution of copper alloys because of complex-salt formation.
4. Sulfur Oxides and Nitrogen Oxides: These are common in industrial and urban areas as effluent gases, e.g., fossil fuelled power stations. These gases dissolve readily, forming strong (e.g., sulfuric and nitric) acids, which are associated with acid rain.

In addition to these gases, there are two other factors that affect atmospheric corrosion: temperature, and chlorides. Temperature is self-explanatory, the warmer the climate, the greater the rate of chemical reactions, including corrosion rate. Chloride is found adjacent to the sea, and the closer a site is to the ocean, the higher the chloride concentration in condensed water. Over 65 years (1940–2005) exposures at the LaQue Center for Corrosion Technology's atmospheric corrosion test facility at Kure Beach, North Carolina, United States, demonstrated that corrosion rates 25 m from the sea were typically much higher than those 250 m from the sea.

Using all these factors, atmospheric sites are generally divided into four main types: rural, urban, industrial, and marine. Rural atmospheres have low chloride and sulfur dioxide levels, but may contain ammonia. The urban atmosphere has some sulfur dioxide (SO_2), while industrial areas have higher sulfur dioxide and nitrogen oxide (NO_x) contents. The marine atmosphere is characterized by high chloride content. It should be noted that atmospheric environments at specific geographic locations may be classified into additional categories such as urban industrial, industrial marine, tropical rainforest, arid desert, and so forth.

The principal cathodic reaction in atmospheric corrosion is usually the reduction of dissolved oxygen and, as stated above, the reaction rates can be high, especially with thin water films. If the dissolved solids and gases content of the water are low, the conductivity will be low, reducing the rate of corrosion, but as the films evaporate the conductivity will increase due to the concentration effect.

11.2 ASSESSING ATMOSPHERIC CORROSIVITY

It is important to be able to assess the corrosivity of atmospheres towards metallic corrosion and this is addressed in ISO 9223[3], which calculates a corrosion index based on the concentration of pollutants such as sulfur dioxide and chlorides, as well as time of wetness, temperature, etc. In addition, the corrosion rate of some standard metals, such as steel, copper, zinc, and aluminum, over a year can be used to modify the corrosion ranking. This produces a ranking from C1 (mild) to C5 (very aggressive). ISO 9226 allows the atmospheric corrosivity assessment to be extended to areas beyond the scope of ISO 9223[4].

While this ranking works for many metals and alloys, it is not a good indicator of the performance of stainless steels in some atmospheres. This is because steel, copper, and zinc corrode uniformly in aggressive atmospheres, while stainless steels tend to suffer predominately localized attack (pitting). For example, an acidic atmosphere that is low in chlorides can be classified as aggressive for steel or zinc coupons, whereas stainless steels may suffer no attack. In marine atmospheres, it is the chloride that is aggressive and this may make some stainless steels pit. However, it will depend on other variables as to how aggressive a particular marine atmosphere is to stainless steel. When testing stainless steels, it is useful to measure these other variables to indicate how these may be affecting alloy performance. In any case, the depth of localized corrosion attack on susceptible stainless steel grades in atmospheric environments is generally quite shallow and may not affect the integrity of the component.

11.3 ATMOSPHERIC EXPOSURE DATA

Over the last 20 years, there have been a number of extended test exposures of a variety of duplex and austenitic stainless steels in atmospheres, mostly in the North Temperate Zone, but some have been in hotter climates.

Fritz et al. exposed 304L, 316L and 6% Mo austenitic stainless steels plus 2205 duplex stainless steel in the 25 m lot at Kure Beach in North Carolina for six months[5]. They tested panels with different finishes to see what effect residual iron from cleaning/handling might have on the corrosion resistance in a very aggressive marine atmosphere. The finishes were mill annealed and pickled, shot-blasted with steel, and shot blasted followed by a pickle in 24% nitric acid at 60°C. Several tests at the Kure Beach site showed it to be a C5 category atmosphere.

After exposure, the panels were examined for pitting and the pit densities were determined, as shown in Table 11.1. The results showed that there was no pitting on the 6% Mo austenitic stainless steel or the 2205 duplex alloy. There was pitting on both 304L and 316L, with the latter being slightly less pitted. All the alloys showed a lot of staining by iron corrosion products in the shot-blasted condition. The 6% Mo and 2205 stainless steel showed no attack under the staining, but the 304L and 316L exhibited pitting, some of which could be attributed to the corrosion of the iron particles. After pickling the shot-blasted samples, some staining of all alloys still occurred, and hence, this method of removal of embedded iron contamination was not satisfactory. The author has always preferred pickling in a nitric/hydrofluoric solution, similar to that used in the mill after annealing, and this is much more effective in removing iron contamination.

TABLE 11.1 Pit densities for some stainless steels exposed at an aggressive marine test site for six months[5].

ALLOY	PIT DENSITY (No./cm^2)
304L	15
316L	12
6% Mo	0
2205	0

Lojewski et. al. carried out atmospheric exposure tests on types 304L, 316L, 2304, and 2205 stainless steels for up to 5 years. In addition, they carried out a one-year exposure on 2202 lean duplex[6]. They used six test sites, including a U.S. subtropical site and five European sites, varying from coastal marine to inland rural, as shown in Table 11.2. The Kure Beach, North Carolina site was closed after two years and the samples were relocated to Daytona Beach, Florida. Because stainless steels are often used for their aesthetic appearance, the samples were assessed for the degree of rust staining using a 10 point logarithmic scale, where 0 is 100% covered by rust, and 10 is 0%, as per ASTM D-610[7]. Intermediate values are such that 2 is 33% rust staining, and 8 is 0.1% staining. The results for a one-year exposure of 2202 lean duplex stainless steel are shown in Figure 11.1 and it is clear that the alloy performed well in rural and urban atmospheres, but suffered increasing staining as the corrosivity of the marine atmosphere increased. The performance of 2202 lean duplex stainless steel was better than 304L and similar to 316L.

TABLE 11.2 Conditions reported by Lojewski et al.[6] at atmospheric exposure test sites.

SITE	TYPE	AV. TEMP. (°C)	WET TIME (%)	CHLORIDE ($mg/m^2/d$)	CORROSIVITY CLASS
Kure Beach	Sub-tropical marine	18.5	31	143	C5+
Daytona Beach*	Sub-tropical marine	22	41	213	C5+
Brest	Temperate marine	12.4	72	215	C4/C5
Lyon	Urban/Industrial	13	36	3.1	C1/C3
Le Creusot	Rural	10	44	1.4	C1/C3
Genoa	Industrial/Urban/ Marine	16.6	27	4.5	C2/C4
Rome	Marine/ Rural	17	54	1.1	C1/C3

* Kure Beach site closed after two years and samples transferred to Daytona Beach.

Figure 11.2 shows the degree of rust staining of the other duplex stainless steels after five years of exposure[6]. None of the alloys was attacked at the rural and urban test sites, but they followed the same pattern as 2202 lean duplex stainless steel with increased staining as the corrosivity of the marine sites increased. It is also clear that 2304 duplex stainless steel suffered more staining than 316L, while 2205 lean duplex stainless steel suffered no rust staining after five years at the Brest, France marine site.

The pit depths were determined after 1 year of exposure at the two most aggressive marine sites, as shown in Figure 11.3[6]. The pitting on all the austenitic and duplex alloys was shallow compared with that on a 12% Cr ferritic stainless steel (UNS S41003). The depth of pitting decreased with increasing Pitting Resistance Equivalent Number (PREN) of the stainless steels (with the exception of 316L) and no pitting was seen on 316L or the two duplex alloys at the Brest site. Such shallow attack would be of no concern for structural items in these alloys in a marine atmosphere.

These tests were extended and Le Manchet and Paul reported the results of 10-year exposures at the same sites, including 2202 lean duplex[8]. Figure 11.4 shows the degree of rust staining for the various alloys, where it can be seen that the lean duplex alloys performed well in milder atmospheres, but in marine atmospheres they were inferior to 316L, which was not as good as 2205. This is clearly shown in Figure 11.5 by the appearance of the stainless steels after 10 years of exposure at the Brest marine site. At the more aggressive Kure/Daytona Beach sites, even 2205 showed staining, but Le Manchet and Paul reported no staining of superduplex stainless steel at this location after four years[8].

Mameng et al. conducted a four-year atmospheric exposure of some stainless steels in the Middle East[9, 10]. One site was right on the seashore of Dubai and another was located 2 km inland where the relative humidity was very low compared with the seashore[11]. The seashore site had higher temperatures, higher salt levels, and lower rainfall. According to testing per ISO 9223, this site was classified as CX category (very aggressive). During the

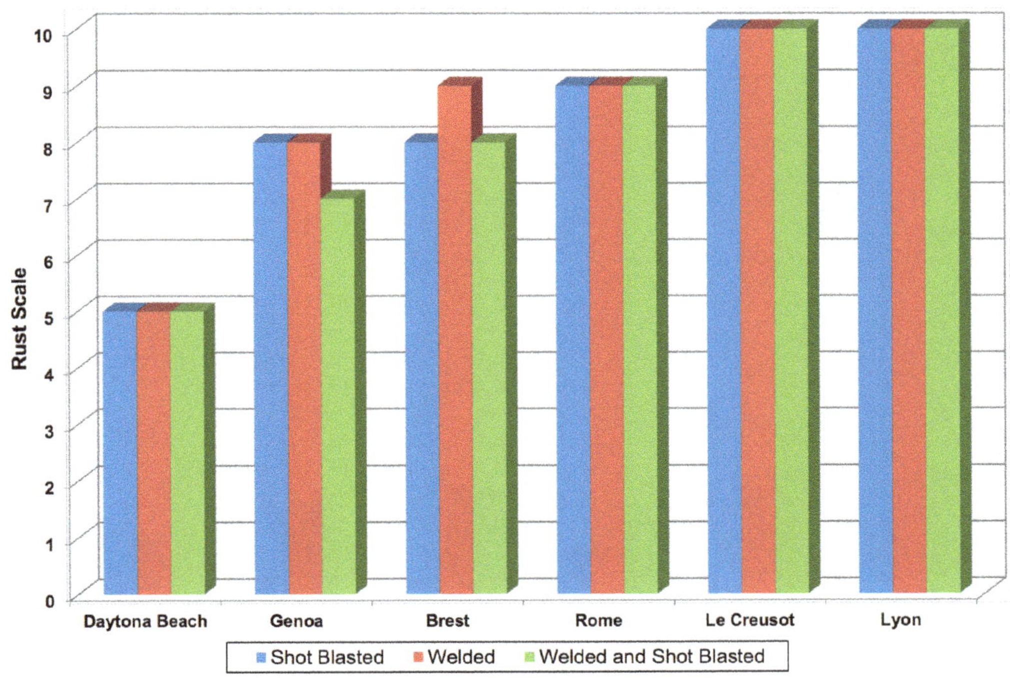

FIGURE 11.1 The degree of rust staining on lean duplex 2202 after one year of exposure[6].

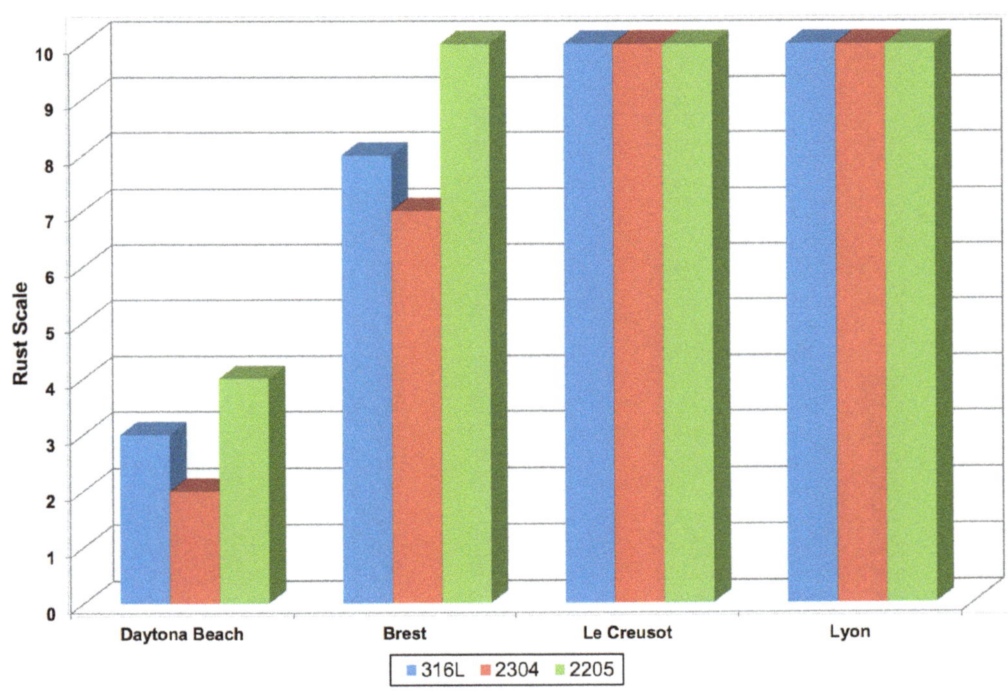

FIGURE 11.2 The degree of rust staining on some stainless steels after five years of exposure at various sites[6].

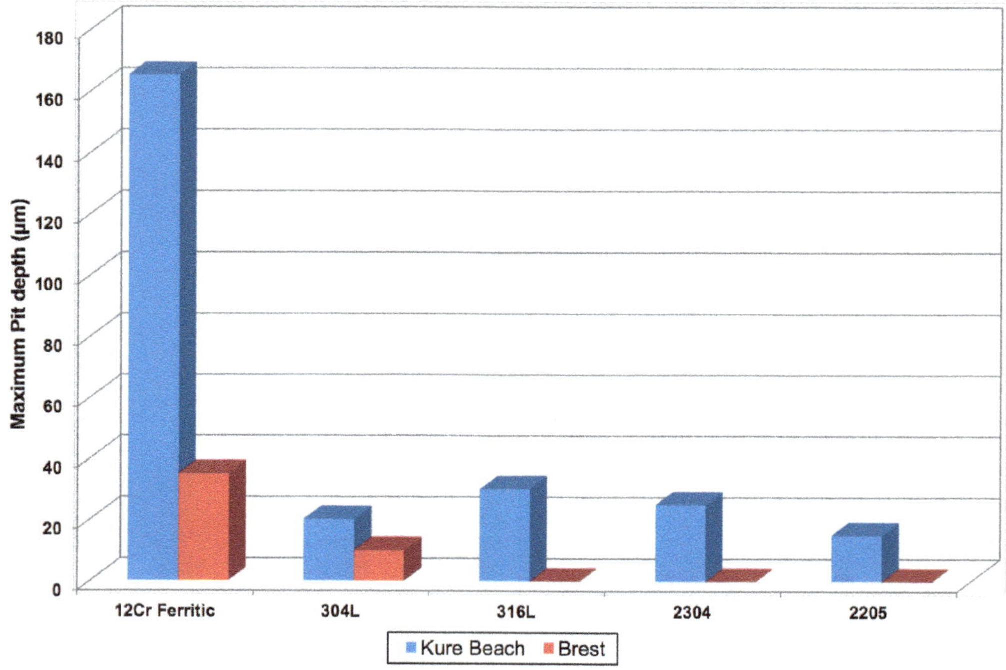

FIGURE 11.3 The maximum pit depth on some stainless steels after one year of exposure at two marine sites[6].

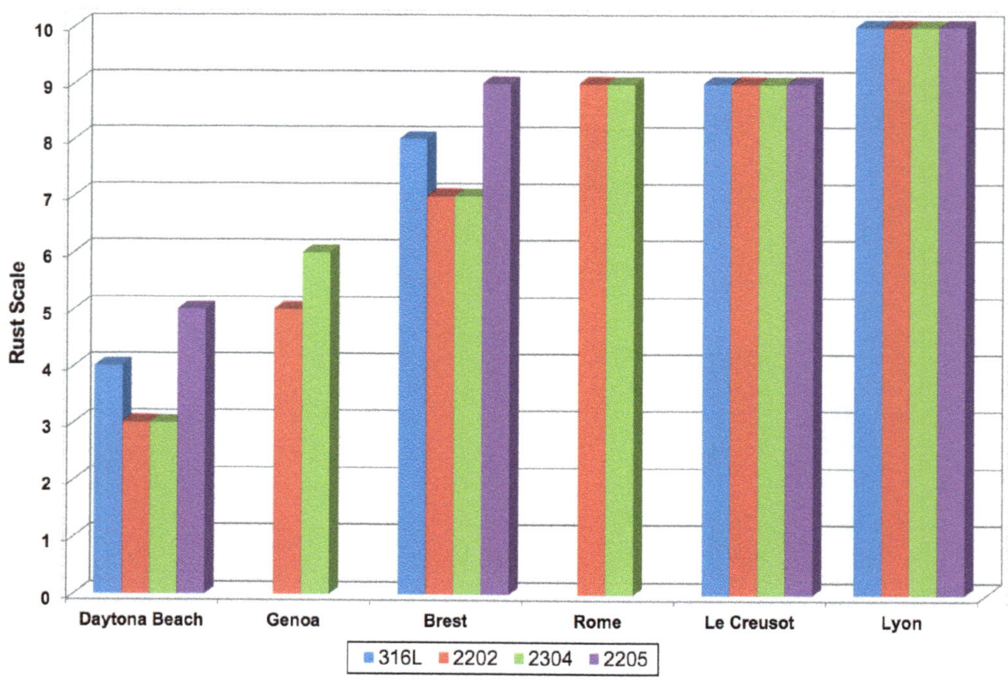

FIGURE 11.4 Degree of rust staining on some stainless steels after a 10-year exposure at various sites[8].

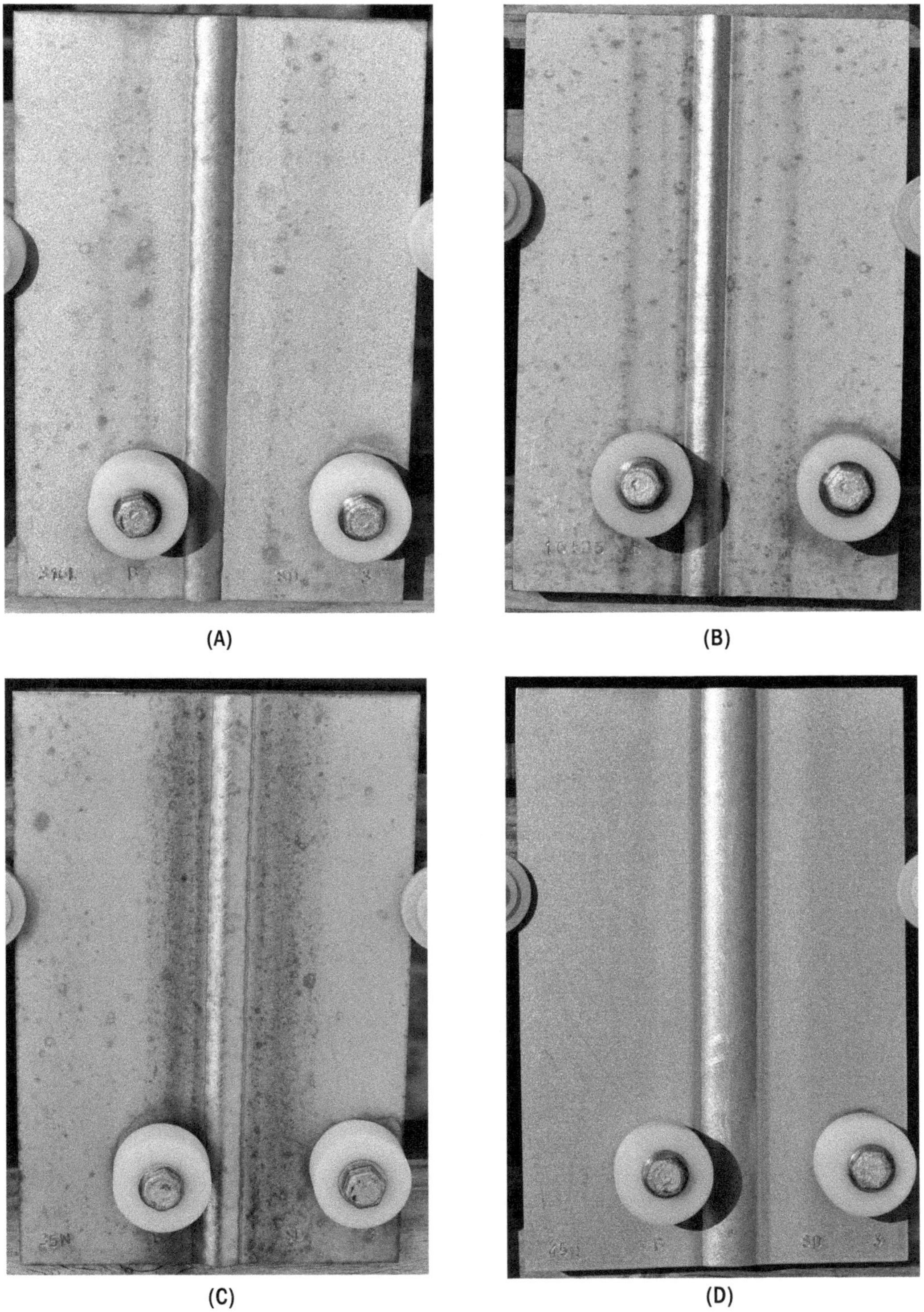

FIGURE 11.5 Appearance of some stainless steel panels after 10 years at the Brest marine site[8]:
A) 316L, B) 2202, C) 2304, and D) 2205.
[Photos courtesy of Industeel]

four-year exposure, the average temperature was 28.8°C, the average chloride deposition rate was 33 $g/m^2/d$ and the annual rainfall was 25 mm. A third site was a rural environment in Riyadh, Saudi Arabia.

Samples were exposed with a variety of surface finishes, as shown in Table 11.3, and some had crevice washers attached. There were also some welded samples that were shot-blasted and pickled after welding. Figure 11.6 shows a test rack at the Dubai inland site, and Figure 11.7 shows the appearance of some of the duplex stainless steels after exposure for four years at the same site.

FIGURE 11.6 Stainless steel exposure panels at the Dubai inland site[9]. [Photo courtesy of Outokumpu Stainless AB]

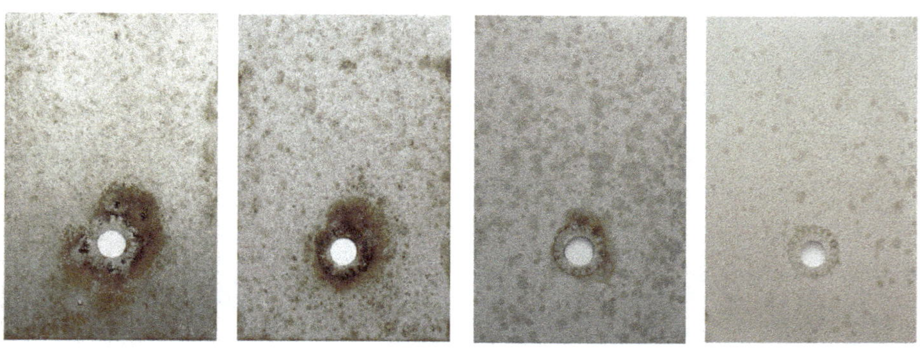

(a) S32101 2B surface | *(b) S32304 2B surface* | *(c) S32205 1D surface* | *(d) S32750 1D surface*

FIGURE 11.7 Appearance of some of the duplex stainless steels exposed for four years at a seaside site in Dubai[10]. [Photo courtesy of Outokumpu Stainless AB]

TABLE 11.3 Condition of 2205 duplex stainless steel samples exposed in the Middle East [9].

SURFACE FINISH	DESCRIPTION	THICKNESS (mm)	TYPE
2M, Patterned	Final rolling is made with an imprinted roll pattern on one side. The pattern was embossed on a bright annealed, 2R finish.	1.2	Plain
1D	Hot rolled, heat treated, mechanically descaled through shot blasting, and pickled.	5.0	Creviced
1D-SP	Original surface 1D, as above, with extra laboratory shot blasting and pickling.	5.0	Welded
2E	Cold rolled, heat treated, mechanically descaled through shot blasting, and pickled.	2.0	Creviced
2R	Cold rolled, annealed in a protective (reducing) atmosphere (bright annealed finish).	1.2	Plain
2E-Brushed	Cold rolled, heat treated, mechanically descaled through brushing, and pickled.	1.0	Plain and creviced
2K	Wet polished to a satin finish	4.0	Plain
Reference	Wet machine grinding with 320 grit SiC paper	3.0	Plain

The results are summarized in Table 11.4 and readers are referred to the original papers [9, 10] for more details. The samples were evaluated both by the maximum pit depth incurred and the degree of rust staining. Pit depths less than 25 μm were not recorded. A Japanese industrial standard developed specifically for stainless steels was used to assess the degree of staining[12]. This is a 10-point scale running from 0–9, with zero being the original surface and 1 being no rust staining but with some dirt or other particles sticking to the surface. The other numbers on the scale represent degree of rust staining as follows: 2 (1% staining), 3 (5%), 4 (12%), 5 (20%), 6 (30%), 7 (50%), 8 (70%), and 9 (100%).

TABLE 11.4 Summary of results from a four-year exposure of 2205 duplex stainless steel in the Middle East[10].

LOCATION	ATMOSPHERIC CORROSIVITY	ALLOY	SURFACE FINISH	MAX. PIT DEPTH (μm)		APPEARANCE*	
				Open	Sheltered	Open	Sheltered
Sea-Side Site, Dubai	CX	304L	2B	290	205	7	6
		316L	1D	185	180	7	6
		2101	2E	106	180	6	4
		2304	2B	130	200	7	6
		2205	2E	95	60	6	3
		2507	1D	<25	<25	4	1
Semi-Marine Site, Dubai	C1	304L	2B	etching	<25	4	3
		316L	1D	etching	<25	3	2
		2101	2E	etching	<25	4	2
		2304	2B	etching	<25	4	2
		2205	2E	<25	<25	3	2
		2507	1D	<25	<25	3	2
Rural Site, Riyadh, Saudi Arabia	C1	304L	2B	<25	<25	3	0
		316L	1D	<25	<25	2	0
		2101	2E	<25	<25	2	0
		2304	2E	<25	<25	2	0
		2205	1D	<25	<25	2	0

* See text for description.

It is clear from the degree of pitting and rust staining compared with the tests of Lojewski et al.[6] that the conditions at the site immediately adjacent to the seawater were extremely aggressive. The results show that there was little difference in the depth of pitting or degree of rust staining with exposure orientation of the surface finish. They also showed that the finer the finish, the better the resistance to atmospheric corrosion. The sheltered samples suffered less attack than those openly exposed. The authors conjectured that this may have been due to sulfate accumulations in the sheltered locations reducing the severity of corrosion, or it may have been that less chloride was deposited in the sheltered regions. Shot blasting and pickling of welds reduced the severity of pitting and the degree of rust staining.

The results from the site 2 km from the coast showed the conditions to be much milder and there was no pitting and little staining at the Riyadh site on all the alloys (Table 11.4).

Gunnarsson et. al. exposed a range of austenitic and duplex stainless steels at eight locations along Swedish roadsides, to determine their corrosion resistance due to atmospheric conditions, road dirt, and salt used to de-ice the roads[13]. They tested a wide range of alloys, but the current discussion will be confined to the duplex alloys 2101, 2304, 2205, and 2507, plus 304L, 316L, and 6% Mo superaustenitic, for comparison. All the samples had a weld and a crevice washer. The sites were characterized using ISO 9223, and steel and zinc coupons, and the results are shown in Table 11.5, along with the road locations and conditions. Site 1 was inland, but all the bridges were located over seawater.

TABLE 11.5 Conditions at eight Swedish roadside atmospheric test sites including the corrosivity per ISO 9223[13].

No.	SITE	TRAFFIC DENSITY (Vehicles/day)	No. OF DE-ICING EVENTS	DISTANCE TO ROAD MARKINGS (m)	CORROSIVITY*
1	Boras, road 40	15,000	900	2.7	C4
2	Gothenberg, Ludbyleden	40,000	700	1.2	C4
3	High Coast Bridge	4,000	1,140	1.5	C4/C5
4	Svartnora Bridge	400	20	0.5	C2
5	The Soderledstunnel	70,000	0+	2.1	C2/C5
6	Oland Bridge	14,000	600	0.3	C5
7	Oresund Bridge, roadside	15,000	290	3.1	C3/C4
8	Oresund Bridge, seaside	-	-	-	C2/C3

* Steel/zinc to ISO 9223
+ No de-icing in the tunnel, but on the roadway just before the entrance. Frequency of de-icing not available.

The results for types 304L, 316L, 2101, and 2304 stainless steels are summarized in Table 11.6. There was no attack on duplex alloys 2205 and 2507 or the 6% Mo austenitic alloy (UNS S31254) at any of the sites. Where there is no mention of a particular type of attack, it means that it was <25 μm. Alloys 2101 and 2304 showed limited corrosion at the more aggressive sites and no corrosion attack at the less aggressive sites. Gunnarsson et al. also observed corrosion on 316L bolts, nuts, and washers used to fasten the samples on the racks, and they suggested the use of 2205 duplex or higher alloyed materials for critical load-bearing components and fasteners at such locations.

TABLE 11.6 Results of five-year roadside atmospheric exposures of some stainless steels in Sweden[13].

SITE	MAXIMUM PIT DEPTH (μm)			
	304L	316L	2101	2304
Boras, road 40	36(P); 60(C)	35(P)	<25	<25
Gothenberg, Ludbyleden	32(P); 51(C)	52(P); 50(C)	<25	35(C)
High Coast Bridge	30(P); 55(C)	27(P); 25(W)	<25	<25
Svartnora Bridge	<25	<25	<25	<25
The Soderledstunnel	32(P): 48(C)	90(P); 172(C)	<25	NT
Oland Bridge	30(P): 61(C)	30(P); 45(C); 25(W)	56(P); 62(C)	<25
Oresund Bridge, roadside	28(P); 45(C)	<30(W)	26(W)	<25
Oresund Bridge, seaside	33(P); 48(C); 35(W)	25(P); 120(C); 28(W)	<25	45(C)

NOTE: P = pitting; C = crevice corrosion
W = weld attack; NT = not tested

When stainless steels are used architecturally and aesthetic appearance is important, there are a number of things that can be done to minimize the risk of staining. This is discussed in detail by, for example, Sussex and Gouch[14], and Houksa[15], and it involves attention both to design and some simple maintenance. It is important that contamination by iron or steel be avoided during cutting, fabrication, transport, and assembly, because corrosion of iron and steel produces unsightly rust and may initiate pitting on susceptible grades of stainless steel. It is important to use a design that avoids entrapment of dirt and debris, which might give rise to localized corrosion. If possible, frequently washing down areas with fresh water can be beneficial where dirt and debris might collect or adhere.

11.4 APPLICATIONS

Duplex stainless steels have been used for a number of applications in the atmosphere where their high strength and minimum maintenance are attractive. Alfonsson describes the first 10 years of applications of 2101 lean duplex under a range of atmospheric conditions[16]. Lean duplex has been used for a number of bridges in Europe in less aggressive atmospheres, where its higher strength offers reduced costs compared to austenitic alloys and the good corrosion resistance offers much reduced lifecycle costs compared to painted carbon steel. Alfonsson also describes the use of 2101 for storage tanks and vessels of non- or mildly corrosive products in moderate atmospheres, where the alloy requires no painting or maintenance compared with carbon steel.

Havn discussed the use of 2101 and 2003 lean duplex alloys offshore, as alternatives to painted carbon steel or 316L austenitic stainless steel[17]. The applications included pipe racks, pipe supports, and piping. He carried out a lifecycle cost analysis and concluded that lean duplex was very cost effective, particularly the 2101 grade.

Duplex alloy 2205 has been widely used in bridges in more aggressive atmospheres, particularly in marine locations. The Stonecutters Bridge in Hong Kong is a large road bridge more than 1 km long over a busy seaway. The bridge towers are reinforced concrete in the lower section, and the top 118 m is a composite sandwich of concrete with 2205 plates on the outside. This provides the stiffness and damping for the stay cables and provides an aesthetically pleasing appearance with minimum maintenance expected over the 120-

year design life. The hot, humid marine conditions would be corrosive to lower alloyed stainless steels, as described previously, and painted carbon steel would require too frequent and difficult maintenance. Figure 11.8 shows a view of the bridge.

Alloy 2205 has also been used in smaller bridge applications in marine locations in, for example, Singapore, Spain, and Australia[18]. Figure 11.9 shows the Marina Bay Bridge in Singapore, constructed entirely of 2205 and exposed to an aggressive marine atmosphere.

Type 316L stainless steel has been widely used for instrument tubing offshore and there have been widespread failures due to external pitting and crevice corrosion at clamps[17]. It is more severe where seawater spray can also contact the tubing. Okerimi and Martin-Thomas report numerous failures of 316L tubing on the starboard side of a Floating Production and Storage (FPSO) vessel, which faced the prevailing wind. There were no failures on the port side[19]. The location was in tropical waters with high humidity and frequent rain showers. Presumably, chlorides on the tubing were not being washed off by the rain showers because of sheltered locations. Openly exposed surfaces—especially those facing skyward—can collect more salts but they may also benefit from washing off if rainfall is frequent at that location. In comparison, ground facing or sheltered surfaces may collect less salt, but susceptible materials could exhibit more corrosion in the long-term because they are much less likely to be washed off by rain. Failures of 316L have also been reported in other locations, such as the Gulf of Mexico and the North Sea. The solution to this can be 316L with a polymeric coating, but this must be removed where the tube enters a fitting. At this area, water/condensation can enter the junction and potentially lead to crevice corrosion. Duplex 2205 would perform better than 316L under marine atmospheric conditions, but could also suffer crevice corrosion if regularly splashed with seawater (see Chapter 5). Superduplex 2507 tubing has been used to solve instrument tube corrosion problems in both the North Sea and the Gulf of Mexico with excellent results.

One of the aspects of restoration of the Statue of Liberty in the mid-1980s was selection of an appropriate, more corrosion-resistant material to replace the original, severely corroded iron armature bars due to galvanic corrosion associated with the mating copper skin and saddle brackets plus self-corrosion of iron. Results of the candidate replacement materials tested at Kure Beach[20]: 316L stainless steel, duplex stainless steel (UNS S32550), 70/30 copper-nickel, aluminum bronze (C61300), and carbon steel control, were reported by Phull et al. Ultimately, type 316L and duplex stainless steel were selected for their corrosion resistance, galvanic compatibility with the copper skin and saddle brackets, and easier fabrication. Type 316L was used for the armature bars (easier fabrication) and duplex stainless steel for support stiffeners (higher strength)[21].

Carbon steel corrosion in aggressive atmospheres typically produces voluminous corrosion products—sometimes referred to as "packout" rusting or "pillowing." The voluminous corrosion products can induce sufficiently high internal stresses between components that can lead to disbondment of coatings, overload, and fracture of fasteners at lap joints, spalling of steel reinforced concrete, and so forth. In contrast, stainless steels are immune to such failures even if they suffer from pitting, crevice corrosion, or rust staining. In other words, attack on stainless steels does not generate voluminous corrosion products. However, susceptible grades of stainless steels can fail by stress corrosion cracking (SCC), under specific environmental conditions, but SCC is very different from the aforementioned voluminous corrosion product overload mechanism that afflicts carbon steels, and duplex stainless steels are very resistant to SCC, as described in Chapter 6.

FIGURE 11.8 Picture of the Stonecutters Bridge in Hong Kong.
[Photo courtesy of Arup]

FIGURE 11.9 Marina Bay Bridge in Singapore constructed entirely of 2205 duplex stainless steel [Photo courtesy of Outokumpu Stainless AB]

REFERENCES

1. E. Mattsson, *Basic Corrosion Technology for Scientists and Engineers*, 2nd ed. (London, United Kingdom: Institute of Materials, Mining and Minerals, 1996): p. 80.
2. L.L. Shreir, R.A. Jarman, and G.T. Burstein, *Corrosion*, 3rd ed. (London, United Kingdom: Butterworth Heinemann, 1994): p. 2:33.
3. ISO 9223, "Corrosion of Metals and Alloys—Corrosivity of Atmospheres—Classification, Determination and Estimation" (Geneva, Switzerland: International Standards Organization).
4. ISO 9226, "Corrosion of Metals and Alloys—Corrosivity of Atmospheres—Determination of Corrosion Rate of Standard Specimens for the Evaluation of Corrosivity" (Geneva, Switzerland: International Standards Organization).
5. J.D. Fritz, J.F. Grubb, B.W. Parks, R.M. Kain, and S.J. Pikul, "The Influence of Iron Contamination on the Corrosion Resistance of Several Stainless Steels," CORROSION 2000, paper no. 00457 (Houston, TX: NACE International, 2000).
6. C. Lojewski, P. Boillot, and J. Peultier, "Atmospheric Corrosion Resistance of Duplex Stainless Steels: Results of a Field Exposure Program," Duplex 2010, Session IV.B.5 (Zutphen, Netherlands: KCI Publishing, 2010).
7. ASTM D610, "Standard Practice for Evaluating Degree of Rusting on Painted Steel Surfaces" (West Conshohocken, PA: ASTM International).
8. S. Le Manchet and D. Paul, "Atmospheric Corrosion Resistance of Stainless Steels for Architecture," CORROSION 2017, paper no. 8880 (Houston, TX: NACE International, 2017).
9. S.H. Mameng, R. Pettersson, L. Wegrelius, and C. Leygraf, "Effect of Surface Finish on the Atmospheric Corrosion of Duplex Grade UNS S32205; Results of a Field Exposure Program in Dubai," CORROSION 2016, paper no. 7231 (Houston, TX: NACE International, 2016).
10. S.H. Mameng and L. Wegrelius, "Atmospheric Corrosion Resistance of Stainless Steel in the Middle East: Result of a Field Exposure Program in Various Environments," CORROSION 2017, paper no. 8964 (Houston, TX: NACE International, 2017).
11. Directorate General for Research and Innovation, "Duplex Stainless Steel in Storage Tank (DUPLEX-TANK)," Final Report, DOI: 10.2777/49448 (Luxembourg: European Commission, 2013).
12. JIS G 0595: "Rating Method of Rust Stain of Atmospheric Corrosion of Stainless Steel" (Tokyo, Japan: Japanese Industrial Standard, 2004).
13. J. Gunnarsson, A. Bergquist, B. Sederholm, and J. Almqvist, "Five Years Exposure of Austenitic and Duplex Stainless Steels in Swedish Road Environments," EUROCORR 2009, European Federation of Corrosion (London, UK: Institute of Materials, Mining and Minerals, 2009).
14. G. Sussex and A. Gouch, "Don't Let Salt Assault Your Stainless Surface," 18th International Corrosion Congress 2011, paper no. 488 (Victoria, Australia: Australasian Corrosion Association, 2011).
15. C. Houska, "Cleaning and Maintaining Stainless Steel," www.ssnews.com (March 16, 2017).
16. E. Alfonsson, "Lean Duplex—The First Decade of Service Experience," Duplex Stainless Steel 2010 (Zutphen, Netherlands: KCI Publishing, 2010).
17. T. Havn, "Lean Duplex Stainless Steels for Structural and Piping Applications," Stainless Steel World America 2008 (Zutphen, Netherlands: KCI Publishing, 2008).
18. www.outokumpu.com/en/stainless-news (Helsinki, Finland: Outokumpu, 2017).
19. A. Okerimi and M. Simon-Thomas, "External Pitting and Crevice Corrosion of 316L Instrument Tubing in Marine Environments and Proposed Solution," CORROSION 2008, paper no. 08254 (Houston, TX: NACE International, 2008).
20. B.S. Phull, R.M. Kain, and S.J. Pikul, "Long-Term Marine Atmosphere Corrosion Testing Experiences," CORROSION 2007, paper no. 07237 (Houston, TX: NACE International, 2007).
21. C. Houska, "Duplex Stainless Steel Revolutionizes Structural Design," Accessible at https://www.constructionspecifier.com/duplex-stainless-steel-revolutionizes-structural-design/3 (Ontario, Canada: CSI, 2015).

12 Galvanic Corrosion

In the past, galvanic corrosion has caused numerous failures of equipment in service, some of them very costly. This chapter describes the causes of galvanic corrosion, the main factors affecting its severity, and some preventative measures that can be adopted. In the vast majority of cases, duplex stainless steels will be the cathode, but the discussion below describes ways that the effect on corrosion can be minimized. This chapter describes three common environments: seawater, fresh water, and the atmosphere. For further information, particularly in complex environments such as acids and alkalis, where the cathode can corrode as well as the anode, the reader is referred to reference 1.

12.1 INTRODUCTION

When a metal is immersed in a conducting liquid, it takes up an electrode potential (also known as the corrosion potential). This is determined by the equilibrium between the anodic and cathodic reactions occurring on the surface and it is usually measured with reference to a standard electrode, such as the saturated calomel electrode (SCE).

Galvanic corrosion occurs when two metals with different potentials are in electrical contact, while immersed in an electrically conducting, corrosive liquid. Because the metals have different natural potentials in the liquid, electrons will flow from the anode (more electronegative) metal to the cathode (more electropositive) in order to equalize the potentials. This will increase the corrosion on the anode (Figure 12.1). This additional corrosion is galvanic corrosion, which is also known as bimetallic corrosion, dissimilar metal corrosion, or contact corrosion. These terms are used interchangeably throughout this chapter.

In general, the reactions that occur under bimetallic corrosion are similar to those that would occur on a single, uncoupled metal, but the rate of attack is increased, sometimes dramatically. With some metal combinations, the change in the electrode potential in the couple compared with the uncoupled potential can induce corrosion that would not have occurred in the uncoupled state, e.g., pitting. In some environments the change in potential of the cathode in the couple, compared to the uncoupled state, can also introduce problems, e.g., hydrogen embrittlement.

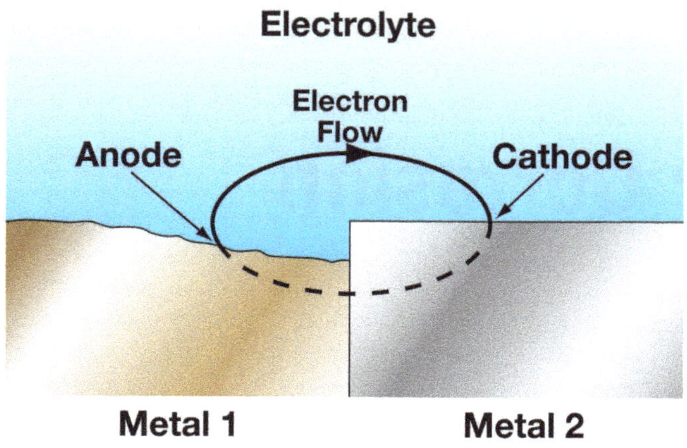

FIGURE 12.1 Schematic diagram of a galvanic corrosion cell.

The effect of coupling two metals together increases the corrosion rate of the anode and reduces, or even suppresses, corrosion of the cathode. Hence, coupling a component to a sacrificial anode can prevent corrosion and this is the principle of cathodic protection.

When a metal is corroding, two processes are occurring. One is the dissolution of metal at the anode, e.g., iron:

$$Fe \rightarrow Fe^{2+} + 2e$$

This must be balanced by a cathodic reaction. Most practical cases of galvanic corrosion occur in solutions containing dissolved oxygen and the primary cathodic reaction is the reduction of dissolved oxygen:

$$O_2 + 2H_2O + 4e \rightarrow 4OH^-$$

Under a few circumstances, the cathodic reaction can be the reduction of hydrogen ions to hydrogen gas:

$$2H^+ + 2e \rightarrow H_2$$

Under uncoupled corrosion conditions, the anodic and cathodic reactions occur at small, local areas on the metal surface. In a bimetallic couple, the cathodic reaction is mostly, or totally, on the electropositive member of the couple and the anodic reaction is mostly, or totally, on the electronegative member of the couple.

12.2 CONDITIONS NECESSARY FOR GALVANIC CORROSION

Here are the basic requirements necessary to cause bimetallic corrosion:

1. An electrolyte bridging the two metals–which may not be aggressive to the individual metals when they are not coupled–and may be in the form of bulk volume of solution, a condensed film, or a damp solid such as soil, salt deposits, or corrosion products.

2. Electrical connection between the two metals. This usually arises from direct physical contact, but it can also arise where electrical continuity is established between two metals, for example, by an insulation-coated conductor, by structural metal work or electrical grounding systems. It is not necessary for the metal junction to be immersed in the electrolyte.
3. A sufficient difference in potential between the two metals to provide a significant galvanic current.
4. A sustained cathodic reaction on the more noble (electropositive) of the two metals by one of the mechanisms described above; in most practical situations, this is the consumption of dissolved oxygen.

12.3 FACTORS AFFECTING GALVANIC CORROSION

There are many factors that affect galvanic corrosion including potential, cathodic efficiency, area ratio, and the nature of the electrolyte. The cathodic efficiency and the area ratio are usually the two most important.

12.3.1 Electrode Potential

The potential, as defined at the beginning of this chapter, is determined by many factors, but the only environment for which this is really well documented is seawater. Unfortunately the potential, even in a single liquid, is not a fixed value, but can vary with changes in temperature, liquid flow rate, the level of aeration, etc. However, the relative ranking of alloys remains largely unaffected by such changes even though the potential may change by several hundred millivolts for some alloys. Figure 12.2 shows the electrochemical series in natural seawater at 10°C, a typical temperature in the North Sea. The difference between the most electropositive and the most electronegative metals is nearly 2 V, and a couple between these materials generates high currents and, hence, high corrosion rates on the anode.

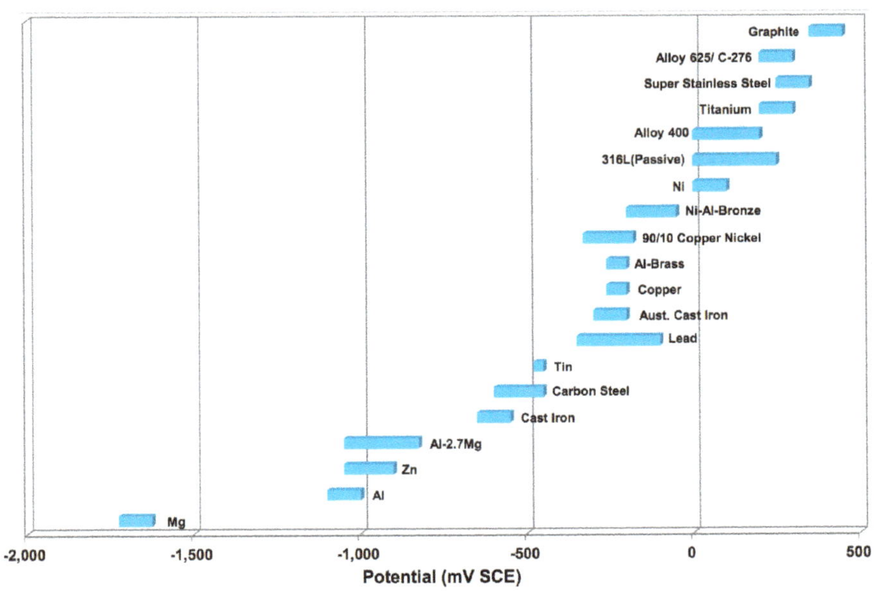

FIGURE 12.2 Potentials of some common metals in seawater at 10°C. [Data courtesy of SINTEF and the author.]

However, the magnitude of the potential difference alone will not necessarily indicate the amount of bimetallic corrosion. For instance, metals with a potential difference of only 50 mV have shown severe bimetallic corrosion problems, while metals with a potential difference of 800 mV have been successfully coupled together. This is because the potential difference gives no information about the kinetics of bimetallic corrosion, which depends on the current flowing between the two metals in the couple.

Figure 12.2 shows the electrochemical series for seawater at 10°C, but the relative position of the metals is very similar for many, near neutral, aerated solutions. Hence, this series has applications outside its initial, anticipated area. It can be seen that stainless steels are in the upper potential regime, and thus are usually the cathodes in a galvanic couple.

12.3.2 Potential Variability

Changes in the corrosion potential of individual metals can result in polarity changes for some couples. A change in potential can be caused by a change in the kinetics of the anodic and/or cathodic reaction or a change in the nature of the reaction e.g., a change from the reduction of dissolved oxygen to the reduction of hydrogen ions as the cathodic reaction. Potential changes can be caused by complexing ions, changes in pH, temperature, or velocity[2], and also by intense aeration, when it enhances film formation on the, hitherto, anodic metal.

Changing the temperature of the seawater can have a significant effect on the potential of some alloys. Figure 12.3 shows the galvanic series for seawater at 40°C. Some metals, such as aluminum, nickel-chrome-molybdenum alloys, high-alloy stainless steels, and titanium show significant changes in potential with temperature, while metals such as zinc and 90/10 copper-nickel (UNS C70600) show only small changes. The reasons for this are mostly connected with the nature of the surface film. For the most electropositive alloys the potential decrease with temperature is due to the absence of a biofilm at 40°C, which forms in natural seawater at ambient temperature. This is described in more detail below.

The high-alloy metals (such as superduplex stainless steels, 6% Mo austenitic stainless steels, high nickel alloys, such as alloy 625 [UNS N06625] and alloy C-276 [N10276]), and titanium can exhibit a wide range of potentials in seawater as shown in Figure 12.4. In natural seawater, at ambient temperature, a biofilm forms over a period of a few days to a few weeks. This film catalyzes the cathodic reduction of dissolved oxygen, giving potentials in the range +250 to +350 mV SCE. Where chlorine or hypochlorite is added to control fouling, small quantities (0.1–0.2 mg/L) prevent the formation of a biofilm and potentials are typically 0 to +100 mV SCE. With higher levels of chlorination (typically 0.5–1.0 mg/L free chlorine) the potential increases to +500 to +700 mV SCE and the reduction of hypochlorite ions to chloride ions has become the main cathodic reaction.

When natural seawater is heated from 25–30°C above ambient temperature, the biofilm ceases to form[3] and the potential is then typically about +100 mV SCE, i.e., "hot" seawater in Figure 12.4. Thierry et al.[4] reported the potentials of a range of stainless steels and copper alloys in natural seawater at 30° and 70°C and showed that all the stainless steels have a similar, lower potential at 70°C and there is hardly any change for the copper alloys.

The above discussion is for fully aerated seawater, but if the dissolved oxygen content decreases, then the potential becomes more electronegative, as shown in Figure 12.4 for stainless steel e.g., –100 mV SCE with 200 ppb dissolved oxygen. In fully deaerated seawater the potential further decreases to about –450 mV SCE. When these high-alloy materials are passive, they draw only a very small current and dissolved oxygen

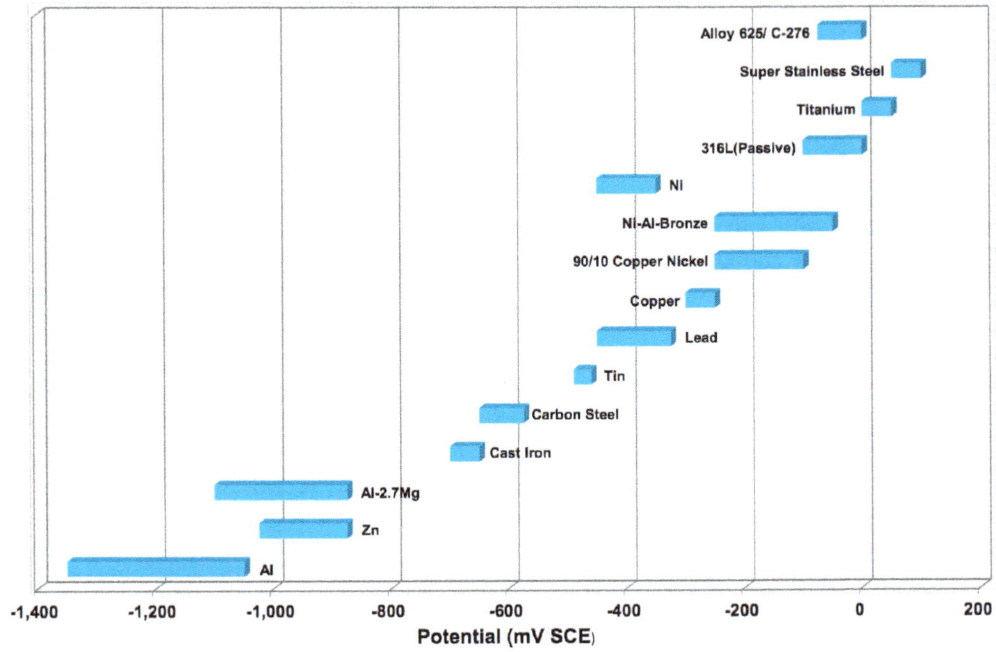

FIGURE 12.3 Potentials of some common metals in seawater at 40°C. [Data courtesy of SINTEF and the author.]

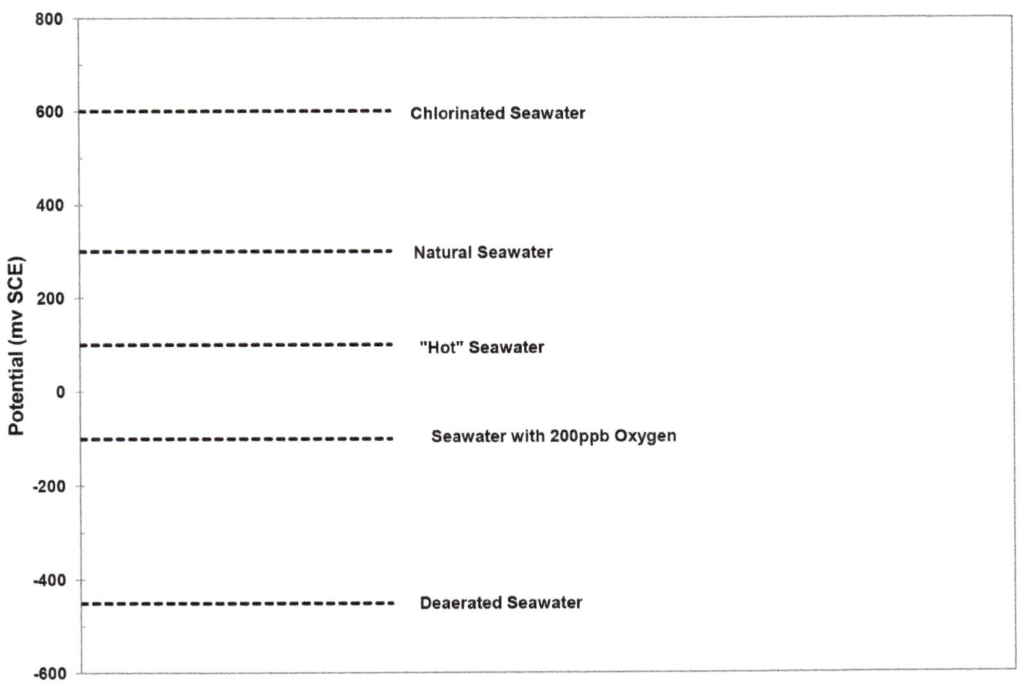

FIGURE 12.4 Potentials of high alloy stainless steels in seawater under a variety of conditions.

contents as low as 20–50 ppb can still give relatively positive potentials. Thierry et al.[4] reported the potentials of some stainless steels and copper alloys over a range of oxygen contents from <10 ppb to fully aerated. All the alloys showed lower potentials at low oxygen contents, but the change was much larger for stainless steels (Figure 12.5).

12.3.3 Electrode Efficiency

When a metal forms part of a galvanic couple, the rate of the anodic or cathodic reaction is not the same for all metals. This is because the kinetics of the reaction vary from metal to metal. The rate of the reaction at the metal surface determines its efficiency as an anode or cathode. In galvanic corrosion, a change of efficiency in the cathodic reaction usually has a more significant effect than one in the anodic reaction.

Some metals, such as titanium, are not very efficient at reducing dissolved oxygen compared with, say, copper alloys. So it is possible for a metal, such as carbon steel, to corrode more when coupled to a copper alloy than to titanium, despite titanium being much more electropositive than copper. All copper alloys are efficient cathodes, including the nickel-copper alloys—alloys 400 and K-500. Stainless steels tend to be less efficient cathodes compared to copper alloys, and may be similar to, or slightly more efficient than, titanium depending on the liquid.

The cathodic efficiency can also change under different conditions. An example is high-alloyed stainless steel in seawater. Figure 12.6 shows the cathodic polarization curves for a 6% Mo stainless steel under three conditions[5]. Curve A shows the behavior in natural seawater after the biofilm has formed. The rest potential is about +300 mV SCE, but under cathodic polarization very large currents are produced for relatively small changes in potential. Hence, the stainless steel with the biofilm is an efficient cathode and corrosion rates of the anode in the couple can be very high. Superduplex stainless steel, titanium, and the nickel-chromium-molybdenum alloys behave in a similar manner in this environment.

Curve B shows the behavior in chlorinated seawater (>0.3 mg/L chlorine). The cathodic reaction is now mainly the reduction of hypochlorite ions to chloride instead of the reduction of dissolved oxygen. Stainless steel in chlorinated seawater is not such an efficient cathode as it is with the biofilm and hence, lower currents are produced at any given potential compared with natural (i.e., unchlorinated) seawater, despite the more electropositive rest potential.

Curve C shows the behavior in seawater when no biofilm forms. The cathodic reaction is the reduction of dissolved oxygen, but without the biofilm, the reaction on the stainless steel is very slow and hence, even lower currents are produced at any given potential than those that occur in chlorinated seawater.

An example that demonstrates this is the use of superduplex, 6% Mo stainless steel or titanium tubes in heat exchangers with nickel-aluminum-bronze (NAB) tube plates. Sometimes, galvanic corrosion of the tube plates has been severe, as shown in Figure 12.7; however, there have been other cases where there has been little acceleration of corrosion of the NAB. The author believes that this is due to chlorination of the seawater. At plants where little or no chlorination has been carried out, the cathodic reaction would not be polarized and severe corrosion could occur. At stations with low-level continuous chlorination, the cathodic efficiency would be much reduced, as shown in Figure 12.6, and corrosion would be less likely.

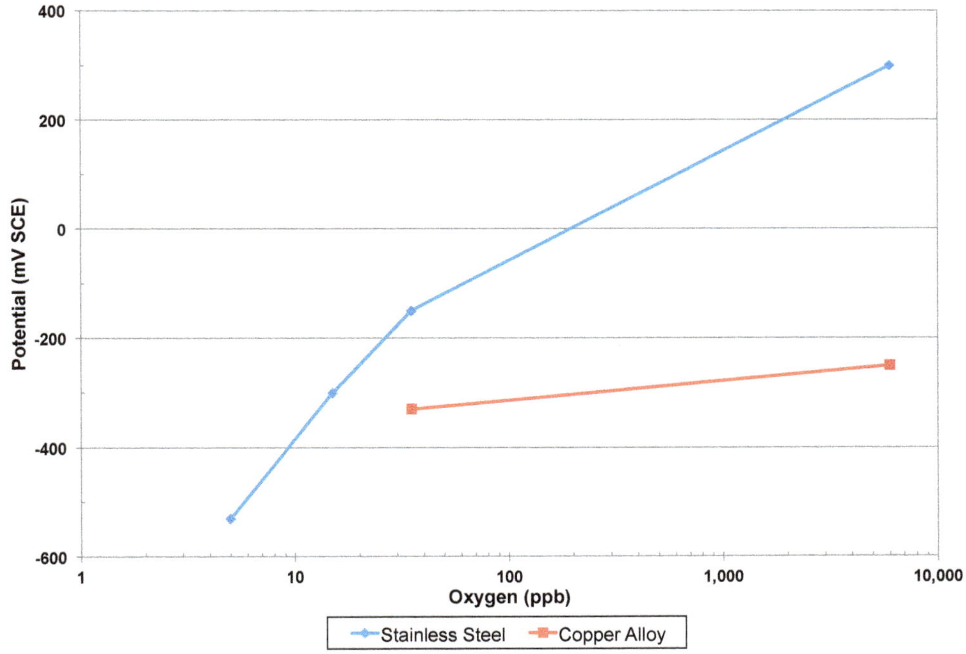

FIGURE 12.5 The potential of stainless steel and copper alloys in seawater as a function of oxygen content[4].

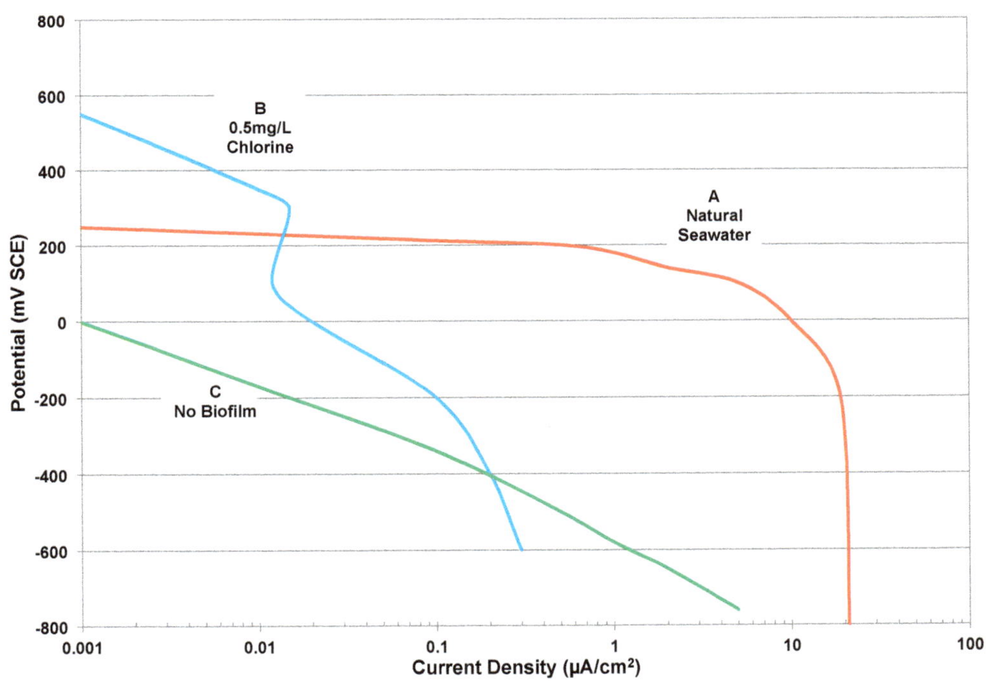

FIGURE 12.6 Cathodic polarization curves for high alloy stainless steels in three types of seawater[5].

FIGURE 12.7 Galvanic corrosion of a NAB tube plate with high alloy stainless steel tubes.

12.3.4 Electrolyte

Electrolyte factors that have a major influence on bimetallic corrosion are composition, pH, and in particular, ionic conductivity, which affects both the intensity and the distribution of corrosion.

The severity of corrosion often increases with increasing ionic conductivity of the electrolyte because in practice, high conductivity is often caused by the presence of aggressive ions such as chloride, or by acid or alkali.

The ionic conductivity of electrolytes can vary widely; typical values are:

Fresh distilled water	0.5–2 μS/cm
Stored distilled water	2–4 μS/cm
Supply water	50–1,500 μS/cm
Sea water	40,000 μS/cm
Saturated sodium chloride solution	250,000 μS/cm
Sulfuric acid	up to 800,000 μS/cm
Molten salts	20,000–10,000,000 μS/cm

Except for very critical items of equipment, such as electronic components, and where it is essential to suppress metal dissolution, bimetallic corrosion is seldom a problem when the conductivity is very low, such as in pure water.

In a solution of low conductivity, or when only a thin film of a highly conducting electrolyte covers the metals, corrosion is confined to an area near the junction between the two metals. While the total amount of corrosion may be low, it is likely to be highly localized and, therefore, may be intense. In a highly conducting solution, corrosion will still be concentrated at the metal junction, but it will be more widespread than in a solution of low conductivity. Intense corrosion is likely to arise even with nominally pure water under condensing conditions if the atmosphere is polluted with acid gases, such as sulfur dioxide, or with salt, e.g., at industrial or sea coast locations.

In common with local cell corrosion, bimetallic corrosion is sensitive to the presence of constituents in the electrolyte that affect stability of the corroded metal ions. In some cases, insoluble corrosion products from the anodic metal can deposit on, and induce crevice corrosion of, the cathodic member of a couple. For instance, corrosion products from an electronegative alloy, such as steel or aluminum, could be deposited on a stainless steel cathode and induce crevice corrosion due to the establishment of an oxygen ion concentration cell between the metal under the deposit and the freely exposed metal. It is also possible to modify the composition of the environment by adding inhibitors to control bimetallic corrosion.

As mentioned above, in some instances the cathodic reaction can become hydrogen evolution. In seawater, this becomes significant at potentials of –850 mV SCE and more electronegative potentials. This means that alloys that are susceptible to hydrogen embrittlement may crack when connected to anodes that produce these low potentials. Duplex stainless steels are susceptible to hydrogen embrittlement and care needs to be taken under situations where hydrogen can be produced to avoid cracking (Chapter 6).

12.3.5 Area Ratio

In a couple, the ratio of the exposed areas of anode-to-cathode materials is very important when considering the likelihood of galvanic corrosion. The larger the cathode compared with the anode, the more oxygen reduction, or other cathodic reaction, can occur and, hence, the greater the galvanic current. This results in more corrosion on the anode.

Adverse area ratios are likely to occur with fasteners and at joints. Weld, braze, or rivet metal should be of the same potential as or, better still, slightly cathodic to the base or parent metal. Duplex stainless steels will usually be the cathodes and so are highly suitable for fasteners or weld filler when coupled to less noble alloys.

The effect of area ratio is not confined to general corrosion, but may also affect other types of corrosion. A failure from a North Sea platform demonstrates this effect. The failure occurred at the junction of a 6% Mo stainless steel pipe and one made of 90/10 copper-nickel. In chlorinated seawater, this would not normally result in any significant acceleration of the corrosion of the 90/10 alloy[1]. However, there was turbulence at the junction due to a protruding gasket and this, combined with the coupling to stainless steel, induced erosion corrosion. This was over a small area and hence, the cathode-to-anode area ratio was very high and resulted in rapid penetration of the copper-nickel[6]. Such an effect would also be expected with a superduplex stainless steel pipe as the cathode.

If the electrical conductance of the electrolyte bridging the bimetallic contact is low, either because the bulk conductivity is low or because the electrolyte is present only as a thin film, the effective areas taking part in bimetallic cell reactions are small and the total amount of corrosion is relatively small, although it may be severe immediately adjacent to the metal junction.

Under immersed conditions in many supply waters, which generally have a relatively low electrical conductivity, adverse effects are uncommon if the contacting metals are of similar area. Thus, stainless steel and copper tubes can usually be joined without problems resulting, but accelerated corrosion of the copper tube is likely to occur if it is attached to a stainless steel tank, as shown in Figure 12.8.

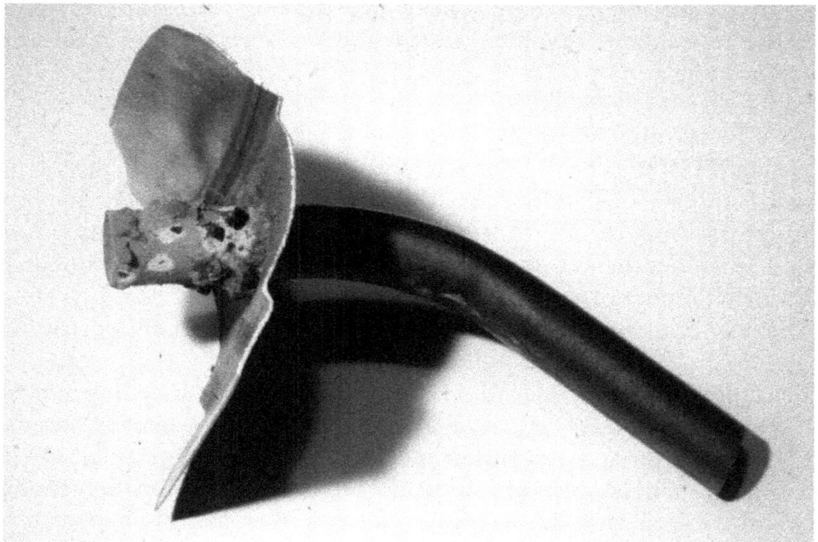

FIGURE 12.8 Severe corrosion of a copper pipe entering a stainless steel hot water cylinder.

Under immersed conditions in a highly conducting electrolyte, such as seawater, effective areas will be greater and severe corrosion may be encountered on small anodic areas of many metals.

By correct selection of area ratio, dissimilar metals have frequently been used successfully together. This is done by making sure that the less noble anode is large, and the more-noble cathode is small. Note that it is not the absolute area of the components that is important, but the wetted area.

12.3.6 Aeration and Flow Rate

The majority of practical situations involving galvanic corrosion arise in aqueous solutions under conditions in which the cathodic reaction is the reduction of dissolved oxygen. As with single metal corrosion, bimetallic corrosion is, therefore, partly dependent on the rate at which oxygen can diffuse to the cathodic surface from the bulk electrolyte.

The galvanic corrosion rate of stainless steels in seawater depends on the flow rate of the water as well as the area ratio. In well-aerated, flowing solutions stainless steels are likely to become passive and corrode less than under stagnant conditions.

In neutral electrolytes complete deaeration will, in many instances, suppress single metal and bimetallic corrosion. However, under such anaerobic conditions, cathodic depolarization and corrosion can occur if sulfate-reducing bacteria (SRB) are present, i.e., an alternative reaction to the reduction of dissolved oxygen is available. In closed, recirculating systems, deaeration is extremely practical and very low, dissolved oxygen concentrations can be maintained with chemical oxygen scavengers. Theoretically, the corrosion of iron and steel in a closed recirculating system should consume the dissolved oxygen fairly quickly. In practice there are often places where small quantities of air can be drawn into the system so corrosion inhibitors and oxygen scavengers become necessary.

In once-through cooling systems, it is difficult to control oxygen levels and 100% deaeration is rarely achieved. In land-based installations, such as multistage flash desalination plants, deaeration to <20 ppb is routinely achieved. However, on offshore oil platforms, where space and weight are at a premium, deaeration of injection water rarely gets below 50 ppb dissolved oxygen and levels of 100 ppb or 200 ppb are not uncommon. Under these conditions corrosion rates are significant for some metals, such as carbon steel, and bimetallic corrosion is also a possibility if coupled to stainless steel.

12.3.7 Metallurgical Condition and Composition

In some cases differences in corrosion potential can exist between metals or alloys of nominally the same composition. Subjection to cold working often tends to make an alloy more anodic. In some alloys heat treatment can produce galvanic differences. With duplex stainless steels, this will not occur unless third phases are precipitated.

In multiphase alloys, it is common for one phase to be anodic to another. Normally these differences are small and do not give rise to problems, but in some instances severe bimetallic corrosion can occur. Correctly-solution treated duplex alloys should have no significant potential difference between the austenite and ferrite, but if the element partitioning between the phases is not at equilibrium, because of an inadequate heat treatment, then significant potential differences might occur in some liquids.

Local changes in composition can also arise at joints made by welding, both in the weld bead and in the heat-affected zone (HAZ) of the parent metal. Problems are usually avoided by selecting appropriate filler metals and welding techniques (see Chapter 3). Because the weld bead has a very small area compared with the parent metal, it is important that it has the same potential as the parent metal or, preferably, is slightly electropositive. Even a reduction of 50 mV in the potential of the weld metal compared to the parent may cause severe galvanic corrosion in some fluids.

12.3.8 Bimetallic Corrosion Without Physical Contact

Localized corrosion of a metal may give rise to soluble corrosion products, which deposit by a displacement reaction onto a less noble metal exposed in the same solution, and form local, intense bimetallic cells. This is particularly evident with the more electropositive metals such as copper, and problems have been encountered in water heating systems as a result of copper dissolving from pipework and depositing onto steel radiators, aluminum alloy pump impellers, and galvanized steel tanks. Insulation of joints between dissimilar metal components is not necessarily a safeguard if cupro-solvent water flows or diffuses from copper to components in susceptible metals. An example is the filling of aluminum saucepans from the hot water tap in an area where the water is slightly cupro-solvent (i.e., a low pH and/or a high free carbon dioxide content). If the copper forms deposits on the aluminum, it can result in localized corrosion that may eventually lead to perforation.

Slightly different is where corrosion products deposit and cause unexpected corrosion. Francis and Byrne described the corrosion of some superduplex stainless steel sprinklers in an offshore firewater system[7]. The original sprinklers were nickel aluminium bronze (NAB) in a superduplex piping system and they corroded rapidly because of galvanic corrosion. However, replacement superduplex sprinkler heads suffered from crevice corrosion in the screwed joint, and this was because copper had precipitated on the pipe adjacent to the heads during corrosion of the NAB. This acted as an efficient cathode under the conditions in the piping and stimulated crevice corrosion in the threads of the superduplex sprinklers[7].

12.3.9 Stifling Effects

In general, the rate of corrosion of coupled and noncoupled metals decreases with time of exposure. This is partly due to the diminishing rate of diffusion of oxygen (or whatever the cathodic reactant) through the electrolyte and through films of corrosion product to cathodic regions, and partly due to the protection afforded to anodic regions by the corrosion product. To some extent, therefore, the bimetallic corrosion rate is affected by the permeability of the corrosion product. In seawater and, to a lesser extent, in hard supply waters, the alkaline conditions produced at the cathode result in the formation of calcareous deposits of calcium and magnesium carbonates, and magnesium hydroxide, which may decrease the rate of bimetallic corrosion. With steel, for example, corrosion is decreased by one order of magnitude after less than a year of exposure. Where corrosion products are water-permeable or hygroscopic, however, their presence can sometimes increase the rate of bimetallic corrosion, either because they are more aggressive than the bulk environment, or because they maintain wet conditions at the metal surface.

12.4 ALLOY GROUPS

The galvanic compatibility of materials in seawater can be divided into four major groups, according to their corrosion behavior, numbered from one to four. Table 12.1 shows the alloy groups and the common alloys found in each class. Group 1 includes all the passive alloys, which by and large, do not usually suffer corrosion in seawater at around ambient temperature. Some definitions are required to identify alloy classes. The nickel-chrome-molybdenum alloy class covers all the nickel alloys with more than 7 wt% molybdenum, such as alloy 625 and alloy C-276. Titanium covers not only grade 2 (commercially pure) but also the common alloys such as Ti-6Al-4V and Ti-0.15% Pd. Super austenitic stainless steels covers all those alloys with a molybdenum content equal to 6 wt% or greater and having a PREN >40 (for which PREN = %Cr + 3.3(%Mo + 0.5x%W) + 16x%N). The superduplex stainless steels covers all duplex alloys with 25% or more chromium and a PREN >40.

TABLE 12.1 Alloy groupings for seawater at ambient temperature.

CATEGORY	TYPE	ALLOY
1	Noble; passive	Nickel-Chrome-Moly Alloys (Mo>7%) 6% Mo Austenitic Stainless Steel Superduplex Stainless Steel Titanium and its alloys
2	Passive; not truly corrosion resistant	Alloy 400/ K-500 904L 22% Cr Duplex Alloy 825 Alloy 20 316L
3	Moderate Corrosion Resistance	Copper Alloys Austenitic Cast Iron
4	Poor Corrosion Resistance	Carbon Steel Cast Iron
		Aluminum Alloys

Group 2 are the passive alloys that are not totally immune to localized corrosion (e.g., crevice corrosion) in seawater. Such alloys are mostly stainless steels and iron-nickel-chromium-molybdenum alloys with PREN values below 40 (e.g., type 316L stainless steel, 2205, and alloy 825). Other metals that fall into this group are the nickel-copper alloys, the best known of which are alloys 400 and K-500. These can suffer crevice corrosion in seawater[8, 9]. The common Group 2 alloys are shown in Table 12.1.

Group 3 covers alloys of moderate corrosion resistance that do not show true passivity. The most common ones are copper and its alloys, and the austenitic cast irons. These cast irons are more noble than conventional cast irons because of their high nickel content.

Group 4 covers alloys with poor corrosion resistance; it has been subdivided into irons and steels, and aluminum alloys. The reason for this is that there can be a substantial potential difference between aluminum alloys and iron and steel in seawater, and so these alloy types need to be treated separately, as shown in Table 12.1.

The principal reason for grouping the alloys as shown in Table 12.1 is that there is ample evidence that bimetallic couples between any of the alloys in a particular group will not generally give rise to any significant bimetallic corrosion. This does not necessarily mean that no corrosion will occur, but no significant additional corrosion will occur. There are a few exceptions to this rule, but they are not common, and even then, with a suitable area ratio, bimetallic corrosion need not be a problem. Hence, when there is any doubt about materials compatibility, the selection of an alloy from the same group will prevent problems. There are obviously cases, in which alloys from different groups can be connected together safely, and some of these are discussed in the following sections, but when in doubt, selecting alloys from the same group is a good rule of thumb.

12.5 GALVANIC CORROSION IN SEAWATER

Superduplex stainless steels are in Group 1 in Table 12.1 and are generally the cathode in many galvanic couples.

Walleen and Andersson[8] examined couples of 6% Mo austenitic stainless steel and 90/10 copper-nickel exposed in natural seawater for 1 year. Although the corrosion rate increases with area of stainless steel, the corrosion only becomes significant at area ratios greater than 1:1. However, in chlorinated seawater systems, the biofilm does not form and the stainless steel is a less efficient cathode. This means that area ratios of 10:1 or more can be tolerated without a significant increase in the corrosion of the copper-nickel[6]. This was demonstrated on a Norwegian offshore platform, where a 90/10 copper-nickel column pipe from a NAB pump was joined to a 6% Mo super-austenitic stainless steel ring main. The seawater had a chlorine residual of ~0.7 mg/L, and after several years of operation there was no significant acceleration of corrosion of the copper-nickel at the junction[10]. This would also be the case with superduplex stainless steel.

Two-phase brasses, such as muntz metal and rolled naval brass are often used for heat exchanger tube plates with copper alloy tubes. The dominant form of corrosion is dezincification and this is countered by making the tube plates thick (35 mm or more). Coupling to other copper alloys does not significantly change the rate of dezincification. However, coupling to 6% Mo super-austenitic or superduplex stainless steel or titanium[11-13], showed increases in the depth of dezincification of from 10–30 times. Hence, an alternative tube plate material or cathodic protection is required in such cases.

Gehring et al.[11-13] carried out studies on 6% Mo stainless steel, and titanium coupled to aluminum bronze, alloy D, in seawater, with an area ratio of stainless steel to bronze of 400:1, as a function of salinity, temperature, and pollution level. The results showed that the corrosion rate of aluminum-bronze (alloy D) is much less than that of muntz metal, particularly at low water temperatures. However, at temperatures of about 20°C or more, the rate of attack of aluminium bronze was unacceptable. As with muntz metal, the solution is to make the tube plate very thick or to protect it cathodically, either by impressed current or by the attachment of soft iron or mild steel anodes.

The author investigated NAB/SS couples when nickel-aluminum-bronze valves were being considered for a superduplex stainless steel seawater piping system[14]. Panels of superduplex and cast NAB were bolted together and exposed in natural seawater or chlorinated seawater (0.7 mg/L chlorine) at about 23°C. Two area ratios were investigated, i.e., stainless steel to NAB of 1:1 and 10:1. The main problem was localized attack. In natural seawater this took the form of both crevice corrosion and pitting on open surfaces. The pitting was along the kappa III phase and hence, was deep, up to ~0.5 mm after 60 days. In chlorinated seawater, there was no open surface pitting, and crevice corrosion only occurred at an area ratio of 10:1.

These results suggested that NAB valves would corrode rapidly in an unchlorinated stainless steel system, and even in a chlorinated system some crevice corrosion might occur. The service experience is that severe corrosion of NAB valves in superduplex piping systems occurred on both the Woodside Goodwyn A and BP Andrew platforms. The valves are usually thick wall castings so the corrosion manifests as the valve being unable to seal properly rather than through-wall penetration. The seawater on Goodwyn A was chlorinated, but the seawater temperature varied from 20–40°C, i.e., higher than in the author's tests. The seawater on BP Andrew was unchlorinated at about 10°C, and severe corrosion occurred within one year. The main valves were replaced in superduplex stainless steel and the others were changed as they failed. Chlorination was also retrofitted because it slows down the rate of attack, because the stainless steel is a less efficient cathode, as discussed above.

The Eastern Trough Area Project (ETAP) in the North Sea also used superduplex stainless steel seawater piping and NAB valves, but with chlorination from the outset. The valves also corroded, but more slowly than on Andrew, and the valves were changed for superduplex stainless steel after about five years. Figure 12.9 shows a corroded NAB butterfly valve from a chlorinated superduplex stainless steel seawater piping system. Some operators have made NAB valves in superduplex stainless steel piping systems work by using internally coated stainless steel spool pieces on either side of each valve, as described later in this chapter. However, the spool pieces need to be five or six pipe diameters long, so there must be space to incorporate them.

Group 1 alloys are not always the cathode in galvanic couples, and corrosion has been reported with superduplex and 6% Mo austenitic stainless steel flanges using graphite-loaded gaskets in chlorinated seawater systems[15-17]. The solution is to use synthetic fiber or neoprene gaskets, and some oil companies have made these mandatory for gaskets in low-pressure stainless steel seawater systems. Figure 12.10 shows corrosion of a Z100 superduplex flange fitted with graphite gaskets.

There is another instance when corrosion of Group 1 alloys can occur in a galvanic couple. This is when they are coupled to a Group 2 alloy that is undergoing crevice corrosion and the solution in the crevice becomes so acidic that it depassivates the Group 1 alloy and causes it to corrode. Shone et al. reported corrosion of superduplex piping under 316L straps holding an electrical cable in place, and which was regularly splashed with seawater[18]. Walleen and Andersson reported attack of a 6% Mo austenitic alloy in a crevice with alloy 400 (UNS N04400) in seawater[8].

FIGURE 12.9 Corroded NAB valve coupled to superduplex stainless steel piping in a seawater system.

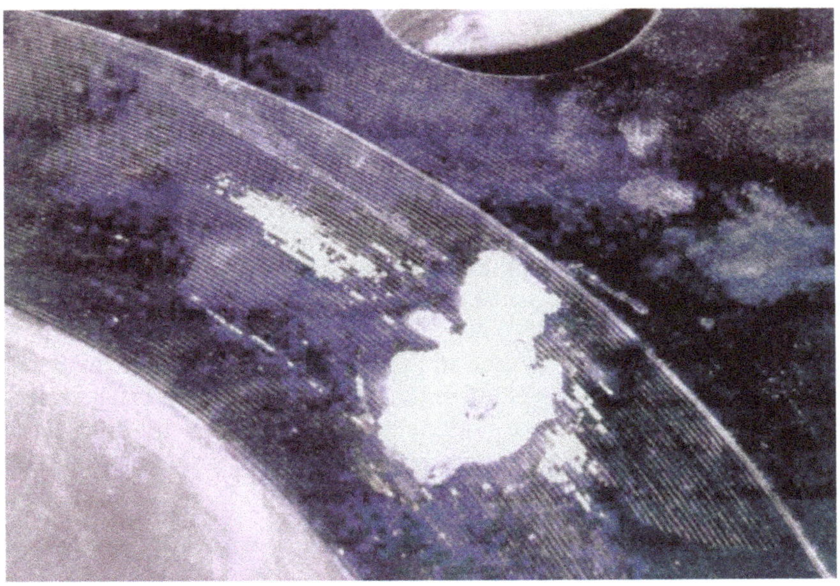

FIGURE 12.10 Corrosion of a superduplex stainless steel flange used with a graphite gasket in a chlorinated seawater system.

Although Group 2 alloys are susceptible to crevice corrosion in seawater, coupling to other alloys need not produce much more attack. Figure 12.11 shows the depth of attack on 2205 duplex coupled to various Group 1 and Group 2 alloys[19]. Although titanium, 6% Mo austenitic, and alloy 625 are all Group 1 alloys, the depth of attack varied, with little additional attack coupled to titanium, and extensive attack coupled to 6% Mo austenitic. This is probably more to do with the amount of biofilm that had formed on these alloys, because in a parallel test they coupled 316L to the same Group 1 alloys and the most severe attack was

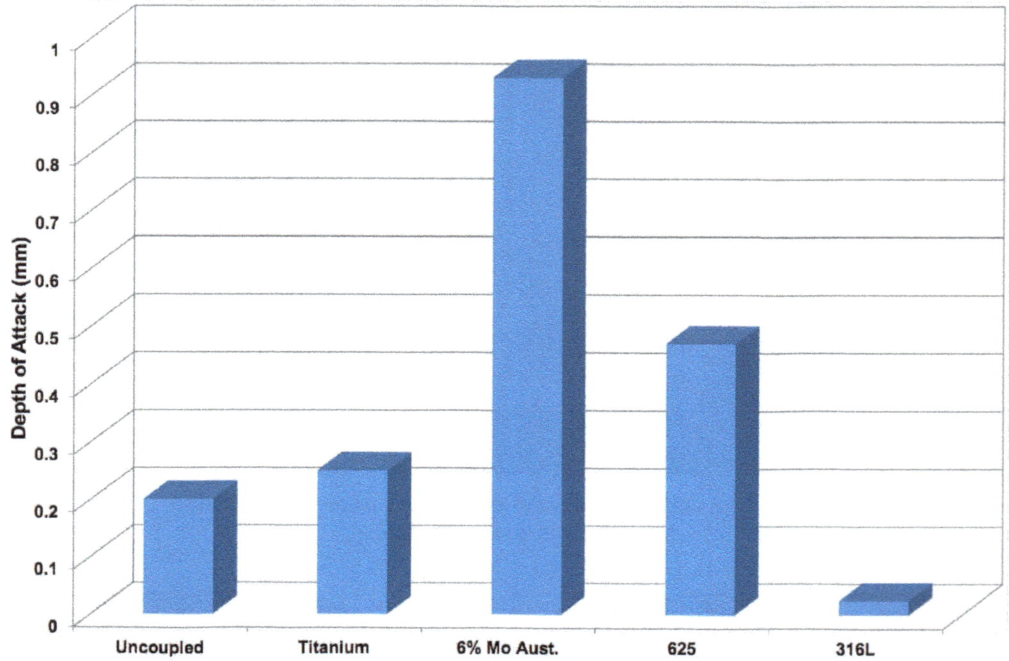

FIGURE 12.11 Depth of crevice corrosion of 2205 in seawater coupled to various alloys (area ratio = 1:1)[19].

seen coupled to titanium, while the couple with 6% Mo austenitic showed the least. It can also be seen that coupling to 316L greatly reduced the attack on 2205, presumably because the 316L corrosion initiated first, delaying the attack of 2205, or causing it because of the acidity in the crevice, as described above.

Alloy 2205 is sometimes used for the tubes in tube-and-shell heat exchangers in seawater or brackish water. It is made to work because the tubes are seal welded to the tube sheets of 316L or 2304, and these are protected by mild steel anodes (carbon ≤ 0.1%). Because there are no crevices on the 2205, this can work provided the anodes are correctly sized to provide adequate protection between maintenance intervals, and the water is chlorinated to avoid fouling.

12.6 GALVANIC CORROSION IN FRESH WATERS

There are two main differences between fresh waters and seawater, apart from the much lower total dissolved solids content: conductivity and hardness. The conductivity of a typical fresh water is about two orders of magnitude less than that of seawater. This means that bimetallic corrosion tends to be confined to the point of contact of the two metals. It also means that galvanic corrosion is often less severe than in seawater, unless the area ratios are very different or the potential difference is very large. The example of copper and stainless steel was discussed above, where copper fittings with stainless steel pipe have been very successful, but a copper pipe entering a stainless steel tank was badly corroded.

Seawater is very hard because of the high magnesium and sulfate contents. Fresh waters are usually subdivided into soft, moderately hard, and hard, for which the measure of hardness is the total hardness, expressed as $CaCO_3$. A soft water is usually from 0–50 mg/L total hardness, a moderately hard water is from 50–150 mg/L, and a hard water has a total hardness greater than 150 mg/L. In fresh waters much of the hardness is often present as calcium bicarbonate, and many hard and moderately hard waters tend to deposit calcium carbonate scale with time. The scale inhibits the access of water to the metal surface, slowing down corrosion. This is the reason that carbon steel, galvanized steel, and cast iron can last for many years in hard waters.

Soft waters do not usually deposit scale, but they also generally have lower dissolved solids contents, including chlorides, so corrosion rates in soft waters are not necessarily much higher than in hard waters. One factor that is important in soft waters is pH. Even if the pH is quite high (e.g., 8.5), there is little buffering action because the bicarbonate content is low and the creation of hydrogen ions, e.g., at a corrosion site, can result in a large local decrease of pH.

Linder and Mattson[20] conducted extensive galvanic corrosion tests both at room temperature (25°C) and at 75°C. Most of the tests were in a water of intermediate hardness (100 mg/L as $CaCO_3$), but additional tests were also carried out in a soft water (hardness = 2.5 mg/L) and a hard water (hardness = 245 mg/L). During these tests they investigated both coupled and uncoupled potentials, as well as galvanic currents for a range of materials commonly used in fresh water systems.

From this data they produced galvanic series for both hot and cold moderately hard water; the one for hot water is shown in Figure 12.12. The heading "stainless steel" covers the four types they tested, i.e., 13% Cr and 18% Cr ferritic, and types 304 and 316 austenitic stainless steels (UNS S30400 and S31600). The lean du-

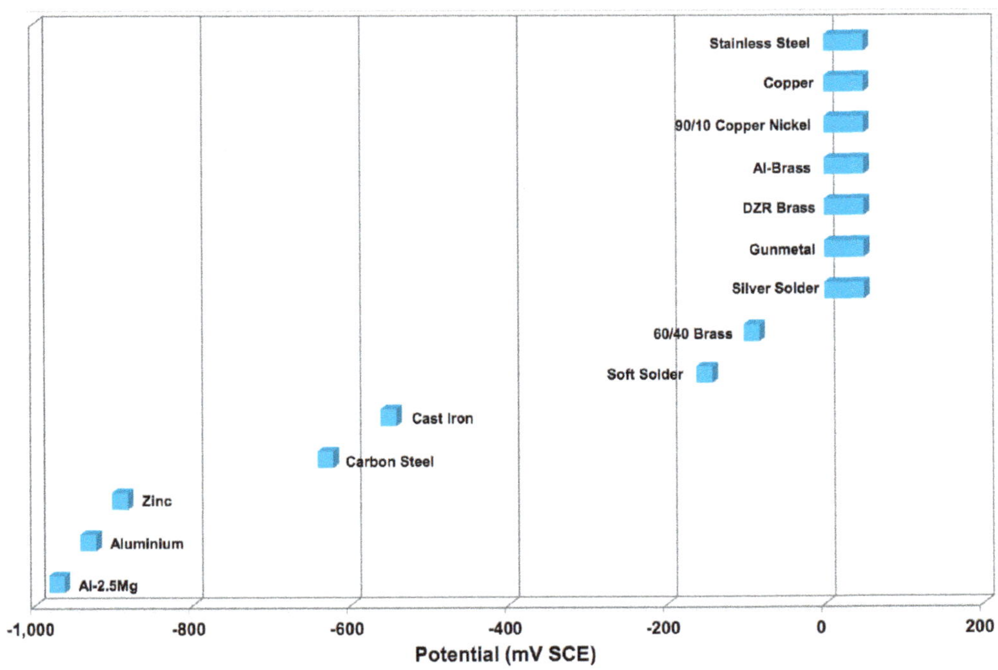

FIGURE 12.12 Galvanic series for some common metals in a moderately hard fresh water at 70°C[20].

plex alloys, such as 2101 and 2304, are regularly used in fresh water systems for items such as hot water tanks and perform similarly to 304 and 316. There were no significant differences between these stainless steels, but there were no tight crevices on the samples and the maximum chloride content was only 75 mg/L, with most of the data generated at 11 mg/L chloride. Higher chloride levels (>100 mg/L) occur in some fresh waters and these might cause crevice corrosion with these alloys, particularly at higher temperatures. The silver solder tested was a copper-silver-phosphorus type, while the soft solder was a tin-silver alloy recommended for potable waters. The names of the other alloys are self-explanatory.

Both series show similar potentials for the stainless steels and most of the copper alloys, with 60/40 brass and the soft solder slightly more electronegative. The order of the other alloys is identical to that in seawater at 25°C, although the potential values are different. However, at 75°C, the relative positions of zinc and aluminum were reversed.

The corrosion currents that were measured did not always reflect the position of the alloys in the two series. From the corrosion data, the metals have been arranged into groups in a similar manner to that done for seawater in Table 12.1. The different groups are shown in Table 12.2 for the moderately hard water at 25°C.

Table 12.2 Grouping of compatible materials in moderately hard fresh water at 25°C[20].

GROUP	TYPE	ALLOYS
1	Passive	Austenitic Stainless Steels (304, 316) Ferritic Stainless Steels (13% Cr, 18% Cr)
2	Good Corrosion Resistance	Copper and its Alloys Silver Solder
		Soft Solder (Sn-Ag)
3	Moderate Corrosion Resistance	Cast Iron Carbon Steel
4	Low Corrosion Resistance	Aluminum Zinc

Although Figure 12.12 shows similar potentials for copper alloys and the stainless steels, it is well known that copper can suffer severe galvanic corrosion when coupled to large areas of stainless steel, as described above. Hence, the copper alloys have been put into a separate group, while acknowledging that under the right circumstances they can be coupled to stainless steel. Linder and Mattsson observed some galvanic currents for the soft solder coupled to copper, although service experience shows few corrosion problems with solders at copper joints. Hence, soft solder has been put into a subsection of Group 2, meaning that problems may arise in some circumstances.

Cast iron and carbon steel have been grouped together, because they show similar potentials. When coupled together at 25°C, a small amount of galvanic corrosion occurred on the cast iron.

Zinc and aluminum have been grouped together because they behaved similarly when coupled to other metals. The potential of aluminum was more electropositive than that of zinc, but when the alloys were coupled together it was the aluminum that became the anode, although the current was small. The reason for this is believed to be that in the couple, the presence of Zn^{2+} ions modifies the potential of the aluminum, moving it electronegatively.

Duplex stainless steels will act as the cathode when coupled to other metals in Table 12.2, particularly if the anode/cathode area ratio is adverse (small anode, large cathode). However, application of one of the methods of prevention in Section 12.8 will prevent most galvanic corrosion problems. Austenitic stainless steels 304L and 316L are often used for freshwater piping and these may be safely coupled to duplex stainless steel tanks.

12.7 GALVANIC CORROSION IN THE ATMOSPHERE

The factors that affect atmospheric corrosion were discussed in Chapter 11, as was the corrosion of duplex stainless steels in various atmospheres. Only in the more aggressive atmospheres, mostly marine, will corrosion of duplex alloys occur, mostly as pitting. In most galvanic couples in the atmosphere, the duplex stainless steels will act as cathodes, although the severity of attack will depend on the couple alloy and the aggressiveness of the atmosphere. Couples with other stainless steels will not usually cause significant galvanic corrosion, unless the atmosphere is very aggressive to only one alloy in the couple.

Kucera and Mattsson[21] published bimetallic corrosion data for a variety of atmospheres. In addition, they made an attempt to categorize the corrosion in terms of engineering risk. Three types of atmosphere are reported: rural, urban, and marine. Some metals suffered little corrosion in any of the environments, such as stainless steel. However, while the corrosion rate of some metals is low in rural atmospheres, it is much greater in urban and/or marine ones, and this could be increased in a galvanic couple with a duplex stainless steel.

Kucera and Mattsson[21] tested many metals in various combinations and the results for marine atmospheres are summarized in Table 12.3. The data from urban and rural areas is similar to that in Table 12.3 and may be found in references 1 and 22. The aim has been to show couples with no significant change in corrosion rate from uncoupled, a slight increase in corrosion rate, and a large increase in corrosion rate. An increase in corrosion rate may still not be much loss of metal and may be acceptable for a particular application. There are no absolutes here and some metal combinations may be acceptable in one case and not in another depending on the criticality of the application. Chromium is included in Table 12.3 because of its widespread use as electroplating in atmospheric applications. It can be seen that apart from magnesium there is only a slight increase in corrosion of most metals coupled to stainless steel. Note that tropical atmospheres are more aggressive and the risk of galvanic corrosion will increase somewhat[1].

TABLE 12.3 The risk of galvanic corrosion in a marine atmosphere for some common metals[21].

CORRODING METAL	COUPLED METAL: Aluminum	Anodized Al	Zinc	Carbon steel	Weathering Steel	Tin	Lead	Copper	Nickel	Stainless Steel	Chromium
Magnesium	No risk	No risk	No risk	Large increase	No data	Some increase	Large increase	Large increase	Large increase	Large increase	Some increase
Aluminum	No risk	No risk	No risk	Large increase	Large increase	Some increase	Large increase	Large increase	Some increase	Some increase	Some increase
Anodized Aluminum	No data	No risk	No data	Large increase	No data	Some increase	Some increase	Large increase	Some increase	Some increase	No risk
Zinc	No risk	Some increase	No risk	Large increase	No data	Large increase	Large increase	Large increase	Large increase	Some increase	Some increase
Carbon Steel	No risk	Some increase	No risk	No risk	No risk	Some increase	No risk	Some increase	Some increase	Some increase	No risk
Weathering Steel	No data	No data	No data	No data	No risk	No data	No data	No data	No data	Some increase	No data
Tin	No risk	No data	No risk	No risk	No data	No risk	No data	Some increase	Some increase	No risk	No risk
Lead	No risk	No risk	No risk	Large increase	No data	Some increase	No risk	Some increase	Some increase	Some increase	Large increase
Copper	No risk	No risk	No risk	No risk	No data	No risk	No risk	No risk	No risk	No risk	No risk
Nickel	No risk	No risk	No risk	No risk	No data	No risk	No risk	No risk	No risk	Some increase	Some increase
Stainless Steel	No risk	No risk	No risk	No risk	No data	No risk	No risk	No risk	No risk	No risk	No risk

- (green) No risk of galvanic corrosion
- (yellow) Some increase in corrosion
- (red) Large increase in corrosion
- (white) No data

It might be thought that coupling copper or one of its alloys to stainless steel may result in accelerated corrosion of the copper. This is not so because the effective area ratio is usually 1:1 and stainless steel is not a very efficient cathode. This was why stainless steel was used to replace corroded steel internal supports of the copper skin on the Statue of Liberty[22] (see Chapter 11).

12.8 METHODS OF PREVENTION

The following sections review the main methods of preventing/combating galvanic corrosion. To do this properly it is necessary to know as much as possible about the environmental conditions. The more information that is available, the more confidence can be placed in the decisions made to mitigate galvanic corrosion.

12.8.1 Materials

Where the design is being completed before any materials are purchased, there is the maximum opportunity to select galvanically compatible metals for the specific application. The following guidelines offer some helpful suggestions.

Use compatible metals wherever possible. Try to select alloys from the same group, as shown in Tables 12.1 and 12.2. Avoid metals that are far apart in the electrochemical series, unless it is well documented that they can be coupled successfully e.g., 90/10 copper-nickel and high-alloy stainless steel in chlorinated seawater.

When welding, pay particular attention to the condition of the weld after fabrication. It is desirable that it be cathodic to the parent metal. When carrying out autogenous welding, choose welding parameters that will achieve this. When using filler metal, select a filler that produces weld metal that has a similar potential or is electropositive to the parent metal. With such a high cathode-to-anode area ratio as occurs in a weld, a weld metal potential only 50 mV negative to that of the parent metal can result in severe corrosion in some fluids[1].

Where compatible materials cannot be used, for whatever reason, one of the following methods of prevention of galvanic corrosion should be considered.

12.8.2 Insulation and Separation

If the anode and cathode can be prevented from coming into electrical contact, then no galvanic corrosion will occur. In some situations this can be done quite conveniently, while in others it is impractical.

Several methods for achieving this when bolting two flat surfaces together were discussed by both Francis[1], and Parkins and Chandler[23], depending on whether it is the bolts or one of the plates that is the anode.

Another method of preventing galvanic corrosion between two adjacent metals could involve the use of a thick coat of nonconducting, insoluble mastic at a joint. The mastic may additionally contain chemicals to inhibit corrosion of the metals as they are leached out.

One method for preventing galvanic corrosion of piping at dissimilar metal joints is the isolating flange. It uses a nonconducting insert between the two metal flanges and nonconducting sleeves and washers for the bolts. This means the two metals are no longer in electrical contact. However, there are two major disadvantages to this technique. First, the sleeves and washers have to be undamaged and correctly assembled for the isolation flange to work properly. On many industrial sites, sufficiently skilled labor is not available and this is difficult to ensure. Second, there is frequently a requirement by the electrical engineers to ground all major items of metal equipment for safety reasons. This has the effect of shorting across the isolation flange and, unfortunately, it is a common industrial practice[24]. Pipe hangers or supports can also act as inadvertent electrical shorting between adjacent pipe runs. Where isolating flanges are used it is important that the resistance between the two components be measured prior to start up, to ensure that isolation has been achieved.

An alternative method of separating the anode and cathode when one is a pipe and the other is a pipe, vessel, valve, etc., is to use an insulated spool piece. This is a length of pipe between the two items that is either internally coated or is made from a nonmetallic, e.g., fiberglass-reinforced plastic (FRP). Note that the coating must extend up onto the flange faces as well as along the pipe bore. The metal pipe being coated must be made of the same metal as that forming the cathode so that defects in the coating do not give rise to rapid localized corrosion. If a coating is being used, it should have a life consistent with the planned maintenance intervals for the plant. A coating that needs replacing at more frequent intervals is often not cost effective and an alternative solution to the galvanic corrosion problem may be required.

The length of the spool piece depends on the electrochemical properties of the two metals and the particular fluid under consideration. An example for seawater serves to demonstrate this principle. When joining a copper alloy (such as 90/10 copper-nickel) to a Group 1 alloy, it is common to use a spool piece approximately six pipe diameters long, providing there is no undue turbulence to cause localized erosion-corrosion of the copper alloy. BP Tankers had an oil tanker with 90/10 copper-nickel piping and a titanium plate heat exchanger. Galvanic corrosion occurred on the copper-nickel adjacent to the heat exchanger, which was solved by fitting a coated titanium spool piece five diameters long. The length of the spool was limited by the available space in the ship, but it prevented further corrosion of the copper-nickel pipe.

Bardal et al.[6] showed that with high alloy stainless steel and 90/10 copper-nickel in chlorinated seawater a spool piece is unnecessary. However, if the chlorination were likely to be turned off at all, a spool piece would be a wise precaution to prevent corrosion at the dissimilar metal joint. In natural seawater, the corrosion rate of 90/10 copper-nickel was ~1.2 mm/y at the junction with a stainless steel pipe.

12.8.3 Painting/Coatings

When trying to prevent a bimetallic corrosion problem it is sometimes possible to use paints or coatings on the affected area. Preferably *both* the anode and cathode should be coated, but it is often sufficient just to coat the cathode. A small defect in the coating will still mean that only a very tiny cathode area is exposed to the seawater. On no account should only the anode be coated. The cathode-to-anode area ratio will be very high at defects in the coating and the rate of corrosion at the anode will usually increase dramatically.

It is important to select coatings that have a life similar to that of the components they are protecting, especially if access is limited or difficult. Otherwise, regular maintenance of the coating to prevent galvanic corrosion will be necessary as the coating breaks down over time.

An alternative method to organic coatings is metallic coatings. The most commonly applied metals are aluminum and zinc, although many other metals can be used as coatings. The most common, and cheapest, method is thermal spraying. However, this produces porous coatings and, for maximum life, polymeric sealers are required after coating. An alternative is to use high velocity oxy fuel (HVOF) spraying, which produces very dense coatings with no need for sealing, although this process is a little more expensive.

Note that where a metallic coating is porous and no sealer is used, it is essential that the coating is anodic to the base metal, otherwise accelerated corrosion will occur at the defects in the coating because of the high cathode to anode area ratio. An example of a poor choice of coating is chromium plating on steel, which is unsuitable in seawater.

For the case of atmospheric corrosion *only*, there are some additional coating measures that can be considered. These all rely on the galvanic corrosion being confined to the joint area with an anode-to-cathode area ratio of approximately 1:1. One solution is to use a nonconducting mastic at the junction, which is effectively increasing the resistance between the anode and cathode. A similar effect can be obtained with polymeric tapes, which may contain inhibitors that can be leached out. The tape is usually applied up to about 25 mm on either side of the bimetallic joint. Rubber coating has also been used successfully to prevent atmospheric galvanic corrosion.

12.8.4 Cathodic Protection (CP)

This method of protection relies on changing the coupled potential of two dissimilar metals. It is not always necessary to suppress corrosion totally, but only to prevent or reduce the corrosion due to the galvanic couple.

There are two methods that are used: impressed current and sacrificial anodes. With the impressed current, a current sufficient to change the potential to the desired value is applied. With a sacrificial anode, the final potential is usually close to the open circuit potential of the anode so the metal chosen for the anodes is important. Not only must the potential be in the desired range, but the anode must also be of a size and have a life and a cost that are practical.

In heat exchangers with duplex or superduplex tubes, a Group 2 or Group 3 alloy may be used for the tube plate. To prevent this corroding, it will require cathodic protection, but if this is done with zinc or aluminum alloy anodes, the potential will be very low and hydrogen embrittlement of any duplex components is likely (see Chapter 6).

The solution is to use soft iron or mild steel (carbon $\leq 0.1\%$) anodes. The potential is now too electropositive for hydrogen evolution, while it is sufficiently low to prevent accelerated corrosion of the tube plates.

A similar principle can be applied in piping systems by fitting a sacrificial spool piece, usually of carbon steel, five to six pipe diameters long between the two components of the bimetallic couple. In seawater the spool piece requires replacement every one or two years, but if access is easy, this can be a cost effective solution. Remember that iron corrosion products from the spool piece will be carried downstream and this must be acceptable in the rest of the system. The corrosion products usually form soft deposits of hydrated iron oxides, 1 mm or more thick, which may not always be desirable.

12.8.5 Inhibitors

Inhibitors are rarely used in once-through seawater systems, although they may be used in recirculating closed systems or in seawater injection systems in the oil and gas industry. Where inhibitors are used, it is important to select one that is compatible with all the metals in the system. With duplex stainless steels, any inhibitor must work on both the ferrite and austenite.

In Section 12.8.3, above, the use of mastics and tapes to prevent atmospheric galvanic corrosion was discussed. In some cases these can be obtained with inhibitors to provide additional protection against galvanic corrosion. As with fully immersed conditions, it is important that suitable inhibitors are used that are compatible with both metals in the couple being protected.

A slightly different method of inhibition is deaeration, which can usually only be applied to recirculating systems. Provided that hydrogen evolution cannot be the cathodic reaction, the addition of oxygen scavengers to prevent the cathodic reduction of dissolved oxygen is a viable method of preventing corrosion, including galvanic corrosion. It is essential to prevent additional entry of oxygen into the system or else regular dosing with a scavenger will be necessary.

12.8.6 Design

If it is addressed early enough in the design stage, galvanic corrosion can be prevented by one of the methods described above. Once construction has started, changes and retrofits become much more difficult and expensive.

There is a range of good practices that should be adopted to minimize the risk of galvanic corrosion. For structures exposed to the atmosphere, one of these is to prevent water accumulation and the entrapment of debris. Designs that allow water to run off freely pose fewer problems than those that create areas where water can collect[1].

Wherever possible, for fully immersed conditions, a design should be chosen to minimize the area of the cathode and/or maximize the area of the anode. If the cathode-to-anode area ratio is kept sufficiently small, additional methods of restricting galvanic corrosion may not be necessary.

Finally, there is one other factor essential to the implementation of good design–the understanding of the workforce carrying out the construction, fabrication, etc. If they are told the importance of what they are doing and why it must be done one particular way then it is more likely to be implemented correctly. Any design to prevent galvanic corrosion must include communication with the workforce of the necessary precautions required during assembly.

REFERENCES

1. R. Francis, *Galvanic Corrosion–A Practical Guide for Engineers*, 2nd edition (Houston, TX: NACE International, 2017).
2. T.S. Lee, E.W. Thiele, and J.H. Waldorf, "The Effect of Seawater Velocity on Corrosion Potentials of Metals," *MP*, 23, 11 (1984): p. 44.
3. G. Alabiso, U. Montini, A. Mollica, M. Beggiato, V. Scotto, G. Marcenaro, and R. Dellepiane, "Marine Corrosion Tests on Metals in Antartica: Preliminary Results, in Marine Corrosion of Stainless Steels—Chlorination and Microbial Effects," EFC Publication No. 10, European Federation of Corrosion (London, UK: Institute of Materials, Mining and Minerals, 1993): p. 36.
4. D. Thierry, C. Leballeur, and N. Larche, "Gavanic Series in Seawater as a Function of Temperature, Oxygen Content and Chlorination," CORROSION 2016, paper no. 7058 (Houston, TX: NACE International, 2016).
5. T. Rogne and U. Steinsmo, "Practical Consequences of the Biofilm in Natural Sea Water and of Chlorination on the Corrosion Behaviour of Stainless Steels, in Sea Water Corrosion of Stainless Steels—Mechanisms and Experiences," EFC Publication No. 19, European Federation of Corrosion (London, UK: Institute of Materials, Mining and Minerals, 1996): p. 55.
6. E. Bardal, R. Johnsen, and P.O. Gartland, "Prediction of Galvanic Corrosion Rates and Distribution by Means of Calculation and Experimental Models," *CORROSION*, 40 (1984): p. 12.
7. R. Francis and G. Byrne, "The Effect of Copper on the Crevice Corrosion of Stainless Steels," *MP*, 54, 3 (2015): p. 70.
8. B. Walleen and T. Andersson, "Galvanic Corrosion of Copper Alloys in Contact with a Highly Alloyed Stainless Steel in Sea Water," in Proc. 10th Scandinavian Corrosion Congress (Stockholm, Sweden: Swedish Corrosion Institute, 1986): p. 149.

9. R. Francis, *Corrosion of Copper and Its Alloys: A Practical Guide for Engineers* (Houston, TX: NACE International, 2010).
10. O. Strandmyr and O. Hagerup, "Field Experience with Stainless Steel Materials in Seawater Systems," CORROSION/98, paper no. 707 (Houston, TX: NACE International, 1998).
11. G.A. Gehring and J.R. Maurer, "Galvanic Corrosion of Selected Tubesheet/Tube Couples Under Simulated Sea Water Condenser Conditions," CORROSION/81, paper no. 202 (Houston, TX: NACE International, 1981).
12. G.A. Gehring and R.J. Kyle, "Galvanic Corrosion in Steam Surface Condensers Tubed with Either Stainless Steel or Titanium," CORROSION/82, paper no. 60 (Houston, TX, USA: NACE International, 1982).
13. G.A. Gehring, "Inlet End Corrosion Problems in Condensers," in Proc. 7th Annual Conference on Materials for Coal Conversion and Utilization (Gaithersburg, MD: National Institute of Standards and Technology, 1982).
14. R. Francis, "Coupling of Superduplex Stainless Steel and Cast Nickel-Aluminum Bronze in Sea Water," *British Corrosion J.*, 34, 2 (1999): p. 139.
15. R. Lye, "Materials Selection Philosophy for Offshore Production Systems," in Proc. Engineering Solutions to Industrial Corrosion Problems, Sandefjord, Norway (Houston, TX: NACE International, 1993).
16. R. Francis, "Experiences with High Alloyed Duplex Stainless Steels in Seawater Systems," in Seawater Corrosion of Stainless Steels—Mechanisms and Experiences, EFC Publication No. 19, European Federation of Corrosion (London, UK: Institute of Materials, Mining and Minerals, 1996): p. 133.
17. O. Strandmyr and O. Hagerup, "Field Experience with Stainless Steel Materials in Sea Water Systems," CORROSION/98, paper no. 707 (Houston, TX: NACE International, 1998).
18. E.B. Shone, R.E. Malpas, and P. Gallagher, "Stainless Steels as Replacement Materials for Copper Alloys in Seawater Handling Systems," *Trans. Inst. Mar. Eng.* 100 (1988): p. 193.
19. E.B. Shone and P. Gallagher, "Galvanic Compatibility of Selected High Alloy Stainless Steels in Seawater," in *Corrosion in Seawater Systems*, ed., A.D. Mercer (London, United Kingdom: Ellis Horwood, 1990): p. 40.
20. M. Linder and E. Mattsson, "Galvanic Corrosion of Combinations of Metals in Fresh Water," in Proc. 7th Scandinavian Corrosion Conference (Oslo, Norway: Royal Norwegian Council for Scientific and Industrial Research, 1975): p. 19.
21. V. Kucera and E. Mattsson, *Atmospheric Corrosion of Bimetallic Structures, in Atmospheric Corrosion*, ed., W.H. Ailor (New York, NY: J. Wiley and Sons, 1982).
22. R. Baboain and E.B. Kliver, *The Statue of Liberty Restoration* (Houston, TX: NACE International, 1990). [See also: *MP*, 35 (1996): p. 5 for a summary.]
23. R. Parkins and M. Chandler, *Corrosion Control in Engineering Design* (Norwich: United Kingdom: HMSO, 1978).
24. R. Johnsen and S. Olsen, "Experience with the Use of S31254 in Seawater Systems: Case Histories from the Field," CORROSION/92, paper no. 397 (Houston, TX: NACE International, 1992).

13 Standards, Quality, and Procurement

Modern duplex stainless steels have been widely used in industry since the 1980s, generally with excellent results when used within the known operating limits of the particular grade. In recent years there have been an increasing number of failures of material because of quality issues (e.g., references 1 to 4). In the 1980s and 1990s and a little beyond, major companies developed their own specifications for corrosion resistant alloys (CRA) to ensure that material supplied would be adequate. However, since the economic crash of 2008, these companies have tended to move QA/QC oversight and expediting lower down the supply chain. Design houses have tended to follow suit and responsibility for material quality has often been moved to equipment suppliers. These companies do not generally have a metallurgist on the staff to review specifications, and there is often little oversight from higher up the supply chain.

Hence, specifications have tended to carry over from job to job without a proper appraisal of what is actually required. There has been a concomitant reliance on standards, such as ASTM, as discussed in the next section. This does not just apply to duplex stainless steels, but to all CRAs. The result has been that CRAs have been supplied to some projects that have been of poor quality, resulting in rapid failure. This has been at great cost to industry. The following sections discuss the use of standards and specifications and the minimum requirements to guarantee good quality duplex stainless steel is supplied.

13.1 STANDARDS

When procuring metals and alloys it is common to use ASTM specifications, such as ASTM A240 for stainless steel plate and sheet. However, these are standards that are largely written by material producers to achieve minimum acceptable requirements. These standards tend to focus on the composition and mechanical properties and may make a brief mention about suitable heat treatment, but do not go into detail about how to achieve the best properties. With the worldwide increase in the number of duplex stainless steel vendors, and increasing project desire to minimize costs, it is essential that additional testing–over and above that required by ASTM–is written into the specification to ensure the supply of good quality material. This principle has been adopted by NORSOK[5], the Norwegian oil and gas standard, and many companies use NORSOK, or incorporate its requirements into their company specifications. ISO 17781 was issued in 2017 and it has adopted or modified NORSOK test methods for duplex stainless steels[6].

One of the most critical stages in the production of duplex stainless steels is the heat treatment. This is not just the temperature, but is also the time at temperature, the heating rate, the transfer time to the quench tank, and the speed of the quench. None of this is covered in ASTM standards and it is critical to obtaining the optimum properties. Francis and Hebdon described this in detail, and also showed the limiting maximum thickness possible with duplex stainless steels[7]. Duplex stainless steels can be produced satisfactorily up to 250 mm thick by some manufacturers, and 300 mm thick sections have been produced by the use of special techniques[7]. Sections thicker than 300 mm cannot be quenched fast enough to avoid the formation of third phases using current best practice. A couple of examples to illustrate this follow in the next paragraphs.

In the first case, some heavy section forgings of superduplex stainless steel were found to give only 6J Charpy impact energy at -46°C and the microstructure was said to be two-phase[7]. The 250 mm thick forging had only been heat treated at 1120°C (solution anneal temperature) for one hour, with the microstructure shown in Figure 13.1. This two-phase structure is austenite and sigma, (the pale phase is austenite) with virtually full transformation of ferrite to sigma. This is an extreme example, but it shows the importance of good heat treatment.

A second case was the purchase of some superduplex fittings by a major oil company for a seawater cooling system. These were thin walled and failed in a matter of months. The problem was sigma phase due to incorrect heat treatment, as shown in Figure 13.2. The pale phase is austenite, the gray phase is ferrite, and the darker particles are sigma phase. This was a high cost failure because of the shut down for temporary repairs, plus the procurement and installation of replacement fittings.

It is not only sigma phase that can cause failures; in Chapter 6 it was pointed out that nitrides can cause sulfide SCC of duplex stainless steels under conditions defined by NACE MR0175/ISO 15156[8] to be satisfactory.

ASTM A923 was written to introduce testing to avoid problems due to sigma phase[9]. Unfortunately it does not address nitrides and it is not conservative in its pass/fail criteria[10].

13.2 TEST REQUIREMENTS

Because of the problems with duplex and superduplex stainless steels described above, Byrne et al. argued that the requirements of ASTM A923 for duplex stainless steel quality are inadequate and that even NORSOK requirements need strengthening[10]. They suggested the following additional tests for 25% Cr superduplex alloys:

1. Impact toughness ≥70 J average at -50°C.
2. ASTM G48A[11] test for 24 h at 50°C (no pitting and weight loss <4 g/m^2).
3. Microsection at X500 (electrolytically etched in oxalic acid, then KOH or NaOH). No third phases to be present. Ferrite to be from 40–60%

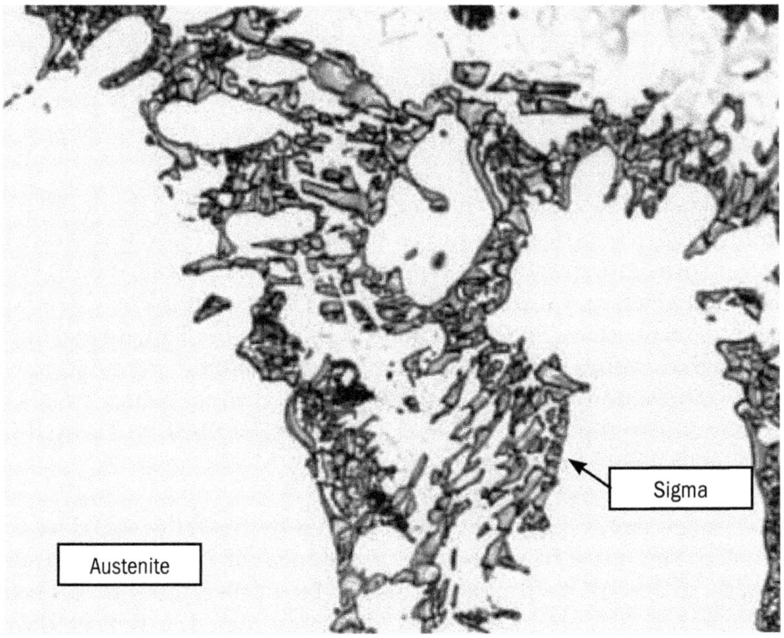

FIGURE 13.1 Duplex microstructure showing almost full transformation of ferrite to sigma phase.

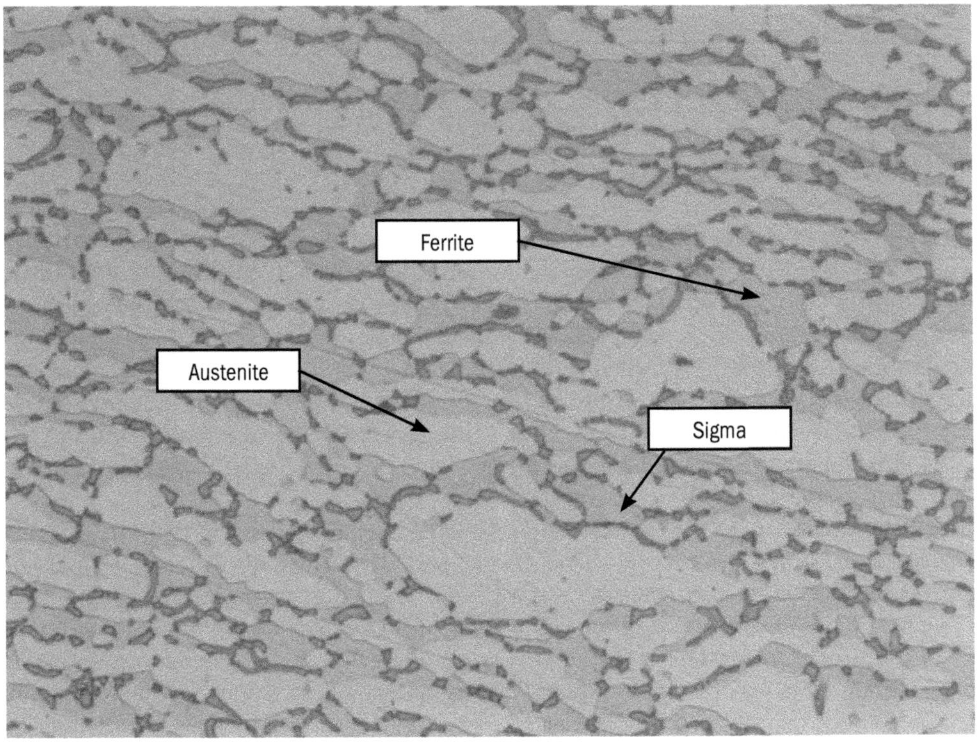

FIGURE 13.2 Microstructure of thin walled superduplex fitting with sigma phase.

It may not be necessary to do all three of these tests, depending on the application. For example, for a seawater application, the corrosion test is probably adequate and the impact toughness test is unnecessary and will not be practicable for thin-walled pipes and fittings.

The requirements for 22% Cr duplex are the same, except the ASTM G48A test temperature is 25°C.

The impact toughness test guarantees that, even after subsequent welding, there will be adequate toughness in the heat-affected zone (HAZ) for most applications. Passing the corrosion test shows that deleterious third phases are not present in sufficient amounts to significantly affect the corrosion resistance. The corrosion test can also detect nitrides, which may cause pitting or just a high weight loss[12]. The weight loss criterion is also useful because it can detect surface chromium depletion due to inadequate pickling after heat treatment and quenching, even though this does not cause pitting. If the chromium denudation is very shallow, pitting does not occur because of the sound metal beneath the denuded zone. A high weight loss may also be due to inadequate scale removal during pickling. It is essential that the finish to be used in service is tested so that the results show the suitability of material as supplied.

The importance of a two-stage etch using oxalic acid followed by NaOH or KOH was discussed in Chapter 2. The microsection enables the phase balance to be determined and can also determine what third phases are present, if any. This is particularly helpful if the material has failed test 1 or 2, because it helps to determine the most appropriate heat treatment to restore the properties. If a duplex alloy contains significant quantities of intermetallic phases, it would be impossible to pass the impact test and/or the corrosion test.

At the time of writing, the test conditions and pass/fail requirements for lean duplex alloys have not been determined.

13.3 IMPORTANCE OF QA/QC

Having specified the tests (described in the previous section), care must still be taken because the certificates may not be true. The author has seen large diameter superduplex bar (420 mm) claiming to meet ASTM properties. However, upon reading the small print on the certificate, or asking the appropriate question, it turns out the test samples were taken from near surface, while ASTM requires them to be at mid-thickness (¼*T*, where *T* is the diameter). Samples from deeper than the surface fail because of the presence of third phases. This emphasizes the need for witnessing of testing–unless the supplier has been fully audited–and the need for QA/QC monitoring throughout. This costs money, but less than that of a major failure. Customers want to lower costs and producers with limited experience may cut corners without realizing the implications.

Many companies rely heavily on NORSOK standards, M-630 for materials and M-650 for vendor approvals[5, 13]. ISO 17782 is currently in development and this is also a vendor approval document[14]. However, just because a vendor claims to be NORSOK approved, be wary. Check when they were last audited, as NORSOK certification lasts five years, but auditing should be done every three or four years, because manufacturers often change things in the production cycle or install new equipment that has not been audited. Auditing can be expensive, but it usually pays for itself when items are needed in a hurry, such as for urgent repairs.

In addition, it must also be checked what the company is approved to supply. This includes both product forms and thickness range. There is usually a good reason if they do not have approval over the full thickness range. It is essential to ensure that any potential supplier is approved for all your material requirements.

Another document that is not as widely used as it should be is EEMUA 218, which gives details on how to obtain good quality duplex stainless steels[15]. Without going into details, its main points follow:

- NORSOK M-650 preferred, but good historic data acceptable
- Audits recommended more frequently than every five years
- UKAS [formerly the United Kingdom Accreditation Service] accredited test laboratories only
- All testing at the center (½*T*)
- All parts of the manufacturing route and heat treatment must be covered in detail
- Surveillance and witnessing during manufacture
- A technical compliance audit required, particularly for new suppliers, *over and above a quality assurance audit*
- Alloy distributors to be treated in the same way as manufacturers, with the same responsibilities
- Positive Material Identification (PMI) and feritscope checks required

In addition to procurement specifications and monitoring during manufacture, it is worthwhile checking material on delivery with some simple nondestructive testing. Hardness tests can sometimes be useful, and Francis et al. described the use of the feritscope to assess the quality of duplex stainless steels[16]. They comment on the factors that affect the readings and show that significant amounts of sigma/chi phase produce a significant reduction in the feritscope reading. It is possible to take a lot of readings very quickly, and looking at the range of the readings suggests when there could be precipitates.

- Readings >40 are OK.
- Readings in the 30s could show problems if inconsistent, and further investigation is required.
- Readings <30 are a strong indication of sigma phase.

Suspect material should then be checked with in situ or destructive metallography, because there are causes other than sigma that can give low feritscope readings, such as poor pickling[16].

The problem fittings in reference 1 passed through five companies–from the manufacturer to the end user–and not one ran a feritscope check, which would have quickly identified the suspect fittings. Francis et al. described a case study where hardness checks on a bar found hard spots and the feritscope confirmed that sigma was present in these areas[16]. This was caused by the use of unsuitable supports in the annealing furnace for the bar, which produced cold spots during heat treatment.

13.4 CASTINGS

The above discussion has principally addressed wrought products, but poor heat treatment can also cause problems with duplex stainless steel castings[17]. The additional tests (impact toughness, ASTM G48[11], and a microsection) described above for wrought products should also be applied to castings. ASTM A995[18] grades 1B and 4A should use the 25°C temperature for the corrosion test, while grades 5A and 6A should use the superduplex temperature of 50°C. Suitable tests for cast lean duplex alloys have yet to be determined.

When procuring duplex castings, it is usual to start with ASTM A995 and add the extra test requirements under the testing options.

When selecting a foundry for duplex stainless steel castings, particularly larger ones, there are some factors to take into account:

- What is their experience? Duplex stainless steels require some skill to cast correctly and it is always best to use a foundry with prior experience.
- Do they have computer modeling? The design of running, gating, and positioning of feeders can affect both the porosity and residual stress in castings. The ability to model this before pouring any metal can save a lot of money and time[17].
- Is the heat treatment furnace adequate? The importance of this is discussed in reference 7.
- Is the quench tank satisfactory? See reference 7.
- Can they do fast transfers? See reference 7.

The author has looked at numerous failures of duplex castings caused by poor quality due to a lack of adequate specifications and a lack of suitable QA/QC.

13.5 WELDING

When duplex stainless steels are put into service it is often necessary to weld them, and some of the most common standards for this are those of ASME International and the American Welding Society (AWS). ASME permits a lot of lea way on thickness from the thickness qualified. With duplex stainless steels it is important to check the welding procedure specification (WPS) and the procedure qualification record (PQR) to ensure they cover the right thickness range. Duplex stainless steels are harder to weld to avoid third phases in thin sections and the qualification thickness should reflect this.

ASME only requires mechanical tests, but for duplex stainless steels these should be supplemented by additional tests, similar to those described above for parent metal. It is also important that these tests are applied both to the PQR and the welder qualifications. It is important that the welder knows what to do to achieve the desired properties.

Below are the tests suggested for superduplex:

1. Impact toughness ≥40J average, 35J minimum single value, at -50°C, in the weld metal, fusion line, fusion line + 2 mm, and fusion line + 5 mm.
2. ASTM G48A test for 24 h at 35°C (no pitting and weight loss <4 g/m^2).
3. Microsection at X500 (electrolytically etched in oxalic acid, then KOH or NaOH). Ferrite to be from 35–65% and no third phases.

Not all of these tests may be required–it depends on the application, as discussed for parent metal above.

The requirements for 22% Cr duplex and CD4MCuN are the same, except the ASTM G48A test temperature is 20°C. Test requirements for lean duplex stainless steels have yet to be determined.

The impact toughness requirements are the same as NORSOK and are perfectly adequate for a duplex weld, although some oil and gas projects may require a lower test temperature. The corrosion test temperature of 35°C for superduplex has been found to be adequate for many applications, such as sour oil and gas fluids.

For demanding oxidizing environments with chlorides, a test temperature of 40°C should be considered. The microsection will help identify any third phases if the weld fails the impact toughness or corrosion tests. It is also important to ensure the ferrite content is not excessive because this can reduce resistance to various types of corrosion, including SCC. If the weld fails any of these tests, the microsection can be used to determine the best modifications to the welding procedure for requalification.

End use will determine the need for specific levels of surface quality, but it must be noted that heat tint from welding will significantly reduce the corrosion resistance of stainless steels in most applications. Tube manufacturers–including seam welded tubing–are moving towards "bright annealing," where the solution annealing and quenching of the finished tube is done under an inert atmosphere to prevent the formation of a surface oxide. If this is not done with precision, a very minor heat tint may result and this can significantly reduce corrosion resistance in the most demanding services. Pickling of the tube (or of the completed heat exchanger) may be necessary. This is recommended especially for sulfuric acid coolers where seawater is the cooling medium on the tube side. Pickled tube has been shown to outperform bright annealed tube in warm seawater service. Surface contamination on stainless alloys can have a marked influence on corrosion resistance, and its removal must be considered whenever equipment is being fabricated. If there is a need to have heat tint removed, the costs for heat tint removal and postweld clean up must be included in cost estimates.

13.6 OVERVIEW

The main points to come out of the discussion in the proceeding sections are as follows:

1. Duplex stainless steels, like all CRAs, require care during manufacture and it is important to purchase to specifications that include all the critical testing to remove the risk of premature failure.
2. It is an advantage to prequalify vendors to ensure that they meet the requisite levels of skill and equipment.
3. It is important that such vendors are re-audited at regular intervals.
4. In addition, there are simple tests that can be carried out upon delivery to ensure that the requirements have been met.
5. All of this requires a robust QA/QC department, which adds an expense, but much less than the cost of an in-service failure, particularly offshore.
6. The test conditions may need modifying for some projects, such as those with very low blow down temperatures.
7. It is important to ensure that duplex stainless steels are procured to adequate quality specifications, and to ensure that testing is carried out as required by the technical specification and witnessed by a third party when appropriate.

REFERENCES

1. G. Rorvik, L.A. Marken, I.M. Kulbotte, M. Aursand, S. Olsen, and K.S. Karlsen, "Influence of Intermetallic Precipitates on the Mechanical Properties and Environmental Cracking Resistance of Duplex Stainless Steel Fittings-A Case History (Part 1)," Duplex Stainless Steels 2010 (Zutphen, Netherlands: KCI Publishing, 2010).
2. S. Mahajanam, H. Rincon, and D. McIntyre, "Metallurgical Examination of Defects in Duplex Stainless Steel Pipe Fittings," *MP*, 49, 4 (2010): p. 56.
3. E. Ryengen and C. Wintermark, "Lessons Learned from Heat Treatment of Components in 22Cr and 25Cr Duplex Stainless Steel (and other materials)," Duplex Stainless Steels 2010 (Zutphen, Netherlands: KCI Publishing, 2010).
4. E. Turbeville, F. Busschaert, S. Benum, and B. Hadsen, A Summary of Recent Experiences on Designing and Fabricating Subsea Oil and Gas Production Systems with 22Cr and 25Cr Piping Materials, Duplex Stainless Steels 2010 (Zutphen, Netherlands: KCI Publishing, 2010).
5. NORSOK M-630, "Material Data Sheets for Piping," Edition 5 (Oslo, Norway: Standards Norway, 2010).
6. ISO 17781, "Petroleum, Petrochemical and Natural Gas Industries–Test Methods for Quality Control of Microstructure of Ferritic/Austenitic (Duplex) Stainless Steels" (Geneva, Switzerland: International Standards Organization, 2017).
7. R. Francis and S. Hebdon, "The Limiting Thickness for Duplex Stainless Steels," CORROSION 2014, paper no. 3651 (Houston, TX: NACE International, 2014).
8. NACE MR0175/ISO 15156, "Petroleum and Natural Gas Industries–Materials for Use in H_2S-Containing Environments in Oil and Gas Production" (Geneva, Switzerland: International Standards Organization, and Houston, TX: NACE International).
9. ASTM A923, "Standard Test Methods for Determining Deleterious Intermetallic Phase in Duplex Austenitic/Ferritic Stainless Steels" (West Conshohocken, PA: ASTM International).
10. G. Byrne, R. Francis, G. Warburton, and M. Maligas, "The Selection of Superduplex Stainless Steel for Oilfield Applications," CORROSION 2004, paper no. 123 (Houston, TX: NACE International, 2004).
11. ASTM G48, "Standard Test Methods for Pitting and Crevice Corrosion Resistance of Stainless Steels and Related Alloys by use of Ferric Chloride Solution" (West Conshohocken, PA: ASTM International).
12. G. Byrne, R. Francis, and G. Warburton, "Variations in Mechanical Properties and Corrosion Resistance of Different Alloys Within the Generic Designation UNS S32760," Duplex America 2000 (Zutphen, Netherlands: KCI Publishing, 2000).
13. NORSOK M-650, "Qualification of Manufacturers of Special Materials, Rev. 4" (Oslo, Norway: Standards Norway, 2011).
14. ISO 17782, "Petroleum, Petrochemical and Natural Gas Industries–Qualification of Manufacturers of Special Materials," in development (Geneva, Switzerland: International Standards Organization).
15. EEMUA Publication 218, "Quality Requirements for the Manufacture and Supply of Duplex Stainless Steels" (London, United Kingdom: Engineering Equipment and Materials Users Association, 2010).
16. R. Francis, G. Warburton, G. Byrne, and J. Wilson, "QA/QC Tools to Ensure the Quality of Duplex Fittings and Other Components," Duplex Stainless Steel 2010 (Zutphen, Netherlands: KCI Publishing, 2010).
17. R. Francis and S. Hebdon, "The Selection of Stainless Steels for Seawater Pumps," CORROSION 2015, paper no. 5446 (Houston, TX: NACE International, 2015).
18. ASTM A995, "Standard Specification for Castings, Austenitic-Ferritic (Duplex) Stainless Steel for Pressure-Containing Parts" (West Conshohocken, PA: ASTM International).

A Nominal Compositions of the Alloys Discussed in the Text

TABLE A.1 The nominal compositions of some commonly used wrought duplex stainless steels.

TYPE	UNS No.	GENERIC NAME	NOMINAL COMPOSITION (wt%)								PREN*
			Fe	Cr	Ni	Mo	N	Cu	W	Mn	
Lean Duplex	S32304	2304	Bal	23	4.8	0.3	0.10	-	-	1.5	24
	S32101	2101	Bal	21	1.5	0.3	0.22	-	-	4.5	26
	S82011	2102	Bal	21	1.5	0.3	0.22	-	-	2.5	26
	S32202	2202	Bal	22	2	0.3	0.2	-	-	1.5	26
	S32003	2003	Bal	20	3.5	1.9	0.17	-	-	1.5	>30
Standard Duplex	S31803	2205	Bal	22	5	3	0.14	-	-	1	34
	S32205	2205	Bal	22	5	3.2	0.17	-	-	1	35
25% Cr Duplex	S32550	Ferralium	Bal	25	6	2.5	0.2	2	-	1	37
Superduplex	S32750	2507	Bal	25	7	3.7	0.25	-	-	0.8	>41
	S32760	Z100	Bal	25	7	3.5	0.25	0.7	0.7	0.7	>41
	S32520	52N+	Bal	25	7	3.5	0.25	1.5	-	0.7	>41
	S39274	DP3W	Bal	25	7	3	0.27	0.5	2	0.7	>41
Hyperduplex	S32707	2707	Bal	27	6.5	4.8	0.4	-	-	1.4	49
	S33207	3207	Bal	32	7	3.5	0.5	-	-	1.4	>50

* PREN =%Cr + 3.3(%Mo + 0.5x%W) + 16x%N
Bal = Balance

TABLE A.2 The nominal compositions of some commonly used wrought stainless steels.

TYPE	UNS No.	GENERIC NAME	NOMINAL COMPOSITION (wt%)									PREN*
			Fe	C^{+}	Cr	Ni	Mo	N	Cu	W	Mn	
Austenitic	S30400	304	Bal	0.08	18	9	-	-	-	-	1.5	18
	S30403	304L	Bal	0.03	18	9	-	-	-	-	1.5	18
	S31600	316	Bal	0.08	17	10	2	-	-	-	1.5	24
	S31603	316L	Bal	0.03	17	10	2	-	-	-	1.5	24
	S31703	317L	Bal	0.03	19	12	3	-	-	-	1.5	29
	S31008	310S	Bal	0.08	25	20	-	-	-	-	1	25
	S32100	321	Bal	0.08	18	10	-	-	-	-	1.5 + Ti	18
	S31265	Saramet/SX	Bal	0.01	17	18	1	-	2	-	1	20
	N08020	Alloy 20	Bal	0.07	20	33	2	-	3.5	-	1 + Nb	27
	N08904	904L	Bal	0.03	20	25	4	-	1.5	-	1	34
	N08028	Alloy 28	Bal	0.03	27	31	3	-	1	-	1.5	37
Super Austenitic	S31254	6% Mo	Bal	0.03	20	18	6	0.2	0.7	-	0.7	43
	N08367	6% Mo	Bal	0.03	20	25	6	0.2	-	-	1	43
	N08926	6% Mo	Bal	0.03	20	25	6	0.2	1	-	1	43
Ferritic	S41003	410	Bal	0.03	12	0.7	-	-	-	-	1	12
Martensitic	S42000	420	Bal	>0.15	13	-	-	-	-	-	0.6	13
Super 13Cr	S41426	Super 13Cr	Bal	0.03	13	5	2	-	-	-	0.5	18

* PREN =%Cr + 3.3(%Mo + 0.5x%W) + 16x%N

Bal = Balance

+ - Maximum

TABLE A.3 The nominal compositions of some commonly used cast stainless steels.

TYPE	$ASTM^{+}$	GENERIC NAME	NOMINAL COMPOSITION (wt%)								PREN*
			Fe	Cr	Ni	Mo	N	Cu	W	Mn	
Austenitic	CF8M	316	Bal	17	10	2	-	-	-	1	24
	CN7M	Alloy 20	Bal	20	28	2.5	-	3.5	-	1	28
	CN3MCu	Mod. Alloy 20	Bal	20	29	2.5	-	3.2	-	1	28
	CK3MCuN	6% Mo	Bal	20	18	6	0.2	0.7	-	0.8	43
Lean Duplex	None	Z21 LCD	Bal	21	4	2	0.16	0.3	0.3	1.5	>30
Standard Duplex	Grade 4A	2205	Bal	22	5.5	3	0.16	0.5	-	1	35
25% Cr Duplex	Grade $1A^{x}$	CD4MCu	Bal	25	5	2	-	3	-	0.7	32
	Grade 1B	CD4MCuN	Bal	25	5	2	0.14	3	-	0.7	34
Superduplex	Grade 5A	2507	Bal	25	7.5	3.7	0.25	-	-	1	>41
	Grade 6A	Z100	Bal	25	8	3.5	0.25	0.7	0.7	0.7	>41

$^{+}$ - ASTM designations taken from A351 for austenitic and A995 for duplex

* PREN =%Cr + 3.3(%Mo + 0.5x%W) + 16x%N

Bal = Balance

X = Grade 1A replaced by grade 1B (see text)

TABLE A.4 Nominal composition of the nickel alloys mentioned in the text.

NAME	UNS No.	NOMINAL COMPOSITION (wt%)									PREN*
		Fe	C+	Ni	Cr	Mo	Cu	Al	Ti	W/Nb	
Alloy 400	N04400	2	0.05	65	-	-	Bal	-	-	-	N/A
Alloy K-500	N05500	1.5	0.05	65	-	-	Bal	3	0.6	-	N/A
Alloy 825	N08825	Bal	0.03	40	21	3	2	-	0.9	-	31
Alloy 625	N06625	3	0.05	Bal	21	9	-	-	-	3.5Nb	51
Alloy C-276	N10276	5	0.01	Bal	15.5	16	-	-	-	3W	72
Alloy C-22	N06022	3	0.015	Bal	22	13	-	-	-	3W	70
Alloy 686	N06686	3	0.01	Bal	21	16	-	-	0.1	3.5W	79
CW12M2	N30002	6.5	0.1	Bal	16	16	-	-	-	4W	73

Bal = Balance
*PREN = %Cr + 3.3(%Mo +0.5x%W) + 16x%N
N/A = Not Applicable
+ = Maximum

TABLE A.5 Nominal composition of the cobalt alloys mentioned in the text.

NAME	UNS No.	NOMINAL COMPOSITION (wt%)								PREN*
		Fe	C	Ni	Cr	Mo	Co	W	N	
Stellite 6	W73006	4	0.10	2	29	0.5	Bal	8	-	44
Ultimet	R31233	3	0.03	9	26	5	Bal	2	0.1	47

Bal = Balance
*PREN = %Cr + 3.3(%Mo +0.5x%W) + 16x%N

TABLE A.6 Nominal composition of copper alloys mentioned in the text.

ALLOY	UNS No.	NOMINAL COMPOSITION (wt%)								
		Cu	Sn	Zn	Pb	Al	Ni	Fe	Mn	As
M-Bronze	C92200	Bal	6	4	1.5	-	-	-	-	-
Muntz Metal	C28000	60	-	Bal	0.5	-	-	-	-	-
Naval Brass	C46400	60	1	Bal	-	-	-	-	-	-
DZR Brass	C35330	62	-	Bal	2	-	-	-	-	0.06
Al-Brass	C68700	78	_	Bal	-	2	-	-	-	0.04
Gunmetal	C83600	Bal	5	5	5	-	-	-	-	-
NAB (wrought)	C63200	Bal	-	-	-	9	4.5	4	-	-
NAB (cast)	C95800	Bal	-	-	-	9	4.5	4	-	-
Al-Bronze	C61300	Bal	0.3	-	-	6.5	-	2.5	-	-
Al-Bronze Alloy D	C61400	Bal	-	-	-	7	-	2.5	-	-
90/10 Cu-N	C70600	Bal	-	-	-	-	10	1.5	1	-
70/30 Cu-Ni	C71600	Bal	-	-	-	-	30	0.8	1	-

Bal = Balance

Index

D

E

F

G

N

O

P

Q

R

S

T

www.ingramcontent.com/pod-product-compliance
Ingram Content Group UK Ltd.
Pitfield, Milton Keynes, MK11 3LW, UK
UKHW051957041026
15590UKWH00009B/2354

9 781575 903699